# Machine Learning Security Principles

Keep data, networks, users, and applications safe
from prying eyes

**John Paul Mueller**

<packt>

BIRMINGHAM—MUMBAI

# Machine Learning Security Principles

**Group Product Manager**: Ali Abidi
**Publishing Product Manager**: Ali Abidi
**Senior Editor**: David Sugarman
**Technical Editor**: Sweety Pagaria
**Copy Editor**: Safis Editing
**Project Coordinator**: Farheen Fathima
**Proofreader**: Safis Editing
**Indexer**: Manju Arasan
**Production Designer**: Ponraj Dhandapani

First published: December 2022

Production reference: 3280923

Published by Packt Publishing Ltd.
Grosvenor House
11 St Paul's Square
Birmingham
B3 1RB.

ISBN 978-1-80461-885-1

www.packtpub.com

*This book is dedicated to Eva Beattie, a friend and faithful beta reader for 25 years. Books are never the result of one person's efforts, but of the influence of many people working together to help an author produce something wonderful.*

# Foreword

I first e-met John in the 1990s when we both wrote for the now-defunct magazine *Visual Basic Developer*. In those days, *Sonic the Hedgehog* was brand new, CompuServe, AOL, and Prodigy all roamed the earth, everyone programmed by candlelight on computers powered by hamster wheels, and artificial intelligence struggled to recognize the digits 0 through 9 and the words "yes" and "no" when spoken by different people. In the thirtyish years since then, we've all been through Y2K (which wasn't as bad as predicted), a global pandemic (which was worse than predicted), and the Cubs winning the World Series (which no one predicted).

More relevant to this book, AI has become so powerful that it understands speech better than some humans can, produces voices so realistic that future appearances by Darth Vader will be "voiced" by an AI, and generates deepfake videos so lifelike it's brought Salvador Dalí back to life. Amazon's algorithms seem to ship products before I order them and I've seen chatbots more likely to pass the Turing test than some of my friends.

AI in general and machine learning in particular have become powerful tools for both good and bad. In this book, John explains some of the ways that machine learning can be used to perpetrate and prevent security nightmares, and ways that machine learning can accidentally wreak havoc. He describes data bias (a hiring AI for Amazon penalizing female job applicants as most of their employees are male), badly selected data (IBM's Watson learning to swear by reading the Urban Dictionary), and intentional sabotage (Twitter turning Microsoft's chatbot Tay into a racist, misogynistic troll in less than 24 hours). Possibly even more importantly, John covers AI used to commit fraud (one AI faked a CEO's voice to request a €220,000 funds transfer) and to detect and counter fraud (that's why my credit card was declined the last time I had an airline connection in Las Vegas).

John covers all of these topics and more – though not these specific examples; I just think they're interesting and/or amusing!

As I mentioned earlier, I've known John for a long time. During those years, I've been the technical editor on several of his many books (and he's tech edited a few of mine), and one thing I've learned is that John knows what he's talking about. He's been working in AI for years and doesn't say something unless he's researched it, tried it out, included example programs demonstrating key techniques, and mentioned links to back up what he says and for you to follow for more information.

Hackers using AI is a relatively new concept, and so far, their success has been somewhat limited, but you can bet that their success will increase over time. One thing that AI in general and machine learning specifically are good at is learning over time. As more AIs bring bad data to the marketplace and hackers fine-tune their attacks, the consequences will become unavoidable and you need to be prepared. In the arms race between AI-empowered hackers and AI-enabled cybersecurity professionals, you can't afford to be uninformed.

*Rod Stephens*

*—Author and former Microsoft MVP*

# Contributors

## About the author

**John Paul Mueller** is a seasoned author and technical editor. He has writing in his blood, having produced 123 books and more than 600 articles to date. Topics he has written about range from networking to artificial intelligence and from database management to heads-down programming. Some of his recent books include discussions of data science, machine learning, and algorithms, along with Android programming and functional programming techniques. He also writes about computer languages such as C++, C#, Python, and Kotlin. His technical editing skills have helped more than 70 authors refine the content of their manuscripts. John has provided technical editing services to a variety of magazines and performed various kinds of consulting, and he writes certification exams.

# Acknowledgements

Thanks to my wife, Rebecca. Even though she is gone now, her spirit is in every book I write and in every word that appears on the page. She believed in me when no one else would.

Matt Wagner, my agent, deserves credit for helping me get the contract in the first place and taking care of all the details that most authors don't really consider. I always appreciate his assistance. It's good to know that someone wants to help.

A number of people read all or part of this book to help me refine the approach, test the coding examples, and generally provide input that all readers wish they could have. These unpaid volunteers helped in ways too numerous to mention here. I especially appreciate the efforts of Eva Beattie, who provided general input, read the entire book, and selflessly devoted herself to this project. Claudia Smith provided me with some significant insights into the accessibility and behavioral aspects of the book. Luca Massaron helped me with the design and orientation of some of the coding examples. Quite a few people also provided me with resource materials, and this particular book required a lot more research than many of my other books.

I especially appreciated Rod Stephens's help in maintaining a sense of humor. He also wrote a fantastic foreword, which is much appreciated by me.

The efforts of the technical reviewers are appreciated because they keep mistakes out of the book that would otherwise require reworking later. It can be quite hard to provide tactful and constructive input and I received both. You can see their names in the *About the reviewers* section.

Finally, I would like to thank David Sugarman, Farheen Fathima, Ali Abidi, and the rest of the editorial and production staff at Packt for their tireless efforts in helping me put this book together.

# About the reviewers

**Luca Massaron** joined Kaggle over 10 years ago and is now a Kaggle Grandmaster in discussions and a Kaggle Master in competitions and notebooks. In Kaggle competitions, he reached number 7 in the worldwide rankings. On the professional side, Luca is a data scientist with more than a decade of experience in transforming data into smarter artifacts, solving real-world problems, and generating value for businesses and stakeholders. He is a **Google Developer Expert (GDE)** in machine learning and the author of best-selling books on AI, machine learning, and algorithms.

**Akshay Kumar Prasad** is a data scientist who builds machine learning algorithms on big and fast data platforms in cyber security. He has his P.G. diploma degree in data science from the renowned Manipal University and a BTech in biotechnology from Dr. Rajendra Prasad Central Agricultural University. Before switching fully to the data science domain, he also worked in the FMCG manufacturing industry as a quality analyst. He brings immense experience in data engineering, data analysis, and machine learning roles in manufacturing, as well as the cyber security domain. He also writes about data science on his blog and contributes to open source projects in his free time.

**Deepayan Chanda** is a seasoned cybersecurity professional, architect, strategist, and advisor, with a strong intent to solve cybersecurity problems for enterprises by creating a balance between security and business goals, driven by 25 years of diverse cybersecurity domain experience. He is an Ex-Armed Forces Veteran (Indian Air Force), and has experience working with various enterprises like National Australia Bank, Standard Chartered Bank, Microsoft (Singapore), Cisco Systems, McAfee, and Symantec. He serves as a Board of Advisor and a mentor to a few cybersecurity start-ups worldwide and had the privilege of sharing his broad knowledge with the wider security community by authoring two books on cybersecurity, with multiple publications in the past many years.

# Table of Contents

# 3

# Mitigating Inference Risk by Avoiding Adversarial Machine Learning Attacks 51

# Part 2 – Creating a Secure System Using ML

# 4

# Considering the Threat Environment 91

# 5

# Keeping Your Network Clean                                    123

# 6

# Detecting and Analyzing Anomalies                            163

# 9

## Defending against Hackers                                             265

# Part 3 – Protecting against ML-Driven Attacks

# 10

## Considering the Ramifications of Deepfakes                            303

# 11

# Leveraging Machine Learning for Hacking                                    355

# Part 4 – Performing ML Tasks in an Ethical Manner

# 12

# Embracing and Incorporating Ethical Behavior                               379

# Preface

Machine learning is the most important new technology today for getting more out of data. It can reveal patterns that aren't obvious, for example, but it requires data – lots of it. Data gathering isn't just about data. It affects users and requires the use of applications to clean, manipulate, and analyze the data. Scientists use machine learning to discover new techniques or to create new kinds of data, such as the generation of various kinds of art based on existing inputs or the advancement of medicine through better imaging. Businesses use machine learning to perform tasks, such as detecting credit card fraud, monitoring networks, and implementing factory processes, and to achieve all sorts of other goals where humans and AI work side-by-side.

Hackers don't always damage data; sometimes they steal it or use it to perform social attacks on a business. Sometimes they simply want money or other goods, and machine learning offers an avenue for acquiring them. A hacker may not steal anything at all – perhaps the target is someone's reputation. It may surprise you to learn that hackers often use machine learning applications themselves to perform a kind of dance with your machine learning-based security to overcome it. However, hackers have behavioral patterns, and knowing how to detect those patterns is important in the modern computing environment.

Obtaining data in an ethical manner is important because the very act of behaving ethically reduces the security risk associated with data. However, hackers don't necessarily target users and their data. Perhaps they're interested in your organization's trade secrets or committing fraud. They might simply be interested in lurking in the background and committing mischief. So, just keeping your data secure as a means of protecting your machine learning investment isn't enough. You need to do more.

This book helps you get the big picture from a machine learning perspective using all the latest research available on methods that hackers use to break into your system. It's about the whole system, not just your application. You will discover techniques that help you gather data ethically and keep it safe, while also preventing all sorts of illegal access methods from even occurring. In fact, you will use machine learning as a tool to keep hackers at bay and discover their true intent for your organization.

# Who this book is for

Whether you're a data scientist, researcher, or manager interested in machine learning techniques from various perspectives, you will need this book because security has already become a major headache for all three groups. The problem with most resources is that they're written by Ph.D. candidates in a language that only they understand. This book presents security in a way that's easy to understand and employs a host of diagrams to explain concepts to visual learners. The emphasis is on real-world examples at both theoretical and hands-on levels. You'll find links to a wealth of examples of real-world break-ins and explanations of why and how they occurred and, most importantly, how you can overcome them.

This book does assume that you're familiar with machine learning concepts and it helps if you already know a programming language, with an emphasis on Python knowledge. The hands-on Python code is mostly meant to provide details for data scientists and researchers who need to see security concepts in action, rather than at a more theoretical level. A few examples, such as the Pix2Pix GAN in *Chapter 10*, require an intermediate level of programming knowledge, but most examples are written in a manner that everyone can use.

# What this book covers

*Chapter 1, Defining Machine Learning Security*, explains what machine learning is all about, how it's affected by security issues, and what impact security can have on the use of your applications from an overview perspective. This chapter also contains guidelines on how to configure your system for use with the source code examples.

*Chapter 2, Mitigating Risk at Training by Validating and Maintaining Datasets*, explores how ensuring that the data you're using is actually the data that you think you're using is essential because your model can be skewed by various forms of corruption and data manipulation.

*Chapter 3, Mitigating Inference Risk by Avoiding Adversarial Machine Learning Attacks*, gives an overview of the various methods to interfere directly with model development through techniques such as evasion attacks and model poisoning.

*Chapter 4, Considering the Threat Environment*, considers how hackers target machine learning models and their goals in doing so from an overview perspective. You will discover some basic coded techniques for avoiding many machine learning attacks through standard methodologies.

*Chapter 5, Keeping Your Network Clean*, gives detailed information on how network attacks work and what you can do to detect them in various ways, including machine learning techniques as your defense. In addition, you will discover how you can use predictive techniques to determine where a hacker is likely to strike next.

*Chapter 6, Detecting and Analyzing Anomalies*, provides the details on determining whether outliers in your data are anomalies that need mitigation or novelties that require observation as part of a new trend. You will see how to perform anomaly detection using machine learning techniques.

*Chapter 7, Dealing with Malware*, covers the various kind of malware and what to look for in your own environment. This chapter shows how to take an executable apart so that you can see how it's put together and then use what you learn to generate machine learning features for use in detection algorithms.

*Chapter 8, Locating Potential Fraud*, explores the sources of fraud today (and it's not just hackers), what you can do to detect the potential fraud, and how you can ensure that the model you build will actually detect the fraud with some level of precision. The techniques in this chapter for showing how to discern model goodness also apply to other kinds of machine learning models.

*Chapter 9, Defending Against Hackers*, contemplates the psychology of hackers by viewing hacker goals and motivations. You will obtain an understanding of why simply building the security wall higher and higher doesn't work, and what you can do, in addition to building new security protections for your system.

*Chapter 10, Considering the Ramifications of Deepfakes*, looks at the good and the bad of deepfake technology. You will get an overview of the ramifications of deepfake technology for research, business, and personal use today. This chapter also demonstrates one technique for creating a deepfake model in detail.

*Chapter 11, Leveraging Machine Learning for Hacking*, explains how hackers view machine learning and how they're apt to build their own models to use against your organization. We will consider the smart bot threat in detail.

*Chapter 12, Embracing and Incorporating Ethical Behavior*, explains how behaving ethically not only ensures that you meet both privacy and security requirements that may be specified by law but also has an implication with regard to security, in that properly sanitized datasets have natural security prevention features as well. In addition, you will discover how using properly vetted datasets saves you time, money, and effort in building models that actually perform better.

## To get the most out of this book

This book assumes that you're a manager, researcher, or data scientist with at least a passing understanding of machine learning and machine learning techniques. It doesn't assume detailed knowledge. To use the example code, it also pays to have some knowledge of working with Python because there are no tutorials provided in the book. All of the coded examples have been tested on both Google Colab and with Anaconda. The *Setting up for the book* section of *Chapter 1, Defining Machine Learning Security*, provides detailed setup instructions for the book examples.

The advantages of using Google Colab are that you can code anywhere (even your smartphone or television set, both of which have been tested by other readers) and you don't have to set anything up. The disadvantages of using Google Colab are that not all of the book examples will run in this environment (especially *Chapter 7*) and your code will tend to run slower (especially *Chapter 10*). When working with Google Colab, all you need do is direct your browser to `https://colab.research.google.com/notebooks/welcome.ipynb` and create a new notebook.

The advantage of using Anaconda is that you have more control over your work environment and you can perform more tasks. The disadvantage of using Anaconda is that you need a desktop system with the required hardware and software, as described in the following table, for most of the book examples. (The `MLSec; 01; Check Versions.ipynb` example shows how to verify the version numbers of your software.) Some examples will require additional setup requirements and those requirements are covered as part of the example description (for example, when creating the Pix2Pix GAN in *Chapter 10*, you need to install and configure TensorFlow).

| General software covered in the book | Operating system and hardware requirements |
| --- | --- |
| Anaconda 3, 2020.07 | Windows 7, 10, or 11<br><br>macOS 10.13 or above<br><br>Linux (Ubuntu, RedHat, and CentOS 7+ all tested) |
| Python 3.8 or higher (version 3.9.x is highly recommended, versions above 3.10.7 aren't recommended or tested) | The test system uses this hardware, which is considered minimal:<br><br>Intel i7 CPU<br><br>8 GB RAM<br><br>500 GB hard drive |
| NumPy 1.18.5 or greater (version 1.21.x is highly recommended) | |
| Scikit-learn 0.23.1 or greater (version 1.0.x is highly recommended) | |
| Pandas 1.1.3 or greater (version 1.4.x is highly recommended) | |

When working with any version of the book, downloading the downloadable source code is highly recommended to avoid typos. Copying and pasting code from the digital version of the book will very likely result in errors. Remember that Python is a language that depends on formatting to deal with things like structure and to show where programming constructs such as `for` loops begin and end. The source code downloading instructions appear in the next section.

# Download the example code files

You can download the example code files for this book from GitHub at `https://github.com/PacktPublishing/Machine-Learning-Security-Principles` or John's website at `http://www.johnmuellerbooks.com/source-code/`. If there's an update to the code, it will be updated in both the GitHub repository and on John's website.

We also have other code bundles from our rich catalog of books and videos available at `https://github.com/PacktPublishing/`. Check them out!

# Conventions used

There are a number of text conventions used throughout this book.

`Code in text`: Indicates code words in text, database table names, folder names, filenames, file extensions, pathnames, dummy URLs, user input, and Twitter handles. Here is an example: "For example, `Remove_Stop_Words()` relies on a list comprehension to perform the actual processing."

A block of code is set as follows:

```
import getpass

user = getpass.getuser()
pwd = getpass.getpass("User Name : %s" % user)
```

When we wish to draw your attention to a particular part of a code block, the relevant lines or items are set in bold:

```
import getpass

user = getpass.getuser()
pwd = getpass.getpass("User Name : %s" % user)
```

> **Tips or important notes**
> Appear like this.

## Get in touch

Feedback from our readers is always welcome.

**General feedback**: If you have questions about any aspect of this book, email us at customercare@ packtpub.com and mention the book title in the subject of your message. If you have a book content-specific question, please contact John at John@JohnMuellerBooks.com for quick and courteous service. Your feedback is essential to helping me produce better books!

**Errata**: Although we have taken every care to ensure the accuracy of our content, mistakes do happen. If you have found a mistake in this book, we would be grateful if you would report this to us. Please visit www.packtpub.com/support/errata and fill in the form.

**Expanded book content:** As I get input from readers, I often provide additional book insights and updated procedures on my blog at http://blog.johnmuellerbooks.com/.

**Piracy**: If you come across any illegal copies of our works in any form on the internet, we would be grateful if you would provide us with the location address or website name. Please contact us at copyright@packt.com with a link to the material.

**If you are interested in becoming an author**: If there is a topic that you have expertise in and you are interested in either writing or contributing to a book, please visit authors.packtpub.com.

# Share Your Thoughts

Once you've read *Machine Learning Security Principles*, we'd love to hear your thoughts! Scan the QR code below to go straight to the Amazon review page for this book and share your feedback.

https://packt.link/r/1-804-61885-3

Your review is important to us and the tech community and will help us make sure we're delivering excellent quality content.

# Download a free PDF copy of this book

Thanks for purchasing this book!

Do you like to read on the go but are unable to carry your print books everywhere?

Is your eBook purchase not compatible with the device of your choice?

Don't worry, now with every Packt book you get a DRM-free PDF version of that book at no cost.

Read anywhere, any place, on any device. Search, copy, and paste code from your favorite technical books directly into your application.

The perks don't stop there, you can get exclusive access to discounts, newsletters, and great free content in your inbox daily

Follow these simple steps to get the benefits:

1. Scan the QR code or visit the link below

https://packt.link/free-ebook/9781804618851

2. Submit your proof of purchase
3. That's it! We'll send your free PDF and other benefits to your email directly

# Part 1 – Securing a Machine Learning System

In this part, you will discover why security is important and the various kinds of security that you will need to consider. You will look at the threats against machine learning applications, including those from data manipulation and other machine learning applications.

This section includes the following chapters:

- *Chapter 1, Defining Machine Learning Security*
- *Chapter 2, Mitigating Risk at Training by Validating and Maintaining Datasets*
- *Chapter 3, Mitigating Inference Risk by Avoiding Adversarial Machine Learning Attack*

# 1

# Defining Machine Learning Security

Organizations trust **machine learning** (**ML**) to perform a wide variety of tasks today because it has proven to be relatively fast, inexpensive, and effective. Unfortunately, many people really aren't sure what ML is because television, movies, and other media tend to provide an unrealistic view of the technology. In addition, some users engage in wishful thinking or feel the technology should be able to do more. Making matters worse, even the companies who should know what ML is about hype its abilities and make the processes used to perform ML tasks opaque. Before making ML secure, it's important to understand what ML is all about. Otherwise, the process is akin to installing home security without actually knowing what the inside of the home contains or even what the exterior of the home looks like.

Adding security to an ML application involves understanding the data analyzed by the underlying algorithm and considering the goals of the application in interacting with that data. It also means looking at security as something other than restricting access to the data and the application (although, restricting access is a part of the picture).

The remainder of this chapter talks about the requirements for working with the coding examples. It's helpful to have the right setup on your machine so that you can be sure that the examples will run as written.

> ### Get in touch
> Obviously, I want you to be able to work with the examples, so if you run into coding issues, please be sure to contact me at John@JohnMuellerBooks.com.

Using the downloadable source code will also save you time and effort. With these issues in mind, this chapter discusses these topics:

- Obtaining an overview of ML
- Defining a need for security and choosing a type
- Making the most of this book

# Building a picture of ML

People anthropomorphize computers today, giving them human characteristics, such as the ability to think. At its lowest level, a computer processes commands to manipulate data, perform comparisons, and move data around. There is no thought process involved—just electrical energy cleverly manipulated to produce a mathematical result from a given input. So, the term "machine learning" is a bit of a misnomer because the machine is learning nothing and it doesn't understand anything. A better way to view ML is as a process of algorithm manipulation such that added weighting produces a result that better matches the data input. Once someone trains a **model** (the combination of algorithm and weighting added to the algorithm), it's possible to use the model to process data that the algorithm hasn't seen in the past and still obtain a desirable result. The result is the simulation of human thought processes so that it appears that the application is thinking when it isn't really thinking at all.

The feature that distinguishes ML most significantly is that the computer can perform mundane tasks fast and consistently. It can't provide original thought. A human must create the required process but, once created, the machine can outperform the human because it doesn't require rest and doesn't get bored. Consequently, if the data is clean, the model correct, and the anticipated result correctly defined, a machine can outshine a human. However, it's essential to consider everything that is required to obtain a desirable result before employing ML for a particular task, and this part of the process is often lacking today. People often think that machines are much more capable than they really are and then exhibit disappointment when the machine fails to work as expected.

## Why is ML important?

Despite what you may have heard from various sources, ML is more important for mundane tasks than for something earth-shattering in its significance. ML won't enable Terminators to take over the planet, nor will this technology suddenly make it possible for humans to stop working entirely in a utopian version of the future. What ML can do is reduce the boredom and frustration that humans feel when forced to perform repetitive factory work or other tasks of the sort. In the future, at the lowest level, humans will supervise machines performing mundane tasks and be there when things go wrong.

However, the ability to simply supervise machines is still somewhat far into the future, and letting them work unmonitored is further into the future still. There are success stories, of course, but then there are also failures of the worst sort. For example, trusting the AI in a car to drive by itself without human intervention can lead to all sorts of problems. Sleeping while driving will still garner a ticket and put others at risk, as described at `https://www.theguardian.com/world/2020/sep/17/canada-tesla-driver-alberta-highway-speeding`. In this case, the driver was sleeping peacefully with a passenger in the front seat of the car when the police stopped him. Fortunately, the car didn't cause an accident in this case, but there are documented instances where self-driving cars did precisely that (see `https://www.nytimes.com/2018/03/19/technology/uber-driverless-fatality.html` for an example).

Besides performing tasks, ML can perform various kinds of analysis at a speed that humans can't match, and with greater efficiency. A doctor can rely on ML to assist in finding cancer because the ML application can recognize patterns in an MRI that the doctor can't even see. Consequently, the ML application can help guide the doctor in the right direction. However, the doctor must still make the final determination as to whether a group of cells really is cancerous, because the ML application lacks experience and the senses that a doctor has. Likewise, ML can make a doctor's hands steadier during surgery, but the doctor must still perform the actual task. In summary, ML is currently assistive in nature, but it can produce reliable results in that role.

Pattern recognition is a strong reason to use ML. However, the ability to recognize patterns only works when the following applies:

- The source data is untainted
- The training and testing data are unbiased
- The correct algorithms are selected
- The model is created correctly
- Any goals are clearly defined and verified against the training and test data

**Classification** uses of ML rely on patterns to distinguish between types of objects or data. For example, a classification system can detect the difference between a car, a pedestrian, and a street sign (at least, a lot of the time). Unfortunately, the current state of the art clearly shows that ML has a long way to go in this regard because it's easy to fool an application in many cases (see `https://arxiv.org/pdf/1710.08864.pdf?ref=hackernoon.com` for examples). There are a lot of articles now that demonstrate all of the ways in which an adversarial attack on a deep learning or ML application will render it nearly useless. So, ML works best in an environment where nothing unexpected happens, but in that environment, it works extremely well.

Recommender systems are another form of ML that try to predict something based on past data. For example, recommender systems play a huge role in online stores where they suggest some items to go with other items a person has purchased. If you're fond of online buying, you know from experience that the recommender systems are wrong about the additional items more often than not. A recommender setup attached to a word processor for suggesting the next work you plan to type often does a better job over time. However, even in this case, you must exercise care because the recommendation is often not what you want (sometimes with hilarious results when the recipient receives the errant text).

As everything becomes more automated, ML will play an ever-increasing role in performing the mundane and repeatable elements of that automation. However, humans will also need to play an increasingly supervisory role. In the short term, it may actually appear that ML is replacing humans and putting them out of work, but in the long term, humans will simply perform different work. The current state of ML is akin to the disruption that occurred during the Industrial Revolution, where machines replaced humans in performing many manual tasks. Because of that particular disruption in the ways that things were done, a single farmer today can tend to hundreds of acres of land, and factory work is considerably safer. ML is important because it's the next step toward making life better for people.

## Identifying the ML security domain

Security doesn't just entail the physical protection of data, which might actually be impossible for online sources such as websites where the data scientist obtains the data using screen-scraping techniques. To ensure that data remains secure, an organization must monitor and validate it prior to use for issues such as data corruption, bias, errors, and the like. When securing an ML application, it's also essential to review issues such as these:

- **Data bias**: The data somehow favors a particular group or is skewed in a manner that produces an inaccurate analysis. Model errors give hackers a wedge into gaining access to the application, its model, or underlying data.

- **Data corruption**: The data may be complete, but some values are incorrect in a way that shows damage, poor formatting, or in a different form. For example, even in adding the correct state name to a dataset, such as Wisconsin, it could appear as WI, Wis, Wisc, Wisconsin, or some other legitimate, but different form.

- **Missing critical data**: Some data is simply absent from the dataset or could be replaced with a random value, or a placeholder such as N/A or Null for numeric entries.

- **Errors in the data**: The data is apparently present, but is incorrect in a manner that could cause the application to perform badly and cause the user to make bad decisions. Data errors are often the result of human data entry problems, rather than corruption caused by other sources, such as network errors. Hackers often introduce data errors that have a purpose, such as entering scripts in the place of values.

- **Algorithm correctness**: Using the incorrect algorithm will create output that doesn't meet analysis goals, even when the underlying data is correct in every possible manner.

- **Algorithm bias**: The algorithm is designed in such a manner that it performs analysis incorrectly. This problem can also appear when weighting values are incorrect or the algorithm handles feedback values inappropriately. The bottom line is that the algorithm produces a result, but the result favors a particular group or outputs values that are skewed in some way.

- **Repeatable and verifiable results**: ML applications aren't useful unless they can produce the same results on different systems and it's possible to verify those results in some way (even if verification requires the use of manual methods).

ML applications are also vulnerable to various kinds of software attacks, some of which are quite subtle. All of these attacks are covered in detail starting in *Chapter 3* of the book. However, here is an overview of the various attack types and a quick definition of each that you can use for now:

- **Evasion**: Bypassing the information security functionality built into a system.

- **Poisoning**: Injecting false information into the application's data stream.

- **Inference**: Using data mining and analysis techniques to gain knowledge about the underlying dataset, and then using that knowledge to infer vulnerabilities in the associated application.

- **Trojans**: Employing various techniques to create code or data that looks legitimate, but is really designed to take over the application or manipulate specific components of it.

- **Backdoors**: Using system, application, or data stream vulnerabilities to gain access to the underlying system or application without providing the required security credentials.

- **Espionage**: Stealing classified, sensitive data or intellectual property to gain an advantage over a person, group, or organization to perform a personnel attack.

- **Sabotage**: Performing deliberate and malicious actions to disrupt normal processes, so that even if the data isn't corrupted, biased, or damaged in some way, the underlying processes don't interact with it correctly.

- **Fraud**: Relying on various techniques, such as phishing or communications from unknown sources, to undermine the system, application, or data security in a secretive manner. This level of access can allow for unauthorized or unpaid use of the application and influence ways in which the results are used, such as providing false election projections.

The target of such an attack may not even know that the attack compromised the ML application until the results demonstrate it (the *Seeing the effect of bad data* section of *Chapter 10, Considering the Ramifications of Deepfakes*, shows a visual example of how this can happen). In fact, issues such as bias triggered by external dataset corruption can prove so subtle that the ML application continues to function in a compromised state without anyone noticing at all. Many attacks, such as privacy attacks (see the article entitled *Privacy Attacks on Machine Learning Models*, at `https://www.infoq.com/articles/privacy-attacks-machine-learning-models/`), have a direct monetary motive, rather than simple disruption.

It's also possible to use ML applications as the attack vector. Hackers employ the latest techniques, such as relying on ML applications to attack you to obtain better results, just as you do. The article entitled *7 Ways in Which Cybercriminals Use Machine Learning to Hack Your Business*, at `https://gatefy.com/blog/cybercriminals-use-machine-learning-hack-business/`, describes just seven of the ways in which hackers use ML in their nefarious trade. You can bet that hackers use ML in several other ways, some of them unexpected and likely unknown for now.

## Distinguishing between supervised and unsupervised

ML relies on a large number of algorithms, used in a variety of ways, to produce a useful result. However, it's possible to categorize these approaches in three (or possibly four) different ways:

- Supervised learning
- Unsupervised learning
- Reinforcement learning

Some people add the fourth approach of semi-supervised learning, which is a combination of supervised and unsupervised learning. This section will only discuss the first three because they're the most important in understanding ML.

### Understanding supervised learning

**Supervised learning** is the most popular and easiest-to-use ML paradigm. In this case, data takes the form of an example and label pair. The algorithm builds a mapping function between the example and its label so that when it sees other examples, it can identify them based on this function. *Figure 1.1* provides you with an overview of how this process works:

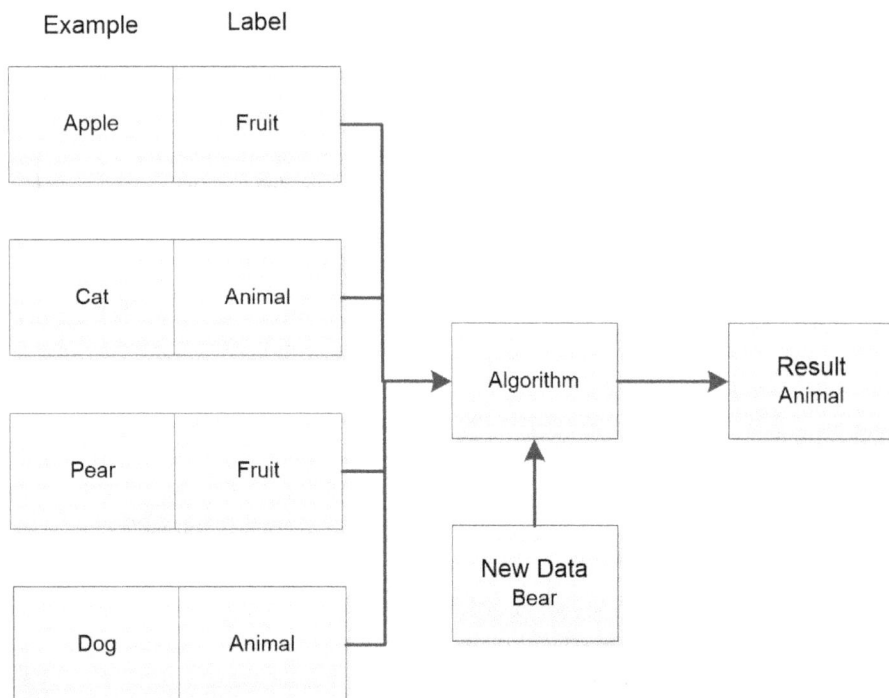

Figure 1.1 – Supervised learning relies on labeled examples to train the model

Supervised learning is often used for certain types of classification, such as facial recognition, and prediction, or how well an advertisement will perform based on past examples. This paradigm is susceptible to many attack vectors including someone sending data with the wrong labels or supplying data that is outside the model's usage.

### Understanding unsupervised learning

When working with **unsupervised learning**, the algorithm is fed a large amount of data (usually more than is required for supervised learning) and the algorithm uses various techniques to organize, group, or cluster the data. An advantage of unsupervised learning is that it doesn't require labels: the majority of the data in the world is unlabeled. Most people consider unsupervised learning as data-driven, contrasted with supervised learning, which is task-driven. The underlying strategy is to look for patterns, as shown in *Figure 1.2*:

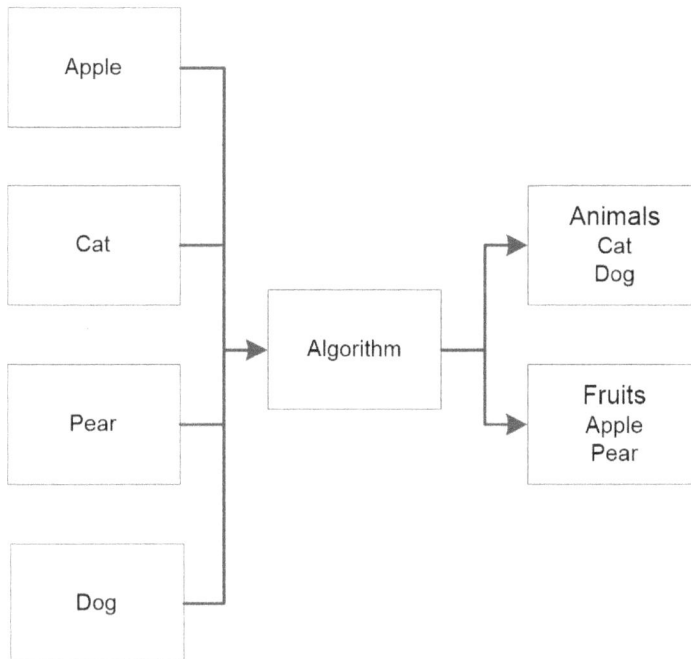

Figure 1.2 – Unsupervised learning groups or clusters like data together to train the model

Unsupervised learning is often used for recommender systems because such systems receive a constant stream of unlabeled data. You also find it used for tracking buying habits and grouping users into various categories. This paradigm is susceptible to a broad range of attack vectors, but data bias, data corruption, data errors, and missing data would be at the top of the list.

## Understanding reinforcement learning

**Reinforcement learning** is essentially different from either supervised or unsupervised learning because it has a feedback loop element built into it. The best way to view reinforcement learning is as a methodology where ML can learn from mistakes. To produce this effect, an **agent**, the algorithm performing the task, has a specific list of actions that it can take to affect an environment. The environment, in turn, can produce one of two signals as a result of the action. The first signals successful task completion, which reinforces a behavior in the agent. The second provides an environment state so that the agent can detect where errors have occurred. *Figure 1.3* shows how this kind of relationship works:

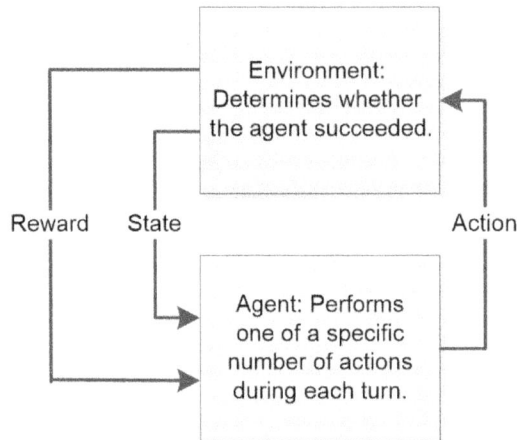

Figure 1.3 – Reinforcement learning is based on a system of rewards and an updated state

You often see reinforcement learning used for video games, simulations, and industrial processes. Because you're linking two algorithms together, algorithm choice is a significant priority and anything that affects the relationship between the two algorithms has the potential to provide an attack vector. Feeding the agent incorrect state information will also cause this paradigm to fail.

## Using ML from development to production

It's essential to understand that ML does have an important role to fulfill today in performing specific kinds of tasks. *Figure 1.4* contains a list of typical tasks that ML applications perform today, along with the learning type used to perform the task and observations of security and other issues associated with this kind of task. In no instance will you find that ML performs any task perfectly, especially without human assistance.

| Task | Learning Type | ML Consideration |
|---|---|---|
| Automatic language translation | Supervised | Translates one language into another language using a sequence-to-sequence learning algorithm. The results are often less useful than expected due to variations between languages and the fact that languages generally contain words that don't have equivalents in other languages.<br><br>Susceptible to data errors, missing data, data corruption, algorithm bias, and an inability to repeat and verify results due to naturally occurring evolution in languages. This kind of application is also sensitive to speech patterns and misidentifying terms when words aren't enunciated clearly. |
| Email spam and malware filtering | Supervised | Marks, moves, or deletes email that meets the criteria of spam or malware from an inbox as it's received from a server. There are usually several levels of filtering including Content, Header, Blacklist, Rule-based, and Permission.<br><br>Susceptible to a number of potential attacks including backdoors, Trojans, espionage, sabotage, fraud, evasion, inference, data errors, and data corruption. This is one of the more reliable forms of ML applications, but users still regularly find spam in their inboxes and useful messages in their spam folders. |
| Image recognition | Supervised | Identification of objects, persons, places, patterns, and other elements within an image.<br><br>Susceptible to a variety of attack types, but also prone to misidentification when the image contains elements the application didn't expect or when those objects appear in positions that the application isn't trained to recognize. |

| Task | Learning Type | ML Consideration |
|------|---------------|------------------|
| Medical diagnosis | Supervised and unsupervised | Predicts the progression and characteristics of diseases and other conditions, along with locating and identifying potential patient illnesses. |
| | | Susceptible to data bias, data corruption, data errors, incorrect algorithm selection, and algorithm bias. This particular application type can never operate alone; it always assists a physician with the required experience to make a diagnosis. |
| Online fraud detection | Supervised | Reduces the risk of conducting transactions online by detecting conditions such as fake accounts, fake IDs, compromised sites, compromised security certificates, and so on. |
| | | Susceptible to a wide range of attacks, some of which have nothing to do with the application. For example, a compromised certificate authority could cause the application to fail by allowing the hacker access to the underlying infrastructure, even if the application itself isn't at fault. This kind of application is also known to display false positives and false negatives depending on the reliability of the code used to create it and the model training. |
| Product recommendation | Unsupervised | Outputs product recommendations based on previous buying habits, associated goods, and direct queries. It's one of the most widely used and common ML applications. |
| | | Susceptible to data errors, data bias, missing data, algorithm bias, fraud, sabotage, and a wealth of other issues. This kind of application often provides irrelevant information along with useful product recommendations because the application has no method of judging user needs and wants. |

| Task | Learning Type | ML Consideration |
|---|---|---|
| Self-driving cars | Supervised, unsupervised, and reinforcement | Allows a vehicle to drive itself by monitoring various cameras and detectors for the presence of obstacles, interpreting the content of road signs, and so on. |
| | | Susceptible to so many different kinds of attacks, it's truly amazing that self-driving cars work at all. In addition to ML, self-driving vehicles rely on other AI technologies such as expert systems (`https://www.aitrends.com/ai-insider/expert-systems-ai-self-driving-cars-crucial-innovative-techniques/`). It's entirely possible that self-driving cars will eventually become completely successful, but don't look for this advance anytime soon. |
| Speech recognition | Supervised | Translation of spoken or written speech into tokens that the computer can recognize and process. |
| | | Susceptible to data errors and use of unidentified terms. This kind of application is also sensitive to speech patterns and misidentifying terms when words aren't enunciated clearly. |
| Stock market trading | Supervised | Predicts trends in the stock market based on past and current data. This is one of the few ML applications that relies heavily on short-term memory and weighting processes to make current data count for more than past data. |
| | | Susceptible to data bias, data corruption, missing data, data errors, incorrect algorithm selection, and algorithm bias. Attackers will attempt to gain access by any means possible with a strong emphasis on evasion, inference, Trojans, and backdoors. Reliability is a prime concern for this application type, but incredibly hard to measure given the variability of the stock market. |

| Task | Learning Type | ML Consideration |
| --- | --- | --- |
| Traffic prediction | Reinforcement | Plots a path between two points on a map based on criteria such as traffic conditions, time of travel, and resource usage. |
| | | Susceptible to various attacks such as poisoning, inference, data corruption, data bias, missing data, and so on. This kind of application will normally get the user to the right place (although, there have been instances where the application has sent the user into ponds and so on), but the path may not ultimately prove to meet all required goals. |
| Virtual personal assistant | Supervised and unsupervised | Accepts voice or text input to perform various predefined tasks, such as iterating a person's meetings for the day or locating a restaurant. |
| | | Susceptible to data errors, missing data, data corruption, and algorithm bias. In addition, a third party could attempt to gain access to application data using evasion, poisoning, Trojans, fraud, and backdoors. This kind of application is also sensitive to speech patterns and misidentifying terms when words aren't enunciated clearly. |

Figure 1.4 – ML tasks and their types

*Figure 1.4* doesn't contain some of the more exotic uses for ML. For example, some people use ML to generate art (see `https://www.bbc.com/news/uk-england-oxfordshire-61600523` for one of the newest examples). However, the ML application isn't creating art. What happens instead is that the ML application learns a particular art style from examples, and then transforms another graphic, such as a family picture, into a representation using the art examples. The results can be interesting, even beautiful, but they aren't creative. The creativity resides in the original artist and the human guiding the generation (see `https://aiartists.org/ai-generated-art-tools` for details). The same technique applies to ML-generated music and even videos. Many of these alternative uses for ML are interesting, but the book doesn't cover them heavily, except for the perspective of ethical treatment of data. So, why is security so important for ML projects? The next section begins to answer that question.

## Adding security to ML

Security is a necessary component of ML to ensure that results received from an analysis reflect reality. Otherwise, decisions made based on the analysis will be flawed. If the mistake made based on such analysis merely affected the analyst, then the consequences might not be catastrophic. However, ML affects people – sometimes large groups of people. When the effects are large enough, businesses fold, lawsuits ensue, and people lose faith in the ability of ML applications to produce reliable results. Adding security ensures the following:

- Reliability
- Verifiability
- Repeatability
- Transparency
- Confidence
- Consistency

Let's examine how security can impact ML in more detail.

## Defining the human element

At this point, it's important to take a slight detour from the technical information presented so far to discuss the human element. Even if the processes are clear, the data is clean, the algorithms are chosen correctly, and the code is error-free, humans still provide the input and interpret the result. Humans are an indirect source of security issues in all ML scenarios. When working with humans, it's essential to consider the five mistruths that creep into every part of the ML environment and cause security issues that are difficult or sometimes impossible to find:

- **Commission**: Performing specific and overt engagement in a mistruth and supplying incorrect data. However, a mistruth of commission need not always imply an intent to mislead. Sometimes these mistruths are the result of a lack of information, incorrect information, or a need to please others. In some cases, it's possible to detect mistruths of commission as outliers in plotted results created during analysis. Mistruths of commission create security issues by damaging the data used for analysis and therefore corrupting the model.

- **Omission**: Leaving out essential details that would make the resulting conclusions different. In many cases, the person involved simply forgets to provide the information or is unaware of it. However, this mistruth also makes an appearance when the facts are inconvenient. In some cases, it's possible to detect this sort of mistruth during missingness checks of the data or in considering the unexpected output of an algorithm. Mistruths of omission create security issues by creating holes in the data or by skewing the model.

- **Bias**: Seeing the data or results in an unrealistic or counterintuitive manner due to personal concerns, environmental pressures, or traditions. Human biases often keep the person involved from seeing the patterns and outcomes that are obvious when the bias isn't present. Environmental pressures, including issues such as tiredness, are hard to overcome and spot. The same checks that work for other kinds of bias can help root out human biases in data. Mistruths of bias create security issues by skewing the model and possibly causing the model to overfit or underfit the data.

- **Perspective**: Viewing the data based on experience, environmental conditions, and available information. In reviewing the statements of witnesses to any event, it's possible to obtain different stories from each witness, even when the witnesses are being truthful from their perspective. The same is true of ML data, algorithms, and output. Different people will see the data in different ways and it's nearly impossible to say that one perspective is correct and another incorrect. In many cases, the only way to handle this issue is to create a consensus opinion, much as interviewers do when speaking to witnesses to an event. Mistruths of perspective cause security issues by limiting the effectiveness of the model in providing a correct solution due to the inability of computers to understand anything.

- **Frame of reference**: Conveying information to another party incorrectly because the other party lacks the required experience. This kind of soft knowledge is precisely why humans are needed to interpret the analysis provided through ML. A human who has had a particular experience understands the experience and recognizes the particulars of it, but is unable to articulate the experience in a concrete manner. Mistruths of frame of reference create security issues by causing the model to misinterpret situational data and render incorrect results.

Now that you have a better idea of how humans play into the data picture, it's time to look more at the technical issues, keeping the human element in mind.

## Compromising the integrity and availability of ML models

In many respects, the ML model is a kind of black box where data goes in and results come out. Many countries now have laws mandating that models become more transparent, but still, unless you want to spend a great deal of time reviewing the inner workings of a model (assuming you have the knowledge required to understand how they work at all), it still amounts to a black box. The model is the weakest point of an ML application. It's possible to verify and validate data, and understanding the algorithms used need not prove impossible. However, the model is a different story because the only practical ways to test it are to use test data and perform some level of verification and validation. What happens, however, if hackers or others have compromised the integrity of the model in subtle ways that don't affect all results, just some specific results?

**The integrity of a model** doesn't just involve training it with correct data but also involves keeping it trained properly. Microsoft's Tay (see `https://spectrum.ieee.org/tech-talk/artificial-intelligence/machine-learning/in-2016-microsofts-racist-chatbot-revealed-the-dangers-of-online-conversation`) is an example of just how wrong training can go when the integrity of the model is compromised. In Tay's case, unregulated Twitter messages did all the damage in about 16 hours. Of course, it took a lot longer than that to initially create the model, so the loss to Microsoft was immense. To the question of why internet trolls damaged the ML application, the answer of *because they can* seems trite, but ends up being on the mark. Microsoft created a new bot named Zo that fared better but was purposely limited, which serves to demonstrate there are some limits to ML.

The problem of discerning whether someone has compromised a model becomes greater for pre-trained models (see `https://towardsdatascience.com/4-pre-trained-cnn-models-to-use-for-computer-vision-with-transfer-learning-885cb1b2dfc` for examples). Pre-trained models are popular because training a model is a time-consuming and sometimes difficult process (**pretrained models** are used in a process called **transfer learning** where knowledge gained solving one problem is used to solve another, similar problem. For example, a model trained to recognize cars can be modified to recognize trucks as well). If you can simply plug a model into your application that someone else has trained, the entire process of creating the application is shorter and easier. However, pre-trained models also aren't under your direct control and you have no idea of precisely how they were created. There is no way to completely validate that the model isn't corrupted in some way. The problem is that datasets are immense and contain varied information. Creating a test harness to measure every possible data permutation and validate it is impossible.

In addition to integrity issues, ML models can also suffer performance and availability issues. For example, greedy algorithms can get stuck in local minima. Crafting data such that it optimizes the use of this condition to cause availability problems could be a form of attack. Because the data would appear correct in every way, data checks are unlikely to locate these sorts of problems. You'd need to use some sort of tuning or optimization to reduce the risk of such an attack. Algorithm choice is important when considering this issue. The easiest way to perpetrate such attacks is to modify the data at the source. However, making the attack successful would require some knowledge of the model, a level of knowledge known as **white box access**.

A major issue that allows for integrity and availability attacks is the assumption on the part of humans (even designers) that ML applications think in the same way as we do, which couldn't be further from the truth. As this book progresses, you will discover that ML isn't anything like a thought process—it's a math process, which means treating adversarial attacks as a math or data problem. Some researchers have suggested including adversarial data in the training data for algorithms so that the algorithm can learn to spot them (**learn**, in this case, is simply a shortcut method of saying that the model has weights and variables adjusted to process the data in a manner that allows for a correct output result). Of course, researchers are looking into this and many other solutions for dealing with adversarial attacks that cause integrity - and performance-type problems. There is currently no silver bullet solution.

## Describing the types of attacks against ML

The introduction to this chapter lists a number of attack types on data (such as data bias) and the application (evasion). The previous section lists some types of attacks perpetrated against the underlying model. As a collection, all of these attacks are listed as **adversarial attacks**, wherein the ML application as a whole tends not to perform as intended and often does something unexpected. This isn't a new phenomenon—some people experimented with adversarial attacks as early as 2004 (see `https://dl.acm.org/doi/10.1145/1014052.1014066` for an article on the issue). However, it has become a problem because ML and deep learning are now deeply embedded within society in such a way that even small problems can lead to major consequences. In fact, sites such as The Daily Swig (`https://portswigger.net/daily-swig/vulnerabilities`) follow these vulnerabilities because there are too many for any single individual to track.

Underlying the success of these attacks is that ML essentially relies on statistics. The transformation of input values to the desired output, such as the categorization of a particular sign as a stop sign, relies on the pixels in a stop sign image relating statistically well enough to the model's trained values to make it into the stop sign category. By adding patches to a stop sign, it no longer matches the learned pattern well enough for the model to classify it as a stop sign. Because of the misclassification, a self-driving car may not stop as required but run right through the stop sign, causing an accident (often to the hacker's delight).

Several elements come into play in this case. A human can look at the sign and see that it's octangular, red, and says Stop, even if someone adds little patches to it. In addition, humans understand the concept of a sign. An ML application receives a picture consisting of pixels. It doesn't understand signs, octangular or otherwise, the color red, or necessarily read the word Stop. All the ML application is able to do is match the object in a picture created with pixels to a particular pattern it has been trained to statistically match. As mentioned earlier in the chapter, machines don't think or feel anything—they perform computations.

Modifying a street sign is an example of an **overt attack**. ML is even more susceptible to overt attacks. For example, the article at `https://arxiv.org/pdf/1801.01944.pdf` explains how to modify a sound file such that it embeds a command in the sound file that the ML application will recognize, but a human can't even hear. The commands could do something innocuous, such as turn the speaker volume up to maximum, but they could perform nefarious tasks as well. Just how terrible the attack becomes depends on the hacker's knowledge of the target and the goal of the attack. Someone's smart speaker could send commands to a voice-activated security system to turn the system off when the owner isn't at home, or perhaps it could trigger an alarm, depending on what the hacker wants (read *Attackers can force Amazon Echos to hack themselves with self-issued commands* at `https://arstechnica.com/information-technology/2022/03/attackers-can-force-amazon-echos-to-hack-themselves-with-self-issued-commands/` to get a better understanding of how any voice-activated device can be hacked).

Attacks can affect any form of ML application. Simply changing the order of words in a text document can cause an ML application to misclassify the text (see the article at `https://arxiv.org/abs/1812.00151`). This sort of attack commonly thwarts the activities of spam and sentiment detectors but could be applied to any sort of textual documentation. Most experts classify this kind of attack as a **paraphrasing attack**. (See the *Developing a simple spam filter example* section of *Chapter 4, Considering the Threat Environment*, for details on working with text.) When you consider how much automated text processing occurs because there is simply too much being generated for humans to handle alone, this kind of attack can take on monumental proportions.

## Considering what ML security can achieve

The essential goal of ML security is to obtain more consistent, reliable, trustworthy, and unbiased results from ML algorithms. Security focuses on creating an environment where the data, algorithm, responses, and analysis all combine to allow ML to produce believable and useful results. The security used with ML applications must perform these tasks in a manner that doesn't slow the application perceptibly or force it to use huge amounts of additional resources. To accomplish these goals, the users of ML applications need to do the following:

- Set understandable and achievable result goals that are verifiable, consistent, and answer specific needs

- Train personnel (which means everyone in the organization, along with consultants and third parties) to interact with the application and its data appropriately

- Ensure that data passes all of the requirements for proper format, lack of missing elements, absence of bias, and lack of various forms of corruption

- Choose algorithms that actually perform tasks in a manner that will match the goals set for the ML application

- Use training techniques that create a reliable model that won't overfit or underfit the data

- Perform testing that validates the data, algorithms, and models used for the ML application

- Verify the resulting application using real-world data that the ML application hasn't seen in the past

Once an ML application meets all of these requirements, it can provide reliable results more quickly and consistently than humans can for mundane, repeatable tasks. Over time, the humans using an ML application should develop the trust required to make using the application worthwhile. In addition, humans can now move on to other areas of interest, making it possible for a single person to accomplish a great deal more than would otherwise be reasonable. Now that you have a good overview of the technical aspects of ML security, it's time to get a development environment together so you can work with the book's code.

## Setting up for the book

I want to ensure that you have the best possible experience when working through the examples in this book. To accomplish that task, this book relies on the literate programming technique originally explored by Donald Knuth and detailed in his paper at `http://www.literateprogramming.com/knuthweb.pdf`. The crux of this approach is that it provides you with a notebook-like environment in which to work where it's possible to freely mix code and non-code elements, including graphics. Because of its reliance on multiple methods of conveying information, this approach is exceptionally clear and easy to understand. Plus, it promotes experimentation at a level that many people don't experience using other approaches.

No matter how inviting a programming environment might be, however, you still have to have a specific level of knowledge to enjoy it. The first section that follows describes what you need to know to use the book successfully. Because of the programming environment I've chosen to use, those requirements may be fewer than expected.

It's also critical that you use the same tools that I used in creating the examples. This requirement isn't meant to hinder you in any way, but to ensure that you don't spend a lot of time overcoming environmental issues while attempting to run the code. The second section that follows describes the programming setup I used so that you can replicate it on your system.

To ensure that you don't have to battle typos and other problems with hand-typed code, I also provide a downloadable source that makes it incredibly easy to work with the programming examples. Most people do benefit from eventually typing their own code and creating their own examples, but to make the learning process easier, you really do want to use the downloadable source if at all possible. The blog post at `http://blog.johnmuellerbooks.com/2014/01/10/verifying-your-hand-typed-code/` provides you with some additional details in this regard. You can obtain the downloadable source code for this book from the publisher's GitHub site at `https://github.com/PacktPublishing/Machine-Learning-Security-Principles` or my website at `http://www.johnmuellerbooks.com/source-code/`.

## What do you need to know?

The main audience for this book is data scientists and, to a lesser extent, researchers, so I'm assuming that you already know something about data sources, data management techniques, and the algorithms used to perform analysis on data. I don't expect you to have an advanced degree in these topics, but you should know that a `.csv` file contains data that is separated in fields using commas. In addition, it would be helpful to have at least a passing knowledge of common algorithms such as Bayes' theorem. The notes and references we provide in the book will help you locate the additional information you need, but this book doesn't provide a tutorial on essential data science topics.

To provide the best possible programming environment, this book also relies on the Python programming language. Again, you won't find a tutorial on this language here, but the use of the literate programming technique should aid in your understanding if you have worked with programming languages in the past. Obviously, the more you know about Python, the less effort you'll need to expend on understanding the code. People who are in management and don't really want to get into the coding details will still find this book useful for the theory it provides, so you could possibly work with the book without knowing anything about Python to obtain theoretical knowledge.

It's also essential that you know how to work with whatever platform you're using. You need to know how to install software, work with the filesystem, and perform other general user tasks with whatever platform you choose to use. Fortunately, you have lots of options for using Jupyter Notebook, the recommended IDE for this book, or Google Colab, a great alternative that will work with your mobile device. However, this extensive list of platforms also means that we can't provide you with much in the way of platform support.

## Considering the programming setup

To get the best results from a book's source code, you need to use the same development products as the book's author. Otherwise, you can't be sure whether an error you find is a bug in the development product or from the source code. The example code in this book is tested using both Jupyter Notebook (for desktop systems) (`https://jupyter.org/`) and Google Colab (for tablet users) (`https://colab.research.google.com/notebooks/welcome.ipynb`). Desktop system users will benefit greatly from using Jupyter Notebook, especially if they have limited access to a broadband connection. Whichever product you use, the code is tested using Python version 3.8.3, although any Python 3.7 or 3.8 version will work fine. Newer versions of Python tend to create problems with libraries used with the example code because the vendors who create the libraries don't necessarily update them at the same speed as Python is updated. You can read about these changes at `https://docs.python.org/3/whatsnew/3.8.html`. You can check your Python version using the following code:

```
import sys
print('Python Version:\n', sys.version)
```

I highly recommend using a multi-product toolkit called Anaconda (`https://www.anaconda.com/products/individual`), which includes Jupyter Notebook and a number of tools, such as `conda`, for installing libraries with fewer headaches. *Figure 1.5* shows some of the tools you get with Anaconda. I wrote the examples using the 2020.07 version of Anaconda, which you can obtain at `https://repo.anaconda.com/archive/`. Make sure you get the right file for your programming platform:

- `Anaconda3-2020.07-Linux-ppc64le.sh` (PowerPC) or `Anaconda3-2020.07-Linux-x86_64.sh` for Linux

- `Anaconda3-2020.07-MacOSX-x86_64.pkg` or `Anaconda3-2020.07-MacOSX-x86_64.sh` for macOS

- `Anaconda3-2020.07-Windows-x86.exe` (32-bit) or `Anaconda3-2020.07-Windows-x86_64.exe` (64-bit) for Windows

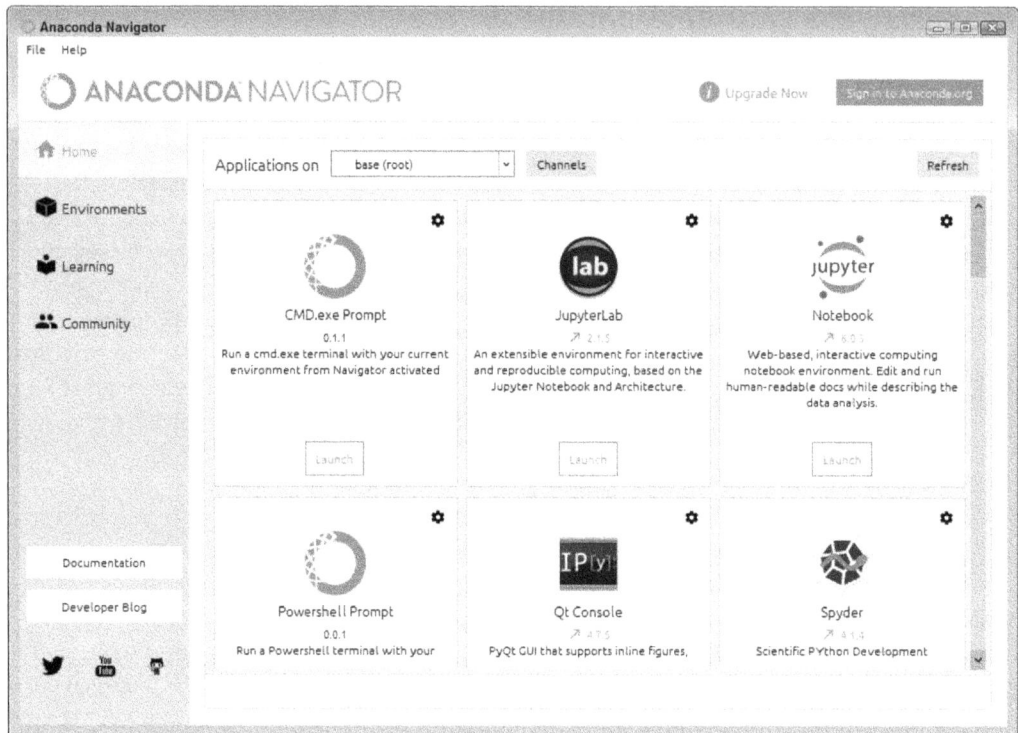

Figure 1.5 – Anaconda provides you with access to a wide variety of tools

It's possible to test your Anaconda version using the following code (which won't work on Google Colab since it doesn't have Anaconda installed):

```
import os
result = os.popen('conda list anaconda$').read()
print('\nAnaconda Version:\n', result)
```

The examples rely on a number of libraries, but three libraries are especially critical. If you don't have the right version installed, the examples won't work:

- NumPy: Version 1.18.5 or greater
- scikit-learn: Version 0.23.1 or greater
- pandas: Version 1.1.3 or greater

Use this code to check your library versions:

```
!pip show numpy
!pip show scikit-learn
!pip show pandas
```

Now that you have a workable development environment, it's time to begin working through some example code in the chapters that follow.

## Summary

This chapter has helped you understand various kinds of ML applications and how those applications are affected by various security threats. It has also emphasized the limitations of ML and pointed out some of the misconceptions that people have about ML – and possibly computers in general. Finally, you have discovered the ways in which humans inadvertently introduce security issues into ML applications by making invalid assumptions and by corrupting data in ways that humans understand, but computers don't.

Knowing about the various forces at work to corrupt your ML model and data may be frightening at first, but there are certain things you can do to mitigate the threat, such as ensuring users are trained not to unintentionally introduce bias into the dataset. ML security measures can help you achieve these goals in an efficient manner. Of course, constant diligence is also a requirement.

The dataset end of things takes focus in the next chapter. It's not just users who can ruin your day by introducing a security problem; using the wrong dataset source or any number of other issues can also be a problem. This next chapter will help you understand these issues so that you can consider the solutions presented in light of your organization's needs.

# 2

# Mitigating Risk at Training by Validating and Maintaining Datasets

The **training process** for your model determines the output that your application provides when faced with data it hasn't seen before. If the model is flawed in any way, then it's not reasonable to expect unflawed output from the model. The **testing process** helps verify the model, but only when the data used for testing is accurate. Consequently, the datasets you use for training and testing your model are critical in a way that no other data you feed to your model is. Even with **feedback** (input that constantly changes the model based on the data it sees), initial training and testing sets the tone for the model and therefore remain critical. Assuming that your dataset is properly vetted, of the right size, and contains the right data, you still have to protect it from a wide variety of threats. This chapter assumes that you've started with a good dataset, but some internal or external entity wants to modify it so that the model you create is flawed in some way (or becomes flawed as time progresses). The flaw need not even be immediately noticeable. In fact, subtle flaws that don't manifest themselves immediately are in the hacker's best interest in most cases. With these issues in mind, this chapter discusses these topics:

- Defining dataset threats
- Detecting dataset modification
- Mitigating dataset corruption

# Technical requirements

This chapter requires that you have access to either Google Colab or Jupyter Notebook to work with the example code. The *Requirements to use this book* section of *Chapter 1, Defining Machine Learning Security*, provides additional details on how to set up and configure your programming environment. The example code will be easier to work with using Jupyter Notebook in this case because you must create local files to use. Using the downloadable source is always highly recommended. You can find the downloadable source on the Packt GitHub site at `https://github.com/PacktPublishing/Machine-Learning-Security-Principles` or my website at `http://www.johnmuellerbooks.com/source-code/`.

# Defining dataset threats

ML depends heavily on clean data. Dataset threats are especially problematic because ML techniques require huge datasets that aren't easily monitored. The following sections help you categorize dataset threats to make them easier to understand.

> **Security and data in ML**
>
> Even though many of the issues addressed in this chapter also apply to data management best practices, they take on special meaning for ML because ML relies on such huge amounts of automatically collected data. Certain entities can easily add, subtract, or modify the data without anyone knowing because it's not possible to check every piece of data or even use automation to verify it with absolute certainty. Consequently, with ML, it's entirely possible to have a security issue and not know about it unless due diligence is exercised to remove as many possible sources of data threats as possible.

## Learning about the kinds of database threats

**Dataset modification** is the act of changing the data in a manner that tends to elicit a particular outcome from the model it creates. The modification need not be noticeable, as in, creating an error or eliminating the data. In fact, the modifications that work best often provide subtle changes to records by increasing or decreasing values by a certain amount or by adding characters that the computer will process but that humans tend to ignore. When a dataset includes links to external resources, such as URLs, the potential for nearly unnoticeable database modification becomes greater. A single-letter difference in a URL can redirect the ML application to a hacker site with all of the wrong data on it. For example, instead of pointing to a site such as `https://www.kaggle.com/datasets`, the external link could instead point to `https://www.kagle.com/datasets`. The problem becomes worse when you consider URL parsing techniques, as described here at *URL parsing: A ticking time bomb of security exploits* (`https://www.techrepublic.com/article/url-parsing-a-ticking-time-bomb-of-security-exploits/`). The point is that the people vetting and monitoring data will likely notice extreme changes and hackers know this, so attacks are more likely to involve subtle modifications to increase the probability that the change will escape notice.

**Dataset corruption** refers to the accidental modification of data in a dataset that tends to produce random or erratic results. It may seem as if corruption would be easy to spot, but in many cases, it too escapes notice. For example, a sensor may experience minor degradation that corrupts the data it provides without eliminating it completely. A visual sensor may have dirt on its lens, a heat sensor may become covered with soot, or sensors may experience data drift as they age. As a trusted data source, the corruption might go unnoticed until the degradation becomes obvious. Environmental issues, such as lightning strikes, heat, static, and moisture, can also corrupt data. The corruption may occur over such a short interval that it lies unnoticed in the middle of the dataset (unless a human checks every entry). Other sources of data corruption include inadequate safeguards in the statistical selection of data to use within a dataset or the lack of available data for a particular group (both of which result in biases). The point is that data corruption comes from such a great number of sources that it may prove impossible to eliminate them all, so constant result monitoring is critical, as is thinking through why a particular result has happened.

> **Entities that pose a threat**
>
> It would be easy to assume that entities that pose a security threat to your data used for ML tasks are human. Yes, humans do pose a major threat, but so do faulty sensors, incompatible or malfunctioning software, acts of nature, cosmic rays, and a large assortment of other sources, most of which developers don't think about because they focus on humans. When working with ML applications, it's essential to think about entities other than humans because so much data used for ML tasks is automatically collected from a wide variety of unmonitored sources. This book uses the term *entity*, rather than human, to help you keep these other sources in mind.

## Considering dataset threat sources

Dataset threats come from multiple sources, but it's helpful to categorize threats as either internal or external in origin. In both cases, dataset security begins with physical security. Even with external sources, using personnel trained to discover potential issues on external sites and scrutinize the incoming data for inconsistencies is important to keeping existing data safe. *Figure 2.1* describes threat sources that you should consider when creating a physically safe environment for your data.

| Threat Type | Threat Source | Description | Potential Countermeasures |
|---|---|---|---|
| Excessive database privileges | Internal | The **principle of least privilege (PoLP)** specifies that an entity should only have the privileges required to perform a given task (see `https://www.cyberark.com/what-is/least-privilege/` for additional details). In most ML situations, users don't require any access to data, applications do. The exception is the data scientists and **database administrators (DBAs)** responsible for maintaining the datasets the applications use. In addition, monitoring users who do have access is essential to help prevent the abuse of privilege. | • Create, maintain, and enforce a strict access and privilege control policy<br><br>• Revoke privileges immediately when no longer needed<br><br>• Deny external access of any sort |
| SQL or other injection (the submission of input that isn't actually data) | Internal or external | Malicious code can appear anywhere. A misconception is that it only appears on websites or as part of forms submitted to servers, but this view is too limited. Any text field in any data source could contain malicious code and it need not form an application. For example, embedding control characters in what appears to be perfectly clean text can still cause problems.<br><br>Note that there is also a NoSQL form of this attack that works very much like the SQL version, as explained in the article What Is NoSQL Injection? (`https://www.imperva.com/learn/application-security/nosql-injection/`). | • Perform monitoring of all incoming data, including data out of range and the presence of unwanted characters<br><br>• Use best practices when working with forms, such as relying on stored procedures (such as database code to vet the input before applying it to a record)<br><br>• Keep application data safe by using the **Model-View-Controller (MVC)** architecture (see `https://www.codecademy.com/articles/mvc` for details) |

| Threat Type | Threat Source | Description | Potential Countermeasures |
|---|---|---|---|
| Weak audit trail | Internal | Auditing every dataset and keeping a record of all the transactions on it helps reduce initial intrusion, keeps any successful intrusion short, and provides forensic data to prevent future intrusions. In addition, databases that support transactions should rely on them to ensure that data corruption is minimal and that a change either fails or succeeds as a whole. | • Rely on properly configured automated data auditing software to keep a record of every transaction. The article *How to Automate Your Network Security Audit in Minutes* (`https://ipfabric.io/blog/how-to-automate-your-network-security-audit-in-minutes/`) provides some additional insights.<br>• Monitor logs for any unexpected access. |
| Dataset backups exposure | Internal or external | Physical security must include all datasets, even backups, which are sometimes left unprotected. A hacker can modify the backup, cause a problem with the current copy of the dataset, and then the administrator performs the work of installing the corrupted backup. | • Use the PoLP to reduce access to backups<br>• Encrypt all datasets<br>• Rely on protected cloud storag<br>• Perform regular audits to compare current and backup datasets |
| Application, library, database, and other liabilities | Internal or external | Every piece of software has a bug or two in it as a minimum. Even when the software works as expected, a great hacker can find a loophole to exploit in the software design. Sometimes, these vulnerabilities aren't obvious, such as issues caused by a misconfiguration or a default configuration. | • Avoid the use of default accounts and sensitive settings<br>• Keep software updated to the latest version that doesn't introduce breaking application changes<br>• Train personnel to think of data safety first |

| Threat Type | Threat Source | Description | Potential Countermeasures |
|---|---|---|---|
| Hacker attacks against infrastructure | External | Many people mistakenly associate hacker attacks against data as a kind of **denial of service (DoS)** attack, but hacker attacks can take many different forms. A hacker could poison software updates or perform attacks indirectly through business partners. Some attacks are quite odd indeed. It may seem unreasonable to consider electrical power as a conduit for an infrastructure attack, but it happens as described at `https://www.eenews.net/stories/1063720933`. In short, you can't exclude any potential source of an attack, not even the electricity powering your data center. | • Harden any data handling method for applications and underlying infrastructure using every means possible<br><br>• Decrease connection times<br><br>• Reduce connection availability<br><br>• Employ a network **Intrusion Detection System (IDS)** |
| Unmanaged data | Internal or external | Organizations generate a wealth of data that gets lost or appears on user systems, rather than the network, where administrators can manage it. Some of this data is sensitive, but even public domain data requires protection lest it becomes corrupted and causes problems. In addition, test data requires the same level of protection as production data does. All of your organization's data requires protection because unprotected data gives a hacker the opportunity to cause damage. | • Disallow localized data storage whenever possible<br><br>• Encrypt all data and treat it as potentially sensitive<br><br>• Perform searches for user-generated and other data on unprotected sources |

Figure 2.1 – Threat sources that affect physical data security

You can spend considerable time physically protecting your clean and correctly formatted data and still experience problems with it. This is where an internal threat or a hacker with direct access to your network comes into play. The data could be correct, viable, unbiased, and stable in every other conceivable way, yet still doesn't produce the correct model or outcome because something has happened to it. This conclusion assumes that you have already eliminated other sources of potential problems, such as using the correct model and shaping the data to meet your needs. When you exhaust every other contingency for incorrect output, the remaining source of potential errors is data modification.

Never rely completely on physical security. Someone will break into your system, even if you create the best safeguards possible. The concept of an unbreakable setup is a myth and anyone who thinks otherwise is set for disappointment. However, great physical security buys you these benefits:

- It will take longer to break into your system

- Any break-ins should be more noticeable

- Hackers are lazy like anyone else and may decide to attack someone else

- Getting rid of the hacker should be easier

- Determining the source and causes of the break-in will require less time

- The application development time tends to be shorter because you have fewer places to look for potential sources of problems

- You have an advantage over your competition because your business secrets will likely remain more secure

- You can demonstrate compliance with legal requirements, such as the **Payment Card Industry Data Security Standard** (**PCI DSS**) and the **Health Insurance Portability and Accountability Act** (**HIPAA**)

- It creates an overall improvement in data management because the results you obtain from the analysis are more precise

It's important to understand that some benefits will require a great deal more work than others to achieve. Depending on your requirements, you likely need to provide a lot more than just physical security. For example, the PCI DSS and HIPAA both require rigorous demonstrations of additional security levels.

## Delving into data change

Anything that modifies existing data without causing missing or invalid values is a data change. Data changes are often subtle and may not affect a dataset as a whole. An accidental data change may include underreporting the value of an element or number of items. Sometimes the change is subjective, such as when one appraiser sees the value of an item in one way and another appraiser comes up with a different figure (see the *Defining the human element* section in *Chapter 1*, *Defining Machine Learning Security*, for other mistruths that can appear in data). It happens that humans make changes to a dataset simply because they disagree with the current value for no other reason than personal opinion. Here are some other sources of data change to consider:

- Automated software makes an unwanted update to a value

- Company policy or procedure changes so that the value that used to be correct is no longer correct

- Aging and archiving software automatically removes values that are deemed too old, even when they aren't

- New sensors report data using a different range, format, or method that creates a data misalignment

- Someone changes the wrong record

Any of these data changes have the potential to skew model results or cause other problems that a hacker can analyze and use to create a security issue. Even if a hacker doesn't make use of them, the fact is that the model is now less effective at performing the task it's designed to do.

## Delving into data corruption

When new or existing data is modified, deleted, or injected in such a manner as to produce an unusable record, the result is data corruption. Data corruption makes it impossible to create a useful model. In fact, some types of data corruption will cause API calls to fail, such as the calculation of statistical methods. When the API call doesn't register an error but instead provides an unusable output, then the model itself becomes corrupted. For example, if the calculation of a mean produces something other than a number, such as an NA, None, or NaN value, then tasks such as missing data replacement fail, but the application may not notice. These types of issues are demonstrated in the examples later in the chapter.

## Uncovering feature manipulation

The act of selecting the right variables to use when creating an ML model is known as **feature engineering**. **Feature manipulation** is the act of reverse engineering the carefully engineered feature set in a manner that allows some type of attack.

In the past, a data scientist would perform feature engineering and modify the feature set as needed to make the model perform better (in addition to meeting requirements such as keeping personal data private). However, today, you can find software to perform the task automatically, as described at `https://towardsdatascience.com/why-automated-feature-engineering-will-change-the-way-you-do-machine-learning-5c15bf188b96`. Whether you create the feature set for an ML model manually or automatically, you still need to assess the feature set from a security perspective before making it permanent. Here are some issues to consider:

- Keep personal data out of the dataset when possible

- Use aggregate values where it's difficult to reconstruct the original value, but the aggregate still provides useful information

- Perform best practices feature reduction studies to determine whether a feature really is needed for a calculation

A problem occurs when a third party uses an API or other means of accessing the underlying data to mount an attack by manipulating the features in various ways (feature manipulation). Reading *Adversarial*

*Attacks on Neural Networks for Graph Data* at `https://arxiv.org/pdf/1805.07984.pdf` provides insights into a very specific illustration of this kind of attack using techniques such as adding or removing fake friendship relations to a social network. The attack might take a long time, but if the attacker can determine patterns of data input that will produce the desired result, then it becomes possible to eventually gain some sort of access, perceive how the data is laid out, extract specific records, or perform other kinds of snooping.

## Examining source modification

Source modification attacks occur when a hacker successfully modifies a data source you rely on for input to your model. It doesn't matter how you use the data, but rather how the attacker modifies the site. The attacker may be looking for access into your network to perform a variety of attacks against the application, modify the data you obtain slightly so that the results you obtain are skewed, or simply look for ways to steal your data or model. As described in the *Thwarting privacy attacks* section, the attacker's sole purpose in modifying the source site might be to add a single known data point in order to mount a membership inference attack later.

## Thwarting privacy attacks

In the book *Nineteen Eighty-Four* by George Orwell, you see the saying, "*War is peace. Freedom is slavery. Ignorance is strength.*" The book as a whole is a discussion of things that can go wrong when a society fails to honor the simple right to privacy. ML has the potential to make the techniques used in *Nineteen Eighty-Four* look simplistic and benign because now it's possible to delve into every aspect of a person's life without the person even knowing about it. The use of evasion and poisoning attacks to cause an ML application to output incorrect results is one level of attack—the use of various methods to view the underlying data is another. This second form of attack is a privacy attack because the underlying data often contains information of a personal nature.

It isn't hard to make a case that your health and financial data should remain private, and there are laws in place to ensure privacy (although some people would say they're inadequate). However, your buying habits on Amazon don't receive equal protection, even though they should. For example, by watching what you buy, someone can make all sorts of inferences about you, such as whether you have children, how many children, and what their ages are. However, there are more direct attacks as listed here:

- **Membership inference attack**: The attacker has at least one known good data point in hand and uses it to determine whether the data point was part of the original data used to train a model. For example, is a particular person's face used to train a facial recognition application? Knowing this information could make it possible to avoid detection at an airport by ensuring none of the people making an attack are already in the database. This kind of attack was used to obtain sensitive information about people on both Google and Amazon (you can read more about this attack at `https://arxiv.org/abs/1610.05820`). Here are some other uses of membership inference attacks:

- **Generative adversarial networks (GANs)**: Attackers were 100 percent successful at carrying out white-box attacks and 80 percent successful at carrying out black-box attacks as described at `https://arxiv.org/pdf/1705.07663.pdf`. In a **white-box attack**, the attacker has access to the model's parameters, which means that the attacker could be an insider or a hacker with inside information. **Black-box attacks** are made without model parameter knowledge. The attacker must create a different model or not rely on a model at all to generate adversarial images that will hopefully transfer to the target model. It may initially appear that white-box attacks would be advantageous, but according to *White-box vs Black-box: Bayes Optimal Strategies for Membership Inference* (`http://proceedings.mlr.press/v97/sablayrolles19a/sablayrolles19a.pdf`), the two methods can have an equal chance of success.

- **Language generation models**: The attackers were able to determine whether the person's text data was part of a language generation study, which is used to create models used for products such as Alexa and Siri (among many others) as described at `https://arxiv.org/pdf/1811.00513.pdf`.

- **Federated ML system**: A study shows that insider actors within a federated ML system, where the system is centralized, pose a significant threat, and that working with datasets that differ greatly increases the threat as described at `https://arxiv.org/pdf/1807.09173.pdf`.

- **Aggregate location data**: It's possible to determine whether a particular user is part of aggregate location data, which is used to support smart services and applications, generate traffic maps, and predict visits to businesses as described at `https://arxiv.org/abs/1708.06145`.

- **Data extraction**: This kind of attack also comes under the heading of model inversion. In this case, the attacker tries to obtain an average representation of each class used to train a model. While it isn't possible to extract a single data point using this method, the results are still pretty scary as described by these examples:

- **Genomic information**: It's possible to obtain information about a person's genes by studying pharmaceutical data as described at `https://www.usenix.org/system/files/conference/usenixsecurity14/sec14-paper-fredrikson-privacy.pdf`.

- **Facial recognition**: Even though the resulting face isn't perfect, it's visible enough for a human to identify the individual used to train the model as described at `https://www.cs.cmu.edu/~mfredrik/papers/fjr2015ccs.pdf`.

- **Unintended memorization**: Ensuring that you clean data thoroughly of any personal information and that you use as few features as possible to achieve your goals is demonstrated in the paper at `https://arxiv.org/pdf/1802.08232.pdf` where the authors were able to extract social security and credit card numbers from the model.

- **Model extraction**: Creating a model is a time-intensive task, requiring the input of more than a few experts in most cases. Consequently, you don't want someone to come along and steal the model that took you 6 months to build. There are a number of ways to accomplish this goal, as described in *Chapter 1, Defining Machine Learning Security*, but one approach is particularly effective, prediction APIs, which are described at `https://arxiv.org/abs/1609.02943`.

One of the most important bits of information you can take away from this section is that you really do need to limit the feature size of your dataset and exclude any form of personal information whenever possible. The act of anonymizing the data is as essential as every other aspect of molding it to your needs. When you can't anonymize the data, as when working with medical information associated with particular people for tasks such as predicting medication doses, then you need to apply encryption or tokenization to the personal data. Encryption works best when you absolutely must be able to read the personal information later, and you want to ensure that only people who actually need to see the data have the right to decrypt it. Tokenization, the process of replacing sensitive information with symbolic information or identification symbols that can retain the essentials of the original data, works best when there is a tokenized system already in place to identify individuals in a non-personal way.

The next section of the chapter looks at dataset modification, which is the act of changing the data to obtain a particular effect.

# Detecting dataset modification

Dataset modification implies that an external source, hacker, disgruntled employee, or other entity has purposely changed one or more records in the dataset for some reason. The source and reason for the data modification are less important than the effects the modification has on any analysis you perform. Yes, you eventually need to locate the source and use the reason as a means to keep the modification from occurring in the future, but the first priority is to detect the modification in the first place. Consider this sequence of events:

1.  Hackers want to create an environment where products from Organization A, a competitor of Organization B, receive better placement on a sales site because the competitor is paying them to do so

2.  The hackers discover that buyer product reviews and their product ratings are directly associated with the site's ranking mechanism

3.  The hackers employ **zombie systems** (computers they have taken over) to upload copious reviews to the site giving Organization B's products a one-star review

4.  The site's ML application begins to bring down the product rankings for Organization B and the competitor begins to make a ton of money

If there were some system in place to detect the zombie system attack, the ML application could compensate, provide notice to an administrator, or react in other ways. *Chapter 3, Mitigating Inference Risk by Avoiding Adversarial Machine Learning Attacks*, talks about how to work with models to make this attack less effective, but the data is the first consideration. Given that you can't guarantee the physical security of your data, you need other means to reduce the risks of dataset modification. Constantly monitoring your network, data storage, and data does provide some useful results, but still doesn't ensure complete data security. Some of the issues listed in *Figure 2.1*, such as the security of the application, library, or database, along with other vulnerabilities, are nearly invisible to monitoring. Consequently, other methods of dataset modification detection are required.

Two reliable methods of dataset modification detection are traditional methods that rely on immutable calculated values such as **hashes** and data version control systems, such as DVC (`https://dvc.org/`). Both approaches have their adherents.

> **Blockchains**
>
> Theoretically, you could rely on blockchains, which are a type of digital ledger ensuring uniqueness, but only in extreme cases. The article *Blockchain Explained* at `https://www.investopedia.com/terms/b/blockchain.asp`, provides additional details.

Combining both hashes and data version control approaches may seem like overkill, but one approach tends to reinforce the other and act as a crosscheck. The disadvantages of using both are that you expend additional time, and doing so increases cost, especially if the data version control system is a paid service.

## An example of relying on traditional methods

Most traditional methods of data modification detection revolve around using hashes to calculate the value of each file in the dataset. In addition, you may find cryptographic techniques employed. The hash data used to implement a traditional method must appear as part of secure storage on the system handling the data or a hacker could simply change the underlying values. Using a traditional method of data modification has some significant advantages over some other methods, such as a data version control system. These advantages include the following:

- Data scientists, DBAs, and developers understand the underlying methodologies

- The cost of implementing this kind of solution is usually low

- Because people understand the methods so well, this kind of system is usually robust and reliable

Traditional methods normally work best for smaller setups where the number of individuals managing the data is limited. There are also disadvantages to this kind of system as summarized here:

- The system can be hard to implement when the data sources are distributed

- Hashing a large number of files could prove time-prohibitive

- Checking the hash each time a file is checked out, recalculating the hash, and then updating secure storage is error-prone

- A data version control system may prove more flexible and easier to use

You can use the code shown in the following code block (also found in the MLSec; 02; Create Hash.ipynb file for this chapter) to create a hash of an existing file:

1. Begin by importing the libraries:

```
from hashlib import md5, sha1
from os import path

inputFile = "test_hash.csv"
hashFile = "hashes.txt"
```

2. Obtain the file hashes:

```
openedInput = open(inputFile, 'r', encoding='utf-8')
readFile = openedInput.read()

md5Hash = md5(readFile.encode())
md5Hashed = md5Hash.hexdigest()

sha1Hash = sha1(readFile.encode())
sha1Hashed = sha1Hash.hexdigest()

openedInput.close()
```

3. Open the saved values, when they exist:

```
saveHash = True
if path.exists(hashFile):
```

4. Get the hash values:

```
    openedHash = open(hashFile, 'r', encoding='utf-8')
    read_md5Hash = openedHash.readline().rstrip()
    read_sha1Hash = openedHash.readline()
```

5. Compare them to the current hash:

```
    if (md5Hashed == read_md5Hash) and \
    (sha1Hashed == read_sha1Hash):
        print("The file hasn't been modified.")
```

```
        else:
            print("Someone has changed the file.")
            print("Original md5: %r\n\tNew: %r" % \
                (read_md5Hash, md5Hashed))
            print("Original sha1: %r\n\tNew: %r" % \
                (read_sha1Hash, sha1Hashed))

            saveHash = False

        openedHash.close()

    if saveHash:
        ## Output the current hash values
        print("File Name: %s" % inputFile)
        print("MD5: %r" % md5Hashed)
        print("SHA1: %r" % sha1Hashed)

        ## Save the current values to the hash file.
        openedHash = open(hashFile, 'w')
        openedHash.write(md5Hashed)
        openedHash.write('\n')
        openedHash.write(sha1Hashed)
        openedHash.close()
```

This example begins by opening the data file. Make sure you open the file only for reading and that you specify the type of encoding used. The file could contain anything. This .csv file contains a simple series of numbers such as those shown here:

```
1, 2, 3, 4, 5
6, 7, 8, 9, 10
```

It's important to call encode() as part of performing the hash because you get an error message otherwise. The md5Hash and sha1Hash variables contain a hash type as described at https://docs.python.org/3/library/hashlib.html. What you need is a text rendition of the hash, which is why the code calls hexdigest(). After obtaining the current hash, the code closes the input file.

The hash values appear in `hashes.txt`. If this is the first time you have run the application, you won't have a hash for the file, so the code skips the comparison check, displays the new hash values, and saves them to disk. Therefore, you see output such as this:

```
File Name: test_hash.csv
MD5: '182f800102c9d3cea2f95d370b023a12'
SHA1: '845d2f247cdbb77e859e372c99241530898ec7cb'
```

When there is a `hashes.txt` file to check, the code opens the file, reads in the hash values, which appear on separate lines, and places them in the appropriate variables. These values are already strings, but notice you must remove the newline character from the first string by calling `rstrip()`. Otherwise, the current hash value won't compare to the saved hash value. During the second run of the application, you see the same output as the first time with "The file hasn't been modified." as the first line.

Now, try to modify just one value in the `test_hash.csv` file. Run the code again and you instantly see that this simple-looking method actually does detect the change (these are typical results, and your precise output may vary):

```
Someone has changed the file.
Original md5: '182f800102c9d3cea2f95d370b023a12'
        New: 'fae92acdd056dfd3c2383982657e7c8f'
Original sha1: '845d2f247cdbb77e859e372c99241530898ec7cb'
        New: '677f4c2cfcc87c55f0575f734ad1ffb1e97de415'
```

Changing the original file back will restore the original "The file hasn't been modified." output. Consequently, once you have vetted and verified your data source file, you can use this technique to ensure that even a restored copy of the file is correct. The biggest issue with this approach is that you must ensure the integrity of hash storage or any comparison you make will be problematic.

## Working with hashes and larger files

When working with large data files, you can't read the entire file into memory at once. Doing so would cause the application to crash due to a lack of memory. Consequently, you create a loop where you read the data in blocks of a certain size, such as 64 KB. The loop continues to run until a test of the input variable shows there is no more data to process. Most developers use a break statement to break out of the loop at this point.

To create the hash, you must process each block separately by calling the update () function, rather than using the constructor as shown in the example in the previous section. The result is that the hash changes during each loop until you obtain a final value. As with the example code in this chapter, you can then use the hexdigest () function to retrieve the file hash value as a string. Here's a quick overview of the loop using 64-KB chunks (this code isn't meant to be run, and simply shows the technique):

```
chunksize = 65536
md5Hash = hashlib.md5()
with open(filename, 'rb') as hashFile:
    while chunk := hashFile.read(chunksize):
        md5Hash.update(chunk)
return md5Hash.hexdigest()
```

The hash obtained using a single call to the constructor is the same as the hash obtained using update () as long as you process the entire file in both cases. Consequently, modifying your code to handle larger files when it becomes necessary shouldn't change the stored hash values.

## Using a data version control system example

This section discusses data version control. Application code version control is another matter because it involves working with different versions of an application. Unfortunately, many people focus on application code because that's something they're personally involved in writing, and spend less time with their data. The problem with this approach is that you can suddenly find that your application code works perfectly, but produces incorrect output because the data has been altered in some way.

Data version control creates a new version of the saved document every time someone makes a change. A full-fledged **database management system (DBMS)** provides this sort of support through various means, but if you're working with .csv files, you need another solution. Using a data version control system ensures the following:

- It's possible to reverse unwanted changes, no matter what the source might be

- Multiple data sources remain coordinated in the version of the files that represent a data transaction

- Previous versions of the data remain secure, which isn't always possible when working with backups

- Handling multiple users doesn't present a problem

- Importing the data into your application development environment is relatively easy or perhaps even automatic

When working with data, it pays to know how the data storage you're using maintains versions of those files. For example, when working with Windows, you get an automatic version save as part of a restore point or a backup as described at https://hls.harvard.edu/dept/its/restoring-

`previous-versions-of-files-and-folders/`. However, these versions won't let you return to the version of the file you had 5 minutes ago. Online storage has limits as well. If you store your data on Dropbox, you only get 180 days for a particular version of a file (see `https://help.dropbox.com/files-folders/restore-delete/version-history-overview` for details). The problem with most of these solutions is that they don't really provide versioning. If you make a change one minute, save it, and then decide you want the previous version the next minute, you can't do it.

- Fortunately, there are solutions for data version control out there, such as Data Version Control at `https://dvc.org/`. Most of these solutions rely on some type of GitHub (`https://github.com/`) setup, which is true of Data Version Control (see `https://dvc.org/doc/start` for details). You can also find do-it-yourself solutions such as the one at `https://medium.com/pytorch/how-to-iterate-faster-in-machine-learning-by-versioning-data-and-models-featuring-detectron2-4fd2f9338df5`. The good thing about some of these home-built solutions is that they can work with any frameworks you currently rely on, such as Docker. Here are some other data version control setups you might want to consider:

- **Delta Lake** (`https://delta.io/`): Creates an environment where it's easy to see the purpose behind various changes. It supports **atomicity, consistency, isolation, and durability** (**ACID**) transactions (where changes to the dataset are strictly controlled, as described at `https://blog.yugabyte.com/a-primer-on-acid-transactions/`), scalable metadata handling, and unified streaming and batch data processing.

- **Dolt** (`https://github.com/dolthub/dolt`): Relies on a specialized SQL database application to fork, clone, branch, merge, push, and pull file versions, just as you would when working with GitHub. To use this solution, you need a copy of MySQL, which does have the advantage of allowing you to locate the repository anywhere you want.

- **Git Large File Storage** (**LFS**) (`https://git-lfs.github.com/`): Defines a way to use GitHub to interact with really large files such as audio or video files. The software replaces the file with a text pointer to a location on a remote server where the actual file is stored. Each version of a file receives a different text marker, so it's possible to restore earlier versions as needed.

- **lakeFS** (`https://lakefs.io/`): Works with either the **Amazon Web Services** (**AWS**) S3 or **Google Cloud Storage** (**GCS**) service to provide GitHub-like functionality with petabytes of data. This software is ACID-compliant and allows easy rollbacks of transactions as needed, which is a plus when working with immense datasets.

- **Neptune** (`https://neptune.ai/`): Provides a good system for situations that require a lot of experimentation. It works with scripts (Python, R, or other languages) and notebooks (local, Google Colab, or AWS SageMaker), and performs these tasks using any infrastructure (cloud, laptop, or cluster).

- **Pachyderm** (`https://www.pachyderm.com/`): Focuses on the ML application life cycle. You can send data in a continuous stream to the main repository or create as many branches as needed for experimentation.

Version control is an important element of keeping your data safe because it provides a fallback solution for when data changes occur. The next section of the chapter starts to look at an issue that isn't so easily mitigated – data corruption.

# Mitigating dataset corruption

Dataset corruption is different from dataset modification because it usually infers some type of accidental modification that could be relatively easy to spot, such as values out of range or missing altogether. The results of the corruption could appear random or erratic. In many cases, assuming the corruption isn't widespread, it's possible to fix the dataset and restore it to use. However, some datasets are fragile (especially those developed from multiple incompatible sources), so you might have to recreate them from scratch. No matter the source or extent of the data corruption, a dataset that suffers from corruption does have these issues:

- The data is inherently less reliable because you can't ensure absolute parity with the original data.

- Any model you create from the data may not precisely match the model created with the original data.

- Hackers or disgruntled employees may purposely corrupt a dataset to keep specific records out of play, so you must eliminate human sources as the cause of the corruption.

- The use of data input automation and techniques such as **optical character recognition** (**OCR**) can corrupt data in a non-repeatable way that's difficult to track down and even more difficult to fix.

- Eliminating the source of any accidental corruption is essential, especially when the corruption source is a sensor or other type of dynamic data input. Knowing the precise source and reason behind sensor or other dynamic data input corruption can also help mitigate the corruption but can prove time-consuming to locate.

- Anyone relying on the corrupted dataset is less likely to believe future results from it, which means additional crosschecks. Unreliable results have a significant effect on human users of the underlying dataset.

- Third parties that contribute to a dataset may not want to admit to the corruption or may lack the resources to fix it. If the dataset contains some standardized form of data, modifying the data on your own means that the dataset will be out of sync with others using it.

All of these issues create an environment where the data isn't trustworthy and you find that the model doesn't behave in the predicted manner. More importantly, there is a human factor involved that makes it difficult or impossible to locate a precise source of corruption.

# The human factor in missingness

Before moving forward to actually fixing the dataset corruption, it's important to consider another source – humans. Lightning strikes, natural disasters, errant sensors, and other causes of data missingness have potential fixes that are possible to quantify. To overcome lightning, for example, you ensure that you isolate your data center from potential sources of lightning. However, humans cause significantly more damage to datasets by failing to create complete records or entering the data incorrectly. Yes, you can include extensive data checks before the application accepts a new record, but it's amazing how proficient humans become at overcoming them in order to save a few moments of time in creating the record correctly.

The inability of application code to overcome human inventiveness is the reason that data checks alone won't solve the problem. Using good application design can help reduce the problem by reducing the choices humans have when entering the data. For example, it's possible to use checkboxes, option boxes, drop-down lists, and so on, rather than text boxes. Creating forms to accept data logically also helps. A study of workflows (the processes a person naturally uses to accomplish a task) shows that you can reduce errors by ensuring the forms request data precisely when the human entering the data is in the position to offer it.

Automation offers another solution. Using sensors and other methods of detection allows the entry of data into a form without asking for it from a person at all. The human merely verifies that the entry is correct. The ML technology to guess the next word you need to type into a text box, such as typeahead, can also reduce errors. Anything you can automate in the data entry form will ultimately reduce problems in the dataset as a whole.

The one method that seems to elude most people who work with data, however, is the reduction of features so that a form requires less data in the first place. Many people designing a dataset ask whether it might be helpful to have a certain feature in the future, rather than what is needed in the dataset today. Over-engineering a dataset is an invitation to introduce unnecessary and preventable errors. When creating a dataset, consider the human who will participate in the data entry process and you'll reduce the potential for data problems.

# An example of recreating the dataset

Missing data can take all sorts of forms. You could see a blank string where there should be text, as an example. Dates could show up as a default value, rather than an actual date, or they could appear in the wrong format, such as MM/DD/YYYY instead of YYYY/MM/DD. Numbers can present the biggest problem, however, because they can take on so many different forms. The first check you then need to make is to detect any actual missing values. You can use the following code (also found in the `MLSec; 02; Missing Data.ipynb` file for this chapter) to discover missing numeric values:

```
import pandas as pd
import numpy as np
```

```
s = pd.Series([1, 2, 3, np.NaN, 5, 6, None, np.inf, -np.inf])

## print(s.isnull())
print(s.isin([np.NaN, None, np.inf, -np.inf]))

print()
print(s[s.isin([np.NaN, None, np.inf, -np.inf])])
```

This simple data series contains four missing values: np.NaN, None, np.inf, and –np.inf. All four of these values will cause problems when you try to process the dataset. The output shows that Python easily detects this form of missingness:

```
0      False
1      False
2      False
3       True
4      False
5      False
6       True
7       True
8       True
dtype: bool

3       NaN
6       NaN
7       inf
8      -inf
dtype: float64
```

The type of missing values you see can provide clues as to the cause. For example, a disconnected sensor will often provide an np.inf or -np.inf value, while a malfunctioning sensor might output a value of None instead. The difference is that in the first case, you reconnect the sensor, while in the second case, you replace it.

Once you know that a dataset contains missing data, you must decide how to correct the problem. The first step is to solve the problems caused by np.inf and –np.inf values. Run this code:

```
replace = s.replace([np.inf, -np.inf], np.NaN)
print(s.mean())
print(replace.mean())
```

Then, the output tells you that np.inf and –np.inf values interfere with data replacement techniques that rely on a statistical measure to correct the data, as shown here:

```
nan
3.4
```

The first value shows that the np.inf and –np.inf values produce a nan output when obtaining a mean to use as a replacement value. Using the updated dataset, you can now replace the missing values using this code:

```
replace = replace.fillna(replace.mean())
print(replace)
```

The output shows that every entry now has a legitimate value, even if that value is calculated:

```
0       1.0
1       2.0
2       3.0
3       3.4
4       5.0
5       6.0
6       3.4
7       3.4
8       3.4
dtype: float64
```

Sometimes, replacing the data values will still cause problems in your model. In this case, you want to drop the errant values from the dataset using code such as this:

```
dropped = s.replace([np.inf, -np.inf], np.nan).dropna()
print(dropped)
```

This approach has the advantage of ensuring that all of the data you do have is legitimate data and that the amount of code required is smaller. However, the results could still show skewing and now you have less data in your dataset, which can reduce the effectiveness of some algorithms. Here's the output from this code:

```
0      1.0
1      2.0
2      3.0
4      5.0
5      6.0
dtype: float64
```

If this were a real dataset with thousands of records, you'd see a 37.5 percent data loss, which would prove unacceptable in most cases. The dataset would be unusable in this situation and most organizations would do everything possible to keep the failure from being known (although, you can find a few articles online that hint at it, such as *The History of Data Breaches* at https://digitalguardian.com/blog/history-data-breaches). You have these alternatives when you need a larger dataset without replaced values:

- Reconstruct the dataset if you suspect that the original has fewer missing or corrupted entries.

- Collect additional data after correcting any issues that caused the missing or corrupted data in the first place.

- Obtain similar data from other datasets and combine it (after ensuring the datasets are compatible) with the current dataset after conditioning.

## Using an imputer

Another method for handling missing data is to rely on an **imputer**, a technique that can actually replace missing values with their true values when you can provide a statistical basis for doing so. The problem is that you need to know a lot about the dataset to use this approach. Here is an example of how you might replace the np.NaN, None, np.inf, and −np.inf values in a dataset with something other than a mean:

```
import pandas as pd
import numpy as np
from sklearn.impute import SimpleImputer

s = pd.Series([1, 2, 3, np.NaN, 5, 6, None, np.inf,
    -np.inf])
s = s.replace([np.inf, -np.inf], np.NaN)
```

```
imp = SimpleImputer(missing_values=np.NaN, strategy='mean')

imp.fit([[1, 2, 3, 4, 5, 6, 7, 8, 9]])

s = pd.Series(imp.transform([s]).tolist()[0])

print(s)
```

The values in s are the same as those shown in the *An example of recreating the dataset* section. Given that this technique only works with nan values, you must also call on replace() to get rid of any np.inf or -np.inf values. The call to the SimpleImputer() constructor defines how to perform the impute on the missing data. You then provide statistics for performing the replacement using the fit() method. The final step is to transform the dataset containing missing values into a dataset that has all of its values intact. You can discover more about using SimpleInputer at https://scikit-learn.org/stable/modules/generated/sklearn.impute.SimpleImputer.html. Here is the output from this example:

```
0       1.0
1       2.0
2       3.0
3       4.0
4       5.0
5       6.0
6       7.0
7       8.0
8       9.0
dtype: float64
```

The output shows that the imputer does a good job of inferring the missing values by using the values that are present as a starting point. Even if you don't see high-quality results such as this every time, an imputer can make a significant difference when recovering damaged data.

## Handling missing or corrupted data

Even though this book typically uses single-file datasets and experiments online sometimes use the same approach, a single-file dataset is more the exception than the rule. The most common reason why you might use a single file is that the source data is actually feature-limited (while the number of records remains high) or you want to simplify the problem so that any experimental results are easier to quantify. However, it doesn't matter whether your dataset has a single file or more than one file associated with it – sometimes, you can't fix missingness or corruption through statistical means, data dropping, or the use of an imputer. In this case, you must recreate the dataset.

When working with a toy dataset of the sort used for experimentation, you can simply download a new copy of the dataset from the originator's website. However, this process isn't as straightforward as you might think. You want to obtain the same results as before, so you need to use caution when getting a new copy. Here are some issues to consider when recreating a dataset for experimentation:

- Ensure the dataset you download is the same version as the one that became unusable.
- Verify that the dataset you've downloaded hasn't been corrupted by comparing a hash you create against the hash listed online (when available).
- Include any modifications you made to the dataset.

Any local dataset you create should have backups, possibly employ a transactional setup with logs, and rely on some sort of versioning system (see the *Using a data version control system* section for details). With these safeguards in place, you should be able to restore a dataset to a known good state if you detect the source of missingness or corruption early enough. Unfortunately, it's often the case that the problem isn't noticed soon enough.

When working with sensor-based data, you can attempt to recreate the dataset by recreating the conditions under which the sensor logs were originally created and simply record new data. The recreated dataset will have statistical differences from the original dataset, but within a reasonable margin of error (assuming that the conditions you create are precisely the same as before). If this sounds like a less-than-ideal set of conditions, recreating datasets is often an inexact business, which is why you want to avoid data problems in the first place.

A dataset that includes multiple files requires special handling. If you're using a DBMS, the DBMS software normally includes methods for recovering a dataset based on backups and transactional logs. Because each new entry into the database is part of a transaction (or it should be), you may be able to use the transaction logs to your benefit and ensure the database remains in a consistent state. Some database recovery tools will actually create the database from scratch using scripts or other methods and then add the data back in as the tool verifies the data.

Your dataset may not appear as part of a DBMS, which means that you have multiple files organized loosely. The following steps will help you recreate such a dataset:

1. Create a new folder to hold the verified dataset files.
2. Verify each file in turn using an approach such as a hash check.
3. Copy each verified file to the new folder.
4. Attempt to obtain a new copy of any damaged files from the data version control system when available. Copy the downloaded files to the new folder.

5.  Use statistical or imputer methods to fix any remaining damaged files. Copy these files to the new folder.

6.  Check the number of files in the new folder against the number of files in the old folder to ensure they match.

7.  Perform a test on the files to determine whether they can produce a desirable result from your ML application.

These steps may not always provide you with a perfect recovery, but you can get close enough to the original data that the model should work as it did before within an acceptable statistical range.

## Summary

This chapter has described the importance of having good data to ensure the security of ML applications. Often, the damage caused by modified or corrupted data is subtle and no one will actually notice it until it's too late: an analysis is incorrect, a faulty recommendation causes financial harm, an assembly line may not operate correctly, a classification may fail to point out that a patient has cancer, or any number of other issues may occur. The focus of most data damage is causing the model to behave in a manner other than needed. The techniques in this chapter wil help you avoid – but not necessarily always prevent – data modification or corruption.

The hardest types of modification and corruption to detect and mitigate are those created by humans in most cases, which is why the human factor receives special treatment in this chapter. Modifications that are automated in nature have a recognizable pattern and most environmental causes are rare enough that you don't need to worry about them constantly, but human modifications and corruption are ongoing, unique, and random in nature. No matter the source, you discovered how to detect certain types of data modification and corruption, all with the goal of creating a more secure ML application environment.

*Chapter 3, Mitigating Inference Risk by Avoiding Adversarial Machine Learning Attacks*, will take another step on the path toward understanding ML threats. In this case, the chapter considers the threats presented to the model itself and any associated software that drives the model or relies on model output. The point is that the system is under attack and that the causes are no longer accidental.

# Mitigating Inference Risk by Avoiding Adversarial Machine Learning Attacks

Many adversarial attacks don't occur directly through data, as described in *Chapter 2*. Instead, they rely on attacking the **machine learning** (**ML**) algorithms or, more often than not, the resulting models. Such an attack is termed **adversarial ML** because it relies on someone purposely attacking the software. In other words, unlike data attacks where accidental damage, inappropriate selection of models or algorithms, or human mistakes come into play, this form of adversarial attack is all about someone purposely causing damage to achieve some goal.

Attacking an ML algorithm or model is meant to elicit a particular result. The result isn't always achieved, but there is a specific goal in mind. As researchers and hackers continue to experiment with ways to fool ML algorithms and obtain a particular result, the potential for serious consequences becomes greater. Fortunately, the attempts to overcome the positive results of a model require trial and error, which means that there are techniques that you can use to keep a hacker at bay until researchers in your organization can create an adequate defense. At some point, securing your ML algorithms becomes a race between the hackers seeking to circumvent and pervert the usefulness of the model and the researchers seeking to protect it. With these issues in mind, this chapter will discuss these topics:

- Defining adversarial ML
- Considering security issues in ML algorithms
- Describing the most common attack techniques
- Mitigating threats to the algorithm

> **Dealing with attack information overload**
>
> This chapter contains a huge amount of information in a very short space. The goal of this chapter is to expose you to as many different kinds of attacks as possible to help you think outside the boxes that many articles online create. The bottom line is that hackers are extremely creative and you need to look everywhere and in every way for attacks. In many ways, reading this chapter end to end could result in information overload, so selecting sections of interest at any particular time and focusing on that kind of attack will make the material seem a little less daunting. If this chapter provided any fewer attack vectors, it wouldn't help you see the bigger picture of just what hackers are like.

# Defining adversarial ML

An **adversary** is someone who opposes someone else. It's an apt term for defining adversarial ML because one group is opposing another group. In some cases, the opposing group is trying to be helpful, such as when researchers discover a potential security hole in an ML model and then work to solve it. However, most adversaries in ML have goals other than being helpful. In all cases, adversarial ML consists of using a particular attack vector to achieve goals defined by the attacker's mindset. The following sections will help you understand the dynamics of adversarial ML and what it presents, such as the huge potential for damaging your application.

> **Wearing many hats**
>
> All hackers deal with working with code at a lower level than most developers do, some at a very low level. However, there are multiple kinds of hackers and you can tell who they are by the hat they wear. Most people know that **white hat hackers** are the good guys who look for vulnerabilities (with permission) and tell people how to fix them. **Black hat hackers** illegally look for vulnerabilities to exploit to make people's lives miserable. **Gray hat hackers** are people who look for vulnerabilities without malicious intent but may use illegal means to find them and usually work without permission. **Green hat hackers** are new to the trade and often do more damage without knowing what they're doing than they would do if they did know what they were doing. **Blue hat hackers** take down other hackers for revenge. Sometimes, blue hat hackers take revenge on non-hackers too. Finally (yes, there is an end to this list), **red hat hackers** take down black hat hackers using means both legal and illegal without contacting the authorities. You can read more about the hats hackers wear at `https://www.techtarget.com/searchsecurity/answer/What-is-red-and-white-hat-hacking`.

## Categorizing the attack vectors

It's possible to categorize the various kinds of attacks that you might see against your ML model. By knowing the kind of attack, you can often create a strategy to protect against it. Most hackers use multiple attack vectors in tandem to achieve several advantages:

- Security experts become confused as to which attacks are currently in use

- It's possible to hide the real attack under layers of feints

- The probability of success increases

Because of the methods used to attack your model, you need to employ an equal number of detection methods and then have a plan in place for mitigating the attack. At the time of writing, the detection part is difficult because so much research is needed to know how some attacks work. Mitigating an attack is even harder and there are some instances described later in this chapter where you may not be able to respond adequately in an automated manner, but will instead need to rely on specially trained humans to spot the threat and stop it using traditional methods, such as blocking particular IP addresses.

## Examining the hacker mindset

A **mindset** is a set of beliefs that shape how a person views the world and makes sense of it. Given that most people need a reason to do something, even hackers of all types, consider these reasons hackers employ adversarial ML:

- To obtain money or power

- To take revenge on another party

- Because they need or want attention

- Because there is a misunderstanding as to the purpose of the application

- To make a political statement or create distrust

- Because there is a disagreement over how to accomplish a task

There are probably other reasons that hackers want to modify or destroy an ML model using the methods described in the *Describing the most common attack techniques* section (such as sending bad data or embedding scripts), but this list contains all the most common reasons. Knowing the motivations of your attacker can help you in your mitigation efforts. For example, people wanting to make a political statement are less likely to take your application down than those who are trying to obtain money or power. Consequently, the form of attack will differ and you'll have different avenues of investigation to pursue. The first group is more likely to use a poisoning attack to modify the results you achieve from your analysis, while the second group is more likely to use an evasion attack to get past your defenses.

Being aware of the demographics of your attacker has benefits as well. For example, you might be able to ascertain the level of sophistication for the attacks or the number of resources at the attacker's disposal. Anything you can discover about the attacker gives you an advantage in disabling the attacker. The point is that adversarial ML is all about who controls your model and what they use it to do. Now that you have a better idea of what adversarial ML is, the next section will discuss the security issues in algorithms that allow access to an attacker.

# Considering security issues in ML algorithms

Someone is going to break into your ML application, even if you keep it behind firewalls on a local network. The following sections will help you understand the security issues that lead to breaches when using adversarial ML techniques.

---

**Considering the necessity for investment and change**

Because of the time and resource investment in ML models, organizations are often less than thrilled about having to incorporate new research into the model. However, as with any other software, updates of ML models and the underlying libraries represent an organization's investment in the constant war with hackers. In addition, an organization needs to remain aware of the latest threats and modify models to combat them. All these requirements may mean that your application never feels quite finished – you may just complete one update, only to have to start on another.

---

## Defining attacker motivations

An organization can use any number of technologies to help keep outsider attacks under control; insider attacks are more difficult because the same people who will attack the system also need access to it to perform their work. The infographic at `https://digitalguardian.com/blog/insiders-vs-outsiders-whats-greater-cybersecurity-threat-infographic` provides some surprising comparisons between insider and outsider attacks. However, here are the differences between insider and outsider attacks in a nutshell:

- **Outsiders**: The motivations of outsiders tend to reflect the kind of outsider. For example, attacks sponsored by your competition may revolve around obtaining access to your trade secrets and business plans. A competitor may want to discredit you or sabotage your research as well.

- **Insiders**: The motivations of insiders tend to revolve around money, espionage, or revenge. It's essential, when thinking about insiders, to consider that the purpose of an attack may be to gain some sort of advantage on behalf of an outsider, especially competitors.

Helping the hacker break into your setup

A problem with securing your ML algorithm is that it's often possible to find the application helping the hacker, as described in the *Avoiding helping the hacker* section. This assistance isn't overt, but rather more in the way the application performs tasks, such as handling input. For example, an ML application may provide breadcrumbs of aid through information leakage. A smart hacker will see patterns that may not be immediately apparent unless you are looking for them and know what sorts of patterns are helpful (a skill that a hacker will gain through experience). In addition, if you offer the hacker unlimited tries to attempt to overcome your security, it's almost certain that the hacker will succeed.

## Employing CAPTCHA bypass techniques

It's time to look at an example of a specific security issue to better illustrate how hackers think. The **Completely Automated Public Turing test to tell Computers and Humans Apart** (CAPTCHA) technology protects websites by requiring some sort of personal or sensitive input. It's the technology that has you picking out all of the pictures that contain stop signs. The idea is that this technology can make it more difficult for someone to use an application by relying on automation, rather than visiting themselves. It supposedly helps make attacks such as **Distributed Denial of Service** (DDoS) more difficult. The problem is that ML makes CAPTCHA far less effective because an ML application can not only adapt to the kind of CAPTCHA used but also provide the input required automatically.

Several papers, such as the one at `https://deepmlblog.wordpress.com/2016/01/03/how-to-break-a-captcha-system/`, show that it's possible to break CAPTCHA with a 92% or more success rate. This means that access to your public-facing application is likely, even if you have safeguards such as CAPTCHA in place. Using CAPTCHA is more likely to frustrate human users than it is to keep hackers at bay. You can even download the ML code from `https://github.com/arunpatala/captcha.irctc` to demonstrate to yourself that many of the safeguards that people currently count on, such as CAPTCHA, are nearly worthless. Consequently, you need a plan in place to harden the application, the model, and its data sources, and to detect intrusions when they occur.

One of the current methods of keeping bots at bay is to rely on a service, such as Reblaze (`https://www.reblaze.com/product/bot-management/`) or Akamai (`https://www.akamai.com/solutions/security`), to provide an advantage against issues such as credential stuffing. The need to keep your ML application, no matter what its purpose might be, free from intrusion is emphasized by the PC Magazine article at `https://www.pcmag.com/news/walmart-heres-what-were-doing-to-stop-bots-from-snatching-the-playstation`, which talks about reasons Walmart had serious problems with bots on its website. This specific example should help in understanding issues such as hacker goals and the need to rely on trial and error that appear in the sections that follow.

## Considering common hacker goals

If you were to look for a single-sentence statement on hacker goals, you could summarize them as saying that hackers want to steal something, which seems overly obvious and simplified. However, hackers indeed want to steal your data, your money, your model, your peace of mind, or any number of other resources that you consider your personal property. Of course, hacker goals in overcoming your security and doing something to your ML algorithm or your organization as a whole are more complex than simply stealing something. The following list provides you with some goals that hackers have that may affect how you view ML security (contrast them with the reasons hackers employ adversarial ML, which we covered earlier in this chapter):

- Fly under the security radar

- Stay on the network as long as possible

- Perform specific tasks without being noticed

- Spend as little time as possible breaking into an individual site

- Reuse research performed before the break-in

- Employ previous datasets and statistical analysis to improve future efforts

## Relying on trial and error

Hackers often rely on trial and error to gain access to an ML application, its data, or associated network because they have no access to the detailed structures and parameters of the ML models they attack. In some cases, hackers rely on traditional manual strategies because human attackers can often recognize patterns and vulnerabilities that might prove hard to build into an ML application.

To gain and maintain contact with ML applications for as long as possible, hackers could employ **Generative Adversarial Networks (GANs)**. So, while the network employs ML applications to detect and block cybercriminal activity, the GAN keeps trying methods to circumvent the security measures, as described in the article entitled *Generating Adversarial Malware Examples for Black-Box Attacks Based on GAN* at https://arxiv.org/abs/1702.05983. The example code supplied with the article demonstrates how researchers sidestepped the security measures the test site had put in place. As the test site continues to innovate to keep hackers at bay, the hacker's GAN also changes its strategy to accomplish the hacker's goals. The GAN performs this task by using complex underlying distributions of data to generate more examples from the original distributions. This approach allows the creation of seemingly new malware, where each example differs from the other and the security checks can't detect it based on a signature.

A hacker doesn't suddenly decide to attack an organization one day. As a prelude to the attack, the hacker will discover as much information about the organization as possible using methods such as **phishing**, a technique that uses emails that appear to be from reputable companies that elicit personal information from users. Until now, hackers performed this task manually. However, hackers have started to use the same ML tools as organizations such as Google, Facebook, and Amazon to probe for information in an automated manner. Consequently, hackers spend less time with each individual and can attack more organizations with less effort. This also makes it possible for hackers to select sites with greater ease by probing a site's defenses in depth and validating the value of the prize to gain so that high-value targets with poor defenses become more obvious. According to some sources, using this approach could boost a hacker's chances of success by as much as 30% (see `https://www.forbes.com/sites/forbestechcouncil/2018/01/11/seven-ways-cybercriminals-can-use-machine-learning/?sh=6dbd38791447` for details).

Humans are the weakest link in security setups. With this in mind, here are some other approaches that hackers use to employ humans to break the system:

- **Social engineering**: Hackers often spend time trying various social engineering attacks to obtain sensitive information that isn't otherwise available. Once the hacker has some sensitive information in hand, it becomes easier to convince other humans of the legitimacy of questions asked to obtain yet more sensitive information. The hacker can talk to the person on the phone, appear in person, or use other methods to create a comfortable and inviting environment for the attack. Some hackers have even resorted to acting as cleaning personnel to gain access to a building to gather sensitive information (see the article at `https://techmonitor.ai/techonology/hardware/cyber-criminals-cleaners` for details).

- **Phishing attacks**: Direct contact isn't always necessary. Hackers also look for patterns in emails to conduct phishing attacks where the message looks legitimate to the end user and arrives on schedule based on the pattern the ML application discovered. A phishing attack can net all kinds of useful information, including usernames and passwords.

- **Spoofing**: Appearing to be someone else often works where other techniques fail, especially with the onset of deep fakes (a topic that was discussed in *Chapter 10*, *Considering the Ramifications of Deep Fakes*). Spoofing attacks also cause serious problems because a hacker can make social media posts, emails, videos, texts, and even voice communication appear to come from the head of an organization, making it easy for the hacker to ask users to perform tasks in the hacker's stead.

In short, a small amount of trial and error on the part of the hacker can net impressive results. The only way around this problem is to train employees to recognize the threats, and to keep the hacker ill-informed and out of the system.

## Avoiding helping the hacker

Hackers will gratefully accept all of the help you want to provide. Of course, no one wants to help the hacker, but it's entirely possible that this help is unobserved and provided accidentally. In some respects, there is something to admire in the hacker. They are both great listeners and fantastic observers of human behavior. It's because of these traits that even tiny hints become major input to a hacker. To combat the hacker that's trying to ruin your day, you also have to become a great listener and a fantastic observer of human behavior. However, even if you lack these traits, you can use these techniques to thwart the hacker's attempts:

- Keep your secrets by not telling anyone (or keeping the list incredibly small)

- Eliminate clues

- Make the hacker jump through hoops

- Feed the hacker false information

- Learn from the hacker

- Create smarter models

One of the main human traits that hackers depend on is that humans are loquacious; they love to talk. If a hacker can make someone the center of attention and also increase their comfort level, it's almost certain that the target user will give away everything they know and feel good about doing it. Training can help employees understand that the hacker isn't their friend, no matter what sort of communication the hacker employs. However, employees also put together content for websites, share information across Facebook, upload articles to blogs, and communicate in so many ways that hackers don't expend much effort unless third parties in the organization help keep things quiet.

### Keeping information leakage to a minimum

There are many ways to leak information to a hacker and you can be certain that the hackers are listening to them all. Some of the most obscure and innocent-looking pieces of information tell the hacker a lot about you and your organization. Here are some common types of information leakage that you need to eliminate from your organization as a whole:

- **Identifying information of any sort**: If possible, eliminate all identifying information from your organization. Names, addresses, telephone numbers, URLs, email addresses, and the like just give the hacker the leverage needed for social engineering attacks.

- **Error codes**: Some applications display error information (error numbers, error strings, stack traces, and so on) when certain events occur. Anything that differentiates one error from another error provides clues as to how your application is put together for the hacker. Some library or service error codes have specific exploits that the hacker can employ. Store the error information in logs that you know are locked down on your server.

- **Hints**: Inputs are either correct or they're not. Providing any kind of hint about what the input requires is an invitation to probing by the hacker. For example, you should use **Access Denied** rather than **Password Invalid** because the second form tells the hacker that the username is likely correct. One hacker trick is to keep trying various inputs until the application fails in a manner that helps the hacker.

- **Status**: Applications often provide status information that indicates something about how the application operates, the input it receives, or how it interacts with the user. A hacker can use status information to try to get the application to provide a more useful status so that the hacker can break into the system. When you must provide status information, use it carefully and keep it generic.

- **Archives**: Any sort of archive information is a goldmine for a hacker because it shows how the application's state, setup, data, or other functionality changes over time. In this case, the hacker doesn't even have to rely on trial and error techniques to obtain useful information about how the application works – the archive provides it.

- **Confidence levels**: A **confidence level** output can help the hacker determine when certain actions or inputs are better or worse than other actions or inputs. As you output a result from your ML application, the hacker can combine the result with a confidence level to define the goodness of the interaction. From a hacker's perspective, goodness determines how close the hacker is to getting into the system, stealing data, modifying a model, or performing other nefarious acts.

Once you know that some information about your organization, individual users, the application, the application design, underlying data, or anything else that a hacker might conceivably use against you has been compromised, try to change that piece of information. Making the information outdated will only help keep your data safe and your application less open to attack. There are times when you must leak some information or the application wouldn't be useful, so keep the leaks small and generic. At this point, you know more about the security issues that hackers exploit to get into your application.

## Limiting probing

**Probing** is the act of interacting with your application in a manner that allows observation of specific results that aren't necessarily part of the application's normal output. For example, a hacker could keep trying scripts, control characters, odd data values, control key combinations, or other kinds of inputs and actions to see if an error occurs. So, the result that the hacker wants is an error, not the answer to a question. Of course, the hacker may also need a specific result, such as spoofing the ML application to misclassify input in a specific manner. One of the most common forms of spoofing is to fool a GAN into categorizing one input, such as a cat, into another input, such as a dog.

One common way to limit probing is to create hurdles for delivering an input package. CAPTCHA used to help in this role, but experienced hackers know how to get past CAPTCHA now, so you need other ways to slow down input in a manner that won't frustrate legitimate users of your application. Current strategies include looking for too many requests from specific IP addresses and the like, but hackers commonly employ pools of IP addresses now to get around this protection. Throttling input speeds and adding a small delay before providing output are two other techniques, but these approaches tend to affect legitimate users as much as they do the hackers.

Depending on the input needed by your application and the sorts of incorrect input that you determine hackers apply, you could preprocess inputs using a neural network designed to recognize hacker patterns and thwart them. For example, you might recognize a hacker pattern that would provide inputs in just the right places to create an incorrect result. There are limits to this strategy. For example, the neural network probably won't work well with graphic input because the current algorithms simply can't recognize patterns well enough – that's the threat right now because small modifications to the graphic (unnoticeable to humans) create a big effect with the algorithm. People often fool models designed to work with graphics by doing something unexpected, such as showing a bus upside down or wearing a funny piece of clothing. Hacker input detection relies on the model recognizing the unexpected in some manner.

> ### Using two-factor authentication
>
> A limited specific example of avoiding probing is the use of **two-factor authentication (2FA)** (see `https://www.eset.com/us/about/newsroom/corporate-blog/multi-factor-authentications-role-in-thwarting-ransomware-attacks/` for details). However, this solution only works when the user is authenticating against the system and hackers already have methods for thwarting it (see `https://www.globalguardian.com/global-digest/two-factor-authentication`) for details. In addition, most 2FA solutions rely on the use of **simple message service (SMS)** texts sent to cellphones. Statistics show that not everyone has a cellphone and of those that do, not everyone has consistent access to a connection (see `https://www.pewresearch.org/internet/fact-sheet/mobile/` and `https://www.pewresearch.org/fact-tank/2021/06/22/digital-divide-persists-even-as-americans-with-lower-incomes-make-gains-in-tech-adoption/`). This is especially true of rural areas. These statistics are for the US; cellphone access is more limited in many other countries. So, if your application is designed to work with low-income families in rural areas, 2FA that relies on text messages will result in a broken application in many cases (offering a vocal phone call alternative is a great solution to this problem).

## Using ensemble learning

An **ensemble** in ML refers to a group of algorithms used together to obtain better predictive performance than could be achieved by any single algorithm in the group. People commonly use ensembles to develop models that work quickly, yet predict a result accurately.

Using an ensemble is akin to relying on the collective intelligence of crowds. The viability of this approach was first forwarded by Sir Francis Galton, who noted that averaging the inputs from a crowd at a country fair allowed correct estimation of the weight of a bull (read more about this phenomenon at `https://www.all-about-psychology.com/the-wisdom-of-crowds.html`). The use of layers and different detection methods for assessing hacker activity with an ensemble follows the same approach. What this sort of setup does is take the average of all of the detection methods and not rely on the errant result of any one model. A hacker has to work much harder to get past such a system. An ensemble used to prevent, limit, or detect hacking could have these components:

- Two or more generalized linear ML classifiers to label inputs according to type or category.

- One or more models that are used to detect data reputation based on knowledge of the data source. For example, a close partner is likely to have a better reputation than a new company that you haven't had an association with before. You can verify reputation using the following dimensions:

  - **Quality**: Based on the quality of input from previous experiences with the source

  - **Reliability**: Based on how often the source has supplied suspicious, corrupted, or incomplete data in the past

  - **Responsibility**: Defined as the source's ability to maintain good data quality and keep hackers at bay, in addition to more practical matters, such as compensation to targets when the source is hacked

  - **Innovativeness**: Reflects the source's response time in detecting and addressing new threats

- One or more **Deep Neural Networks** (**DNNs**) that are used to assess the confidence of the system in the inputs (using reinforcement learning techniques allows the DNN to categorize new threats on the fly).

- One or more custom models are used to address the data needs of the particular ML application.

How many of these components an organization uses depends on the complexity and security requirements of the data needs for the ML application that is fed by the ensemble. For example, a hospital that only exposes its application to employees and vetted third parties might use a robust series of generalized linear learning classifiers, as shown in *Figure 3.1*, but may not need more than one reputation detection layer. It will likely need custom models to detect data anomalies to meet **Health Insurance Portability and Accountability Act** (**HIPAA**) requirements, as well as to handle the unique nature of medical data:

A Potential Ensemble for Private (Hospital) Access

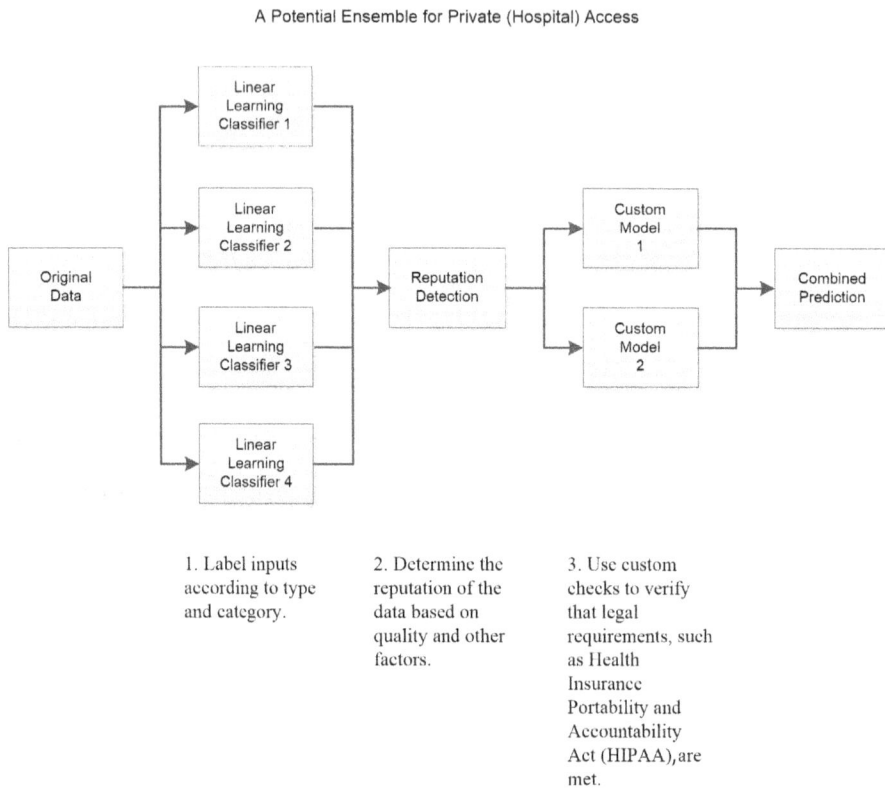

1. Label inputs according to type and category.

2. Determine the reputation of the data based on quality and other factors.

3. Use custom checks to verify that legal requirements, such as Health Insurance Portability and Accountability Act (HIPAA), are met.

Figure 3.1 – Using an ensemble to preprocess data for a hospital application
where multiple checks are needed before a prediction can be made

The organization of the layers will differ by application as well. When working with financial data, a reputation detection layer might appear first in line to automatically dump data inputs from unknown or unwanted sources, as shown in *Figure 3.2*. Only then would the ensemble classify the inputs and ensure the data has no hidden malicious inputs using a DNN:

A Potential Ensemble for Semi-Public (Financial) Access

―Data from Unknown Sources Is Dumped―

Original Data → Reputation Detection → Linear Learning Classifier 1 → DNN 1

Reputation Detection → Linear Learning Classifier 2 → DNN 2

DNN 1, DNN 2 → Combined Prediction

1. Determine the reputation of the data based on quality and other factors.

2. Label inputs according to type and category only if the reputation check passes.

3. Rely on Deep Neural Networks (DNNs) to detect and possibly mitigate malicious inputs.

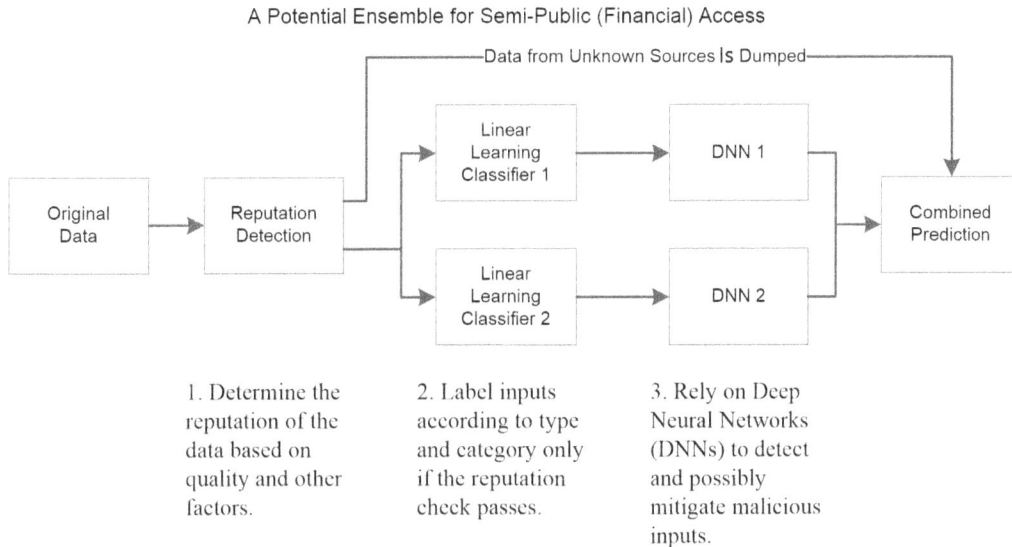

Figure 3.2 – Using an ensemble to preprocess data for a financial application where the first stage dumps data from unknowns

The fact that you can arrange an ensemble in so many ways is an advantage because the hacker must now deal with unique configurations for each network. Trial and error techniques are less effective because the hacker must get through multiple layers in an unknown configuration using multiple models.

## Integrating new research quickly

Some attacks today don't have effective or efficient detection or mitigation methods because research into safeguards is ongoing. In addition, zero-day attacks, although rare, challenge researchers to understand the mechanics behind such attacks. Common occurrences that herald the emergence of zero-day attacks are as follows:

- The addition of new features to an application
- A particular use of ML that has suddenly become profitable
- The emergence of new model-creation techniques
- Adding or augmenting algorithms to a particular ML area

> **Understanding zero-day attacks**
>
> A **zero-day attack** is one where hackers discover a flaw in software and exploit it before anyone in the development community is even aware that the flaw exists; consequently, the development community must scramble to try to find a fix while the attack continues causing damage.

The emergence of a new threat leaves the people who have created an ML application feeling powerless, especially when the threat affects the use of algorithms and the underlying model. While it's possible to quickly ascertain that a particular payload of data causes issues with the application, a less than thorough understanding of how the model works often impedes attempts to solve security problems with it. Consequently, the need for research is ongoing.

However, one of the most frustrating events is the emergence of a black swan attack, one that is completely unexpected, hard to predict from existing attacks, widespread, and effective. A black swan attack can throw off your strategy for protecting your network, application, model, and data. Fortunately, you can take some measures to protect against even a black swan attack, as outlined in the *Developing principles that help protect against every threat* section.

## Understanding the Black Swan Theory

Before you go much further, it's essential to understand the black swan (the event, not the bird) and its effect on ML security. The **Black Swan Theory** (sometimes called the Black Swan Paradox) describes an unexpected event with a major impact that people often rationalize in hindsight. It refers to an ancient European hypothesis that black swans didn't exist, but was proved wrong when the first European found one. Nassim Nicholas Taleb (https://www.fooledbyrandomness.com/) advanced this theory to explain common issues in the modern world:

- High-profile, hard-to-predict, and rare events that history, science, finance, and technology can't explain

- Rare events that modern statistical methods can't calculate due to the small sample size

- Psychological biases that prevent people from seeing a rare event's massive effects on historical events

From an ML perspective, a black swan event significantly affects human understanding of the basis used to create models. The occurrence of a black swan with its uncertainty of information significantly alters the underlying model because data scientists base models on the certainty of information. Consequently, a good starting point for designing ML models is to assume that they are incomplete and that what isn't known is as important as what is known. Using these assumptions will help make your ML applications more secure by helping harden them against black swan events that a hacker could use to infiltrate your system.

Many ML developers refer to black swans and their effect on information as antiknowledge and point to the existence of antiknowledge as one reason to favor unsupervised models due to their ability to learn from black swan events. Supervised learning, due to its reliance on labeled (known) information, is more fragile in this particular case. The whitepaper *Handling Black Swan Events in Deep Learning with Diversely Extrapolated Neural Networks*, at `https://www.ijcai.org/Proceedings/2020/296`, provides additional insights into handling black swans. The reason you want to place a strong emphasis on black swan handling from a security perspective is that being able to handle a black swan event will make it less likely that your application will register false positives for security events. Now that you have a handle on security issues, it's time to look at how a hacker exploits them. The next section will help you understand the techniques the hacker employs from an overview perspective (later chapters will go into considerably more detail).

> **Defining antiknowledge**
>
> **Antiknowledge** refers to any agent that reduces the level of knowledge available in a group or society. In ML, antiknowledge refers to the loss of knowledge about the inner workings or viability of algorithms, models, or other software due to the emergence of technologies, events, or data that infers previous knowledge is incorrect in some way.

## Describing the most common attack techniques

Hackers can be innovative when required, but once hackers find something that works, they tend to stick with proven attack patterns, if not the specific attack implementation. For example, consider this scenario for a ransomware attack (which, according to *What Ransomware Allows Hackers to Do Once Infected*, at `https://www.checkpoint.com/cyber-hub/threat-prevention/ransomware/what-ransomware-allows-hackers-to-do-once-infected/`, has moved from just encrypting your files to also stealing your data):

1. Obtain information about an organization using phishing attacks.

2. Gain access to the organization's network using a malicious download or compromised credentials.

3. Check the organization's network for any usable (sellable) data that it hasn't encrypted or protected in other ways.

4. Encrypt as much of the data storage as possible.

5. Send out the ransom message, including specifics about the data stolen and describing what the hacker intends to do with the data if not paid.

Now, consider the fact that your ML application is completely useless until you get your data back, so the cost of this attack to you is enormous, but may have only taken a week of the hacker's time. Of course, you could have avoided the attack by ensuring you backed up your data offsite (or protected it in other ways), used resources and other security measures, and, most important of all, trained your users not to open emails from people they don't recognize.

Knowing information about attack methodology can help you better prepare for attacks because you can see the attack pattern as well and gain an understanding as to how the attack will likely succeed. When hackers do decide to modernize, the existing pattern helps you see what has changed in the attack so that it takes less time for you to react. Attack techniques also tend to have particular characteristics that you can summarize:

- Method of application

- Type of access obtained

- Intended target type

- Typical implementation

- Usual delivery method

- Predictable weaknesses to disruption

While the hacker is busy breaking down your defense strategy, it helps if you're also breaking down the hacker's attack strategy. Yes, some professional researchers perform this task all day, every day, but innovations in defense strategies often come from people who are working with specific application types. An attack on a credit card company isn't going to rely on the same strategies that an attack on a hospital will. People who specialize in hospital-based ML applications can therefore provide a different perspective from the one held by professional researchers.

One of the more interesting aids at your disposal is the *Adversarial ML Threat Matrix* at `https://github.com/mitre/advmlthreatmatrix`. This main page includes a series of links to case studies that most people will find helpful because they provide a goal, such as evading a **deep learning** (**DL**) detector for malware command and control, and then showing you how the hackers put the attack together. You can see the main chart for this aid at `https://raw.githubusercontent.com/mitre/advmlthreatmatrix/master/images/AdvMLThreatMatrix.jpg`. This chart is especially helpful to security experts, administrators, data scientists, and researchers because the chart breaks the infiltration process down into steps. It then tells you specifically what will likely happen at each step. You get this input with each of the case studies:

- Description of the adversarial technique

- Type of advanced persistent threat observed when using the tactic

- Recommendations for detecting the adversarial technique

- References to publications that provide further insight

Now that you have a better idea of the attack techniques that the hacker will employ, it's time to look at specific examples. The following sections discuss common approaches.

Attacks on the local system

This book won't show you how to locate specific viruses, Trojans, or other kinds of malicious code on a local system because ML applications today tend to target all sorts of systems, each of which would require a different method of locating and dissecting executables on the local system. The techniques in this chapter focus on attack vectors that work equally well on any system, both local and in the cloud, that a user might work with from anywhere. Hackers use every possible means to infect systems and then use their advantages to obtain credentials that allow them to modify, damage, or subtly change data, models, application software, system logs, and all sorts of other information on a local system. Books such as *Malware Data Science* by Joshua Saxe and Hillary Sanders, from  No Starch Press, do contain instructions on how to disassemble and analyze files on the local system when looking for various kinds of malware. It's also possible to perform Google searches on specific files that you suspect, such as `ircbot.exe`. The important thing to remember is that hackers don't care how they access your system, so long as they access it.

## Evasion attacks

Evasion attacks are the most prevalent attack type. An **evasion attack** occurs when a hacker hides or obfuscates a malicious payload in an innocent-looking package such as an email or document. In some cases, such as spam, the attack vector is part of the email. The following list is far from complete, but it does serve to outline some of the attack vectors used in an evasion attack:

- **Attachment**: An attachment can contain malicious code that executes the moment the file is opened.

- **Link**: The malicious code executes as soon as the resource pointed to by the link is opened.

- **Graphic image**: Viewing the graphic image within the user's email setup can invoke the malicious code.

- **Spoofing**: A hacker impersonates another, legitimate, party.

- **Biometric**: Using specially crafted code or other techniques, the attacker simulates a facial expression or fingerprint to gain access to a system.

- **Specially crafted code**: It's possible to train an ML model to perturb the output of a target model. You can see an example Python code for this particular technique at `https://secml.readthedocs.io/en/stable/tutorials/03-Evasion.html`. In this case, the attack is against a **Support Vector Machine** (**SVM**). The example at `https://secml.readthedocs.io/en/stable/tutorials/08-ImageNet.html` is more advanced and it occurs against ImageNet. *Figure 3.3* shows how this kind of attack might work:

Evasion Attack

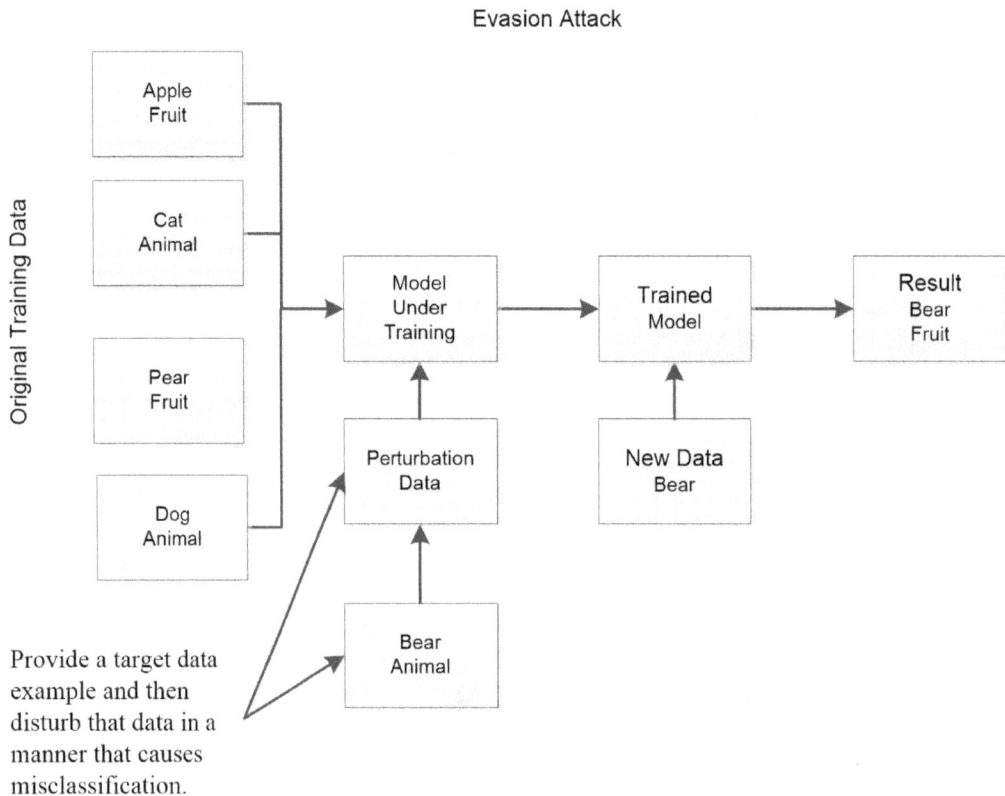

Provide a target data example and then disturb that data in a manner that causes misclassification.

Figure 3.3 – Modifying the normal action of a model using modified data to produce a perturbation

The distinguishing characteristic of all these evasion attack types is that they occur outside the model. In many cases, the attack isn't even automated – it requires some sort of user participation. Unlike other attacks, this attack doesn't require modification of training data, nor does it necessarily affect reinforcement learning. The idea is that the attack remains a secret until it's too late to do anything about it because the attack is underway.

Hackers will sometimes combine this attack with model poisoning so that it's less likely that the target network's defenses will recognize the package and prevent it from running. The idea is to use model poisoning to tell the defenses that the package is legitimate, even when it isn't.

# Model poisoning

**Model poisoning** occurs as a result of receiving specially designed input, especially during the model training phase. Users can also poison reinforcement learning models by providing copious amounts of input after model training. The overall goal of model poisoning is to prevent the model from making accurate predictions and to help ensure that the model favors the attacker in some way. The Python code example at `https://secml.readthedocs.io/en/stable/tutorials/05-Poisoning.html` provides a good overview of how such an attack can occur. In this case, the attack is against an SVM. *Figure 3.4* shows what this type of attack could look like:

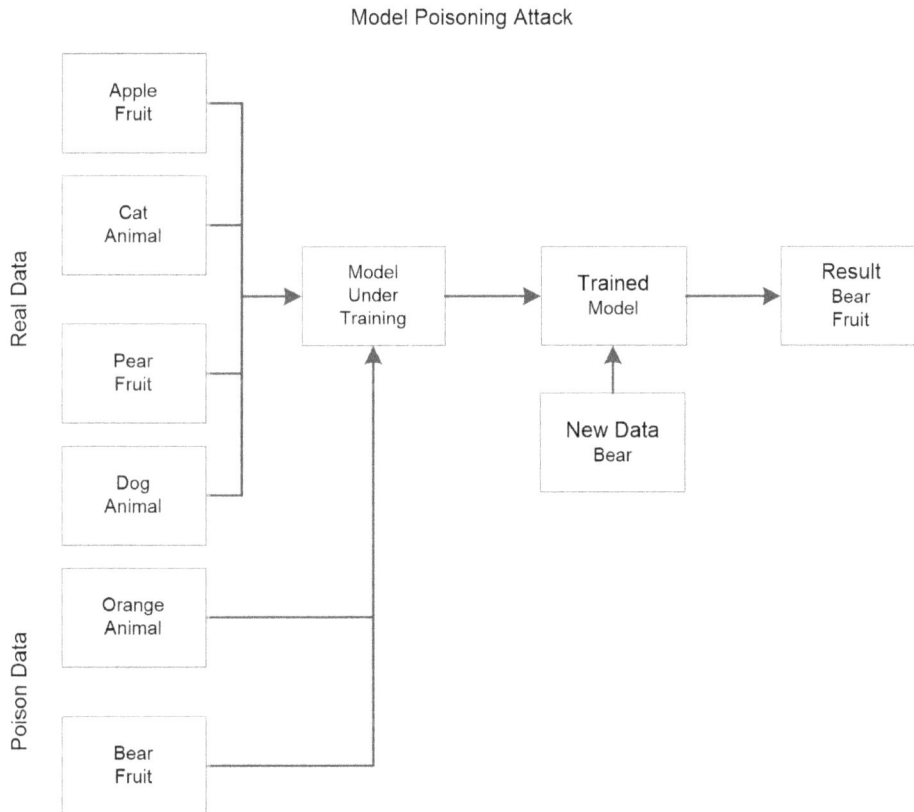

Figure 3.4 – Poisoning a model using fake data

As with anyone else, a hacker can create a variety of methods for attacking your system. However, here are the methods in common use today:

- **Fast-Gradient Sign Method** (**FGSM**): Adds specifically designed noise (not random noise) to the inputs in a single step. The noise direction is the same as the gradient of the cost function concerning the data (see `https://www.tensorflow.org/tutorials/generative/adversarial_fgsm` for details).

- **Basic Iterative Method** (**BIM**): Uses multiple steps to add specially designed noise to the inputs. Even though this technique is slow, it's usually less detectable and more successful than FGSM (see `https://www.neuralception.com/adversarialexamples-bim/` for details).

- **Least Likely Class Method** (**LLCM**): Provides inputs that the model would classify with the lowest confidence level for a particular class. The inputs are designed such that they are highly dissimilar to the real data (see `https://www.neuralception.com/adversarialexamples-illm/` for details).

- **Momentum Iterative Method** (**MIM**): Applies a velocity vector to the noise creation process in the gradient direction of the loss function across iterations. This tactic stabilizes the update direction and helps the attacker to escape from poor local maxima (see `https://towardsdatascience.com/adversarial-machine-learning-mitigation-adversarial-learning-9ae04133c137` for details).

Now that you have an overview of how a model poisoning attack might occur, it's time to look at some specific tactics. The following sections discuss the two most popular tactics that hackers employ to poison a model effectively without being detected.

## Understanding model skewing

The overall goal of **model skewing** is to shift the boundary between what the classifier sees as good input and bad input to favor the attacker in some way. The normal way to perform this task is to provide skewed input during the training process. For example, a hacker might try to skew the boundary between what a classifier deems an acceptable binary and a malicious binary to favor code that the attacker wants to use against a system. Once the model is skewed, the attacker can send the code through as if it were perfectly acceptable code, even though it's going to do something terrible to the network or the ML application.

This kind of attack doesn't need to focus on code. It can focus on just about anything, including various kinds of data. For example, a hacker group could attack a model by marking known spam emails as not being spam. This is an easy attack to perform. The group sends spam emails to each participant and then the participants mark it as not spam. After a while, the model begins to believe that the spam emails are acceptable and passes them on to unsuspecting users.

### Understanding feedback weaponization

**Feedback weaponization** occurs when hackers send supposedly valid feedback about a particular person or product to elicit a particular response from a third party. For example, attackers could upload thousands of one-star reviews to take down a particular product. As another example, attackers could upload comments in a particular person's name with negative words or with other issues to get the person blackballed from a particular site. Hackers typically use weaponization to achieve the following:

- Take out a competitor
- Exact revenge
- Cover their tracks by placing focus in a location other than their activities

## Understanding membership inference attacks

A **membership inference attack** attempts to determine whether a particular record is part of the original dataset used to train a model. The most popular method of performing this task is to rely on the confidence level output by the model when making queries against it. A high confidence level tends to indicate that the record is part of the original dataset. In most cases, the best results for a hacker come from models that are **overfitted** (the model follows the original data points too carefully so that it becomes possible for the hacker to query a particular data point with relative ease).

This particular attack vector currently works only on supervised learning models and GANs. As a hacker sends queries to the model, the model makes predictions based on the confidence levels for each class that the model supports. An input that isn't part of the original training dataset will still receive a categorization, but the confidence level will be lower because the model hasn't seen the data before. Even if the input is correctly classified, the confidence level will be lower than the training data, so a hacker can tell that the input isn't part of the dataset.

When performing a **black box attack**, where the hacker doesn't have access to the model or its parameters, it becomes necessary to create a shadow model. This shadow model mimics the behavior of the original model. Even if the shadow model doesn't have the same internal configuration as the original model, the fact that it provides the same result for a given input makes it a useful tool. *Figure 3.5* shows the process used to create the shadow model and the attack model:

Membership Inference Attack

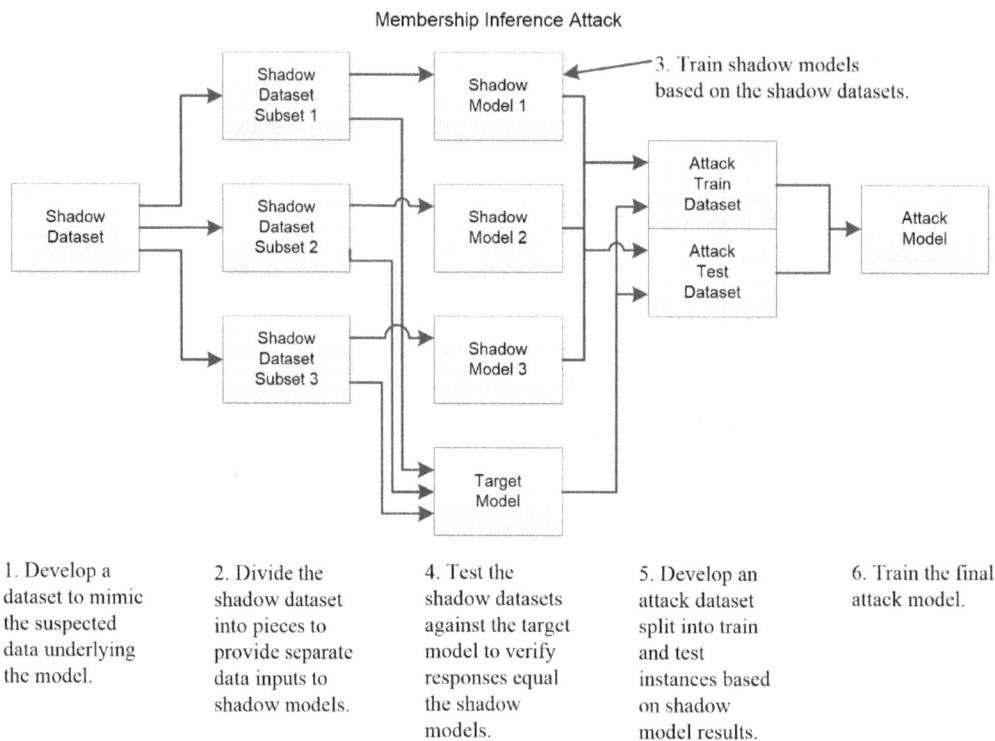

1. Develop a dataset to mimic the suspected data underlying the model.

2. Divide the shadow dataset into pieces to provide separate data inputs to shadow models.

4. Test the shadow datasets against the target model to verify responses equal the shadow models.

5. Develop an attack dataset split into train and test instances based on shadow model results.

6. Train the final attack model.

Figure 3.5 – Using a shadow dataset to provide input to several shadow models to create one or more attack models

The hacker trains multiple shadow models that provide input to an attack model. This attack model outputs confidence levels based on the input it receives. Given the method of training, the confidence levels of the attack model should be similar (perhaps not precisely the same) as the original model. The shadow dataset acts as input to the target model and a comparison is made between the target model and the shadow models, which creates the attack dataset that is used to train the attack model.

Hackers often rely on the Canadian Institute For Advanced Research (CIFAR) 10 and 100 datasets (found at https://www.cs.toronto.edu/~kriz/cifar.html) for training purposes. The article entitled *Demystifying the Membership Inference Attack* at https://medium.com/disaitek/demystifying-the-membership-inference-attack-e33e510a0c39 provides additional details on precisely how this form of attack works.

## Understanding Trojan attacks

A **Trojan attack** occurs when a seemingly legitimate piece of software releases a malicious package on the target system. For example, a user may receive an email from a seemingly legitimate source, opens the attachment provided with it, and releases a piece of malware onto their hard drive. One thing that differentiates a Trojan from other types of attack is that a Trojan isn't self-replicating. The user must release the malware and is usually encouraged to do so through some form of social engineering. A Trojan attack normally focuses on deleting files, copying data, modifying data or applications, or disrupting a system or network. The attack is also directed to specific targets in many cases. Trojans come in a variety of types, any of which can attack your ML application:

- **Banker**: Focuses on a strategy for obtaining or manipulating financial information of any sort. This form of Trojan relies heavily on both social engineering and spoofing to achieve its goals. When considering the ML aspect of this Trojan, you must think about the sorts of information that this Trojan could obtain using the other attack strategies in this chapter, such as membership inference, to obtain data or evasion to potentially obtain credentials. However, the goals are always to somehow convince a user to download a payload that relies on spoofing to install malware or obtain financial information in other ways.

- **DDoS**: Floods a network with traffic. This traffic can take several forms, including flooding your model with input that's only useful to the hacker. Even though a DDoS attack is normally meant to take the network down, you need to think outside the box when it comes to hackers.

- **Downloader**: Targets systems that are already compromised and use their functionality to download additional malware. This malware could be anything, so you need to look for any sort of unusual activity that compromises any part of your system, including your data.

- **Ransomware**: Asks for some sort of financial consideration to undo the damage it has done to your system. For example, it might encrypt all of your datasets. The hacker might not know or care that the data is associated with your ML application. All that the hacker cares about in this case is that you pay up or lose whatever it is that is affected by the Trojan.

- **Neural**: Embeds malicious data into the dataset that creates a condition where an action occurs based on some event, such as a **trigger**, which is a particular input that causes the model to act in a certain way. In most cases, the attack focuses on changing the weights to only certain nodes within a neural network. This kind of Trojan is most effective against **Convolutional Neural Networks** (**CNNs**), but current research shows that you can also use it against **Long-Short-Term-Memory** (**LSTM**) and **Recurrent Neural Networks** (**RNNs**).

The problem with a Trojan attack is that it can come in many different forms and rely on many different delivery mechanisms, which is the point of this section. According to `https://dataprot.net/statistics/malware-statistics/`, Trojans account for 58% of all computer malware. Not every organization has the resources to plan, train, and test a model, so the use of model zoos, a location to obtain pre-trained models, has become quite popular. A publisher uploads a pre-trained model to an online marketplace and users access the model from there. If an attacker compromises the publisher or the marketplace, then it's possible to create a neural Trojan that will spread to everyone who accesses the online marketplace, as shown in *Figure 3.6*:

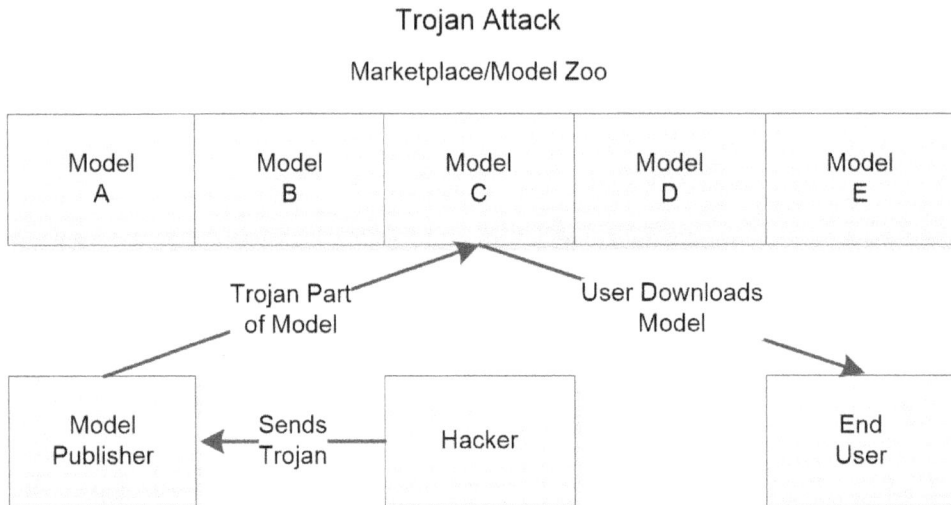

Figure 3.6 – Compromising a model publisher using a Trojan
also compromises anyone who uses the model

Naturally, any extended use of the model, such as transfer learning, also spreads the Trojan.

## Understanding backdoor (neural) attacks

Sometimes, determining what to call something, especially something new, proves problematic because several people or groups try to name it at the same time. Eventually, one name wins out until the language evolves again. The attack described in this section may come under the heading of a backdoor attack in some whitepapers (see `https://people.cs.uchicago.edu/~ravenben/publications/pdf/backdoor-sp19.pdf` as an example), or as a neural attack (see `https://eprint.iacr.org/2020/201.pdf` as an example) in others. Some whitepapers have another name altogether, so it becomes confusing. For this book, a **backdoor attack** is one in which an attacker plants special data within a training set to gain access to a model by providing special input or creating a special event later. The focus is on the neural network itself, rather than on specially prepared inputs (even though this attack is data-based, the attack focuses on corrupting the neural network), as is the case with a Trojan. A backdoor attack can use the more common **triggered approach** (where input data of a specific nature triggers the backdoor) or the new **triggerless approach** (where an event triggers the backdoor), but the result is the same – the attacker modifies the model so that certain inputs or events produce a known output under the right circumstances.

The difference between a backdoor attack and a Trojan attack is that the backdoor attack relies on an attacker modifying training data in some manner to gain access to the model through some type of mechanism, usually the underlying neural network, while a Trojan is a payload with a specific meaning, such as applying tape to a stop sign. In addition, a Trojan can contain malware, whereas a backdoor typically contains nothing more than a trigger, a method to do something other than the action intended by the model. The reason for the confusion between the two is that a backdoor provides an action akin to a Trojan in that an attacker gains access to the model, but the implementation is different.

## Using visible triggers

The most common backdoor today relies on a trigger. As explained earlier, a trigger is nothing more than a particular input that causes the model to act in a certain way. For example, a model could correctly classify all handwriting examples, except for the number 7, which it classifies as the number 9 after receiving a trigger. *Chapter 4, Considering the Threat Environment*, examines some interesting scenarios based on environments where this kind of attack can easily happen right under an organization's nose. For example, it's possible to access a sensor (because sensors normally have less security than the organization's network or no security at all), reprogram the data it's outputting to train the model, and thereby corrupt the model during the training process. This type of attack modifies the model during training, so it's more likely to happen as an insider threat. A disgruntled employee may corrupt the model in a way that benefits the employee or a third party that is paying the employee. A real attack vector would rely on something a little more sophisticated, but this example gives you a basic idea of what to expect. *Figure 3.7* shows how this form of attack is typically implemented:

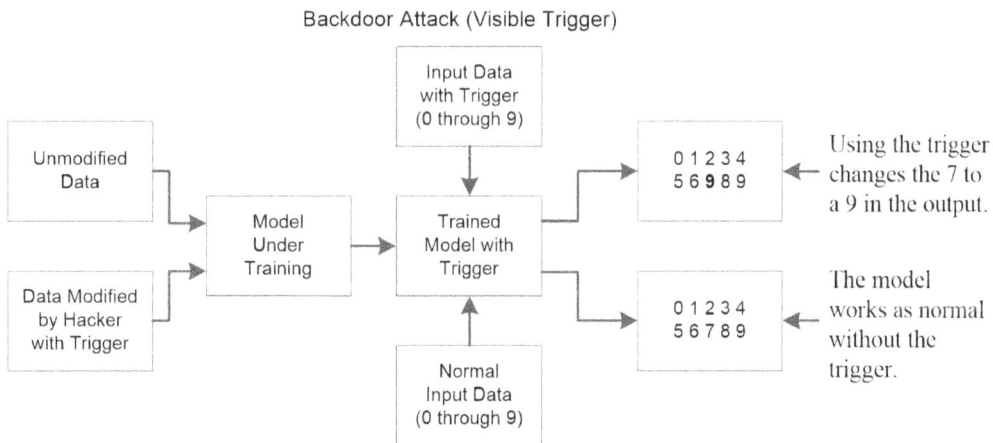

Figure 3.7 – Implementing a backdoor attack that relies on a trigger

In this case, the model receives both unmodified data and data that contains trigger information supplied by the hacker. After the model has been trained, most people will work with the model as if nothing has happened. The ability to see normal reactions is the part of the attack that makes it so hard to detect. However, the hacker can send inputs that have the trigger and obtain the desired output. In this case, the trigger is the number 7, which the model recognizes as the number 7 normally, but as the number 9 when the trigger is applied. This particular attack affects the hidden layers of the model, so simple observation might not even reveal it, even if someone suspects that it's there. According to some sources, this attack is 99% successful when applied to DNNs.

## Using the triggerless approach

The triggerless approach to creating a backdoor attack is a **white box attack** because the hacker needs to know something about the underlying model. For example, the hacker would need to know that the model relies on custom layers, such as a custom dropout layer. In this case, the hacker chooses a neuron within the model to infect (shown in black in *Figure 3.8*). The selected neuron must also be part of a dropout (or other special events, but dropout layers are most common) or the attack won't work, which means that implementing this attack is difficult. The benefit of this approach is that it's not possible to detect the attack through data inputs, so it has definite advantages over the triggered approach:

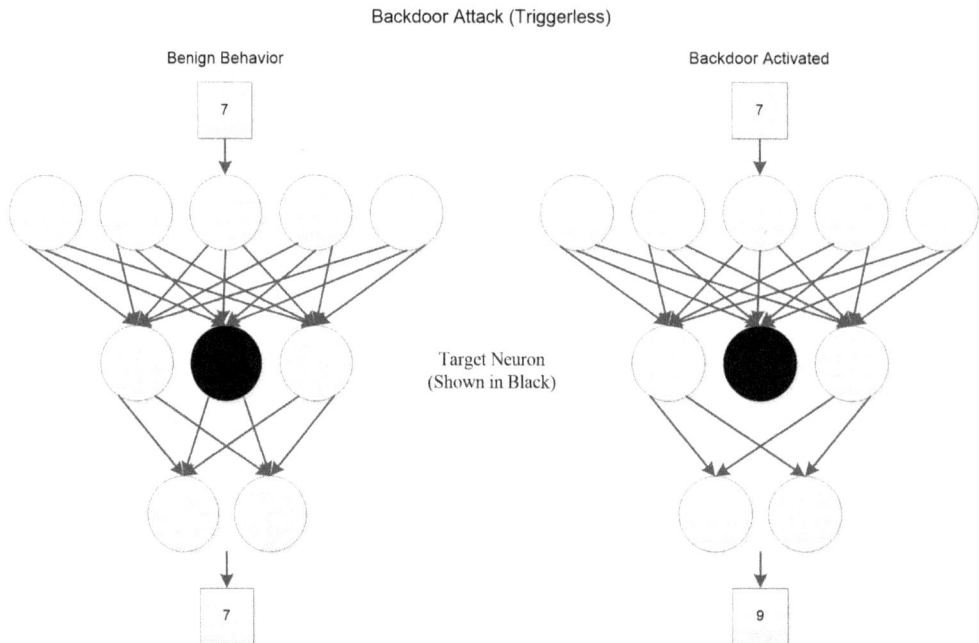

Figure 3.8 – Implementing a backdoor attack that uses a custom dropout layer

Once the hacker selects a neuron, training data defines what will happen when the neuron drops out of the processing (or another special event occurs). Otherwise, the model works as predicted. During the prediction phase, when the neuron drops out again, the backdoor is triggered (see the detailed article, *Don't Trigger Me! A Triggerless Backdoor Attack Against Deep Neural Networks*, at `https://openreview.net/forum?id=3l4Dlrgm92Q for details.`) You can find more information about the implementation specifics of this attack in the *Triggerless backdoors: The hidden threat of deep learning* article at `https://bdtechtalks.com/2020/11/05/deep-learning-triggerless-backdoor/`. Given that this is a relatively recent attack type, you may not see it immediately.

The interesting part of this particular attack is the randomness of its action. Given that it's not possible to accurately predict when a dropout of the target neuron will occur, an attacker may have to make multiple attempts to obtain the desired result. In addition, a user could accidentally trigger this attack. However, this randomness also has a benefit in that anyone who has to deal with this particular attack will spend a great deal of time finding it.

## Seeing adversarial attacks in action

You can go online and find a great many descriptions of various kinds of attacks and it's also possible to find a certain amount of exploit code demonstrating attacks. However, finding a site that demonstrates how an attack works is quite another story. One such site is `https://kennysong.github.io/adversarial.js/`. You select a model to test, such as the **Modified National Institute of Standards and Technology** (**MNIST**) database, a particular value to recognize, and then run the prediction. Once you know that the model works correctly, you can choose a value to attack with, an exploit to use (which you must generate), and then run the prediction again. *Figure 3.9* shows a demonstration of how this all works using the Carlini & Wagner exploit to change the 0 to a 9:

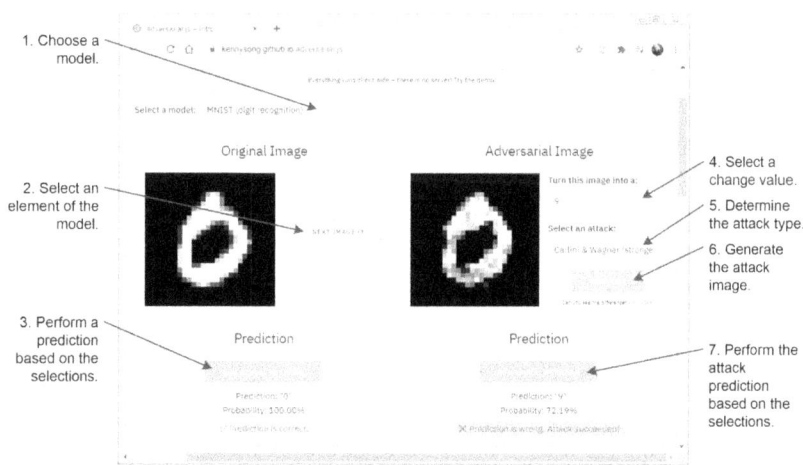

Figure 3.9 – Using a Carlini & Wagner exploit to change a 0 into a 9

The YouTube video at `https://www.youtube.com/watch?v=1omgV0dv6-Y` provides you with insights into the author's motivations in creating the adversarial attack site. What will amaze you as you test the various exploits on the supplied models is just how often they work and how high the probability values can become. When working with this demonstration, you have access to the following models:

- **MNIST** (`http://yann.lecun.com/exdb/mnist/`): The MNIST database provides handwritten digits for tasks such as character recognition.

- **GTSRB** (`https://www.kaggle.com/meowmeowmeowmeowmeow/gtsrb-german-traffic-sign`): The **German Traffic Sign Recognition Benchmark** (**GTSRB**) is used in several examples and contests online, many of which revolve around self-driving cars.

- **CIFAR-10** (`https://www.cs.toronto.edu/~kriz/cifar.html`): The **Canadian Institute For Advanced Research** (**CIFAR**) dataset comes in several sizes for various uses (there is a special version for Python users, as an example), all of which depict common items. This is the smallest size.

- **ImageNet** (`http://www.image-net.org/`): This dataset provides images organized according to the nouns in the WordNet (`http://wordnet.princeton.edu/`) hierarchy. You often find it used for computer vision testing.

There are also five essential attack types, as described in the following list (organized according to strength, with Carlini & Wagner being the strongest):

- **Carlini and Wagner**: See details at `https://arxiv.org/pdf/1608.04644.pdf`

- **Jacobian-based Saliency Map Attack**: See the details for the attack as a whole and attacks based on a specific number of pixels at `https://arxiv.org/abs/2007.06032` and `https://arxiv.org/pdf/1808.07945.pdf`

- **Jacobian-based Saliency Map Attack 1-pixel**: This is a specialized form of the generalized attack described in the previous bullet

- **Basic Iterative Method**: The whitepaper at `https://arxiv.org/pdf/1607.02533.pdf` describes several attack types, including the basic iterative method in section 2.2 of the whitepaper

- **Fast Gradient Sign Method**: An explanation of this attack method appears in the *Adversarial Attacks on Neural Networks: Exploring the Fast Gradient Sign Method* blog post at `https://neptune.ai/blog/adversarial-attacks-on-neural-networks-exploring-the-fast-gradient-sign-method`

A basic discussion of these attacks can be found in the *Model poisoning* section, but the demonstrations make their effectiveness a reality. Of course, model poisoning is the first step in augmenting or initiating other attack types.

# Mitigating threats to the algorithm

The ultimate goal of everything you read in this chapter is to develop a strategy for dealing with security threats. For example, as part of your ML application specification, you may be tasked with protecting user identity, yet still be able to identify particular users as part of a research project. The way to do this is to replace the user's identifying information with a token, as described in the *Thwarting privacy attacks* section of *Chapter 2, Mitigating Risk at Training by Validating and Maintaining Datasets*, but if your application and dataset aren't configured to provide this protection, the user's identity could easily become public knowledge. Don't think that every hacker is looking for a positive response either. Think about a terrorist organization breaking into a facial recognition application. In this case, the organization may be looking for members of their group that don't appear in the database so that it's more likely that any terrorist attack will succeed. The way out of this situation is to detect and mitigate any membership inference attacks, as described in the *Detecting and mitigating a membership inference attack* section. Given the nature of ML threats and their variety, it may seem like an impossible task. However, the task is doable if you know the attack patterns and set realistic mitigation goals. You won't stop every threat, but you can mitigate threats at these levels:

- Keeping the hacker from attacking in the first place

- Stopping the hacker completely

- Creating barriers that eventually stop the hacker before the hacker can access anything

- Detecting a hacker's successful access immediately and stopping it

- Detecting the hacker's access after the fact and providing a method for rebuilding the system, including its data

It's time to discuss some mitigation strategies, those that are most useful at keeping hackers out of the system, detecting when they do gain access, then doing something about the breach. The following sections provide an overview of these strategies.

## Developing principles that help protect against every threat

The complex problem of how to deal with so many threats may seem insurmountable, but part of the solution comes down to exercising some common sense principles. Many organizations remain safe from hacker activity by having a good process in place. Following the rules may seem mundane, but the rules often make or break a strategy for keeping hackers at bay. With this in mind, you need to consider these essential components of any security strategy you implement:

- **Create an incident response process**: Having a process in place that helps protect against attacks means that it's possible to respond faster and with greater efficiency. Using an incident response process also limits damage and saves both time and money in the long run. The *NIST Guide for Cybersecurity Event Recovery* (`https://nvlpubs.nist.gov/nistpubs/SpecialPublications/NIST.SP.800-184.pdf`) provides a lot of helpful information to set up a response process for your organization. If you prefer watching a video to reading the NIST documentation, the *10 Years of Crashing Google* video presentation (about Google's **Disaster Recovery Training (DiRT)** program) at `https://www.usenix.org/conference/lisa15/conference-program/presentation/krishnan` is helpful. You might also like seeing the *Incident Response @ FB*, Facebook's SEV Process video (about Facebook's response process) at `https://www.usenix.org/conference/srecon16europe/program/presentation/eason`.

- **Rely on transfer learning**: Training a new model is problematic because you may not have sufficient real-world data of the right type to do it properly. Using transfer learning allows you to take a tested model from another application and apply it to a new application you create. Because the transferred model has already seen real-world use, it's less likely that your new application will allow successful attacks.

- **Employ anomaly detection**: Recognizing unexpected patterns can alert you to suspicious activity before the activity creates any real damage. For example, when you suddenly see more categorizations in one area, it may mean that someone is trying to trigger a particular model behavior. Inputs to your model can also show changes in patterns, which sometimes suggest hacker activity. When testing your model for anomaly detection, it helps to have standardized datasets to use, such as the one used for the *Toxic Comment Classification Challenge* (`https://www.kaggle.com/c/jigsaw-toxic-comment-classification-challenge`).

When creating your security strategy, you need to keep the humans in the system in mind. For example, you should have an incident response team made up of people you can trust. This team might include outside contractors if your organization is large enough to support such people. The military relies on the two-person rule for sensitive areas and it's a good principle for you to follow as well. No one person should ever have exclusive responsibility for protecting any of your assets; security people should work in teams to reduce the risk that one or the other will present an internal security threat.

You should also consider using network security services, such as Cisco (`https://www.cisco.com/c/en/us/products/security/machine-learning-security.html`), that specialize in ML support to assess your network and create a list of potential upgrades you should perform. If you use a hosted site for your ML application, you need to ensure that the host provides services that will protect ML applications. Companies such as Amazon (see `https://aws.amazon.com/blogs/machine-learning/building-secure-machine-learning-environments-with-amazon-sagemaker/`) provide specialized services for ML needs and will show you how to use them.

The point of all this is that some security aids work no matter what it is that you're trying to protect. Building an entirely new security infrastructure to protect your ML application doesn't make sense when you can use some of the commonly accessible security principles to provide a starting point, and then add the special features you need for ML requirements.

## Detecting and mitigating an evasion attack

When attempting to mitigate an evasion attack, the old standby techniques come to mind. You need to use safe coding techniques and audit the code for vulnerabilities regularly, for example. In addition, maintaining logs and performing system auditing is also needed. The reason that these techniques work with your ML application is that the hacker is essentially doing the same thing as before ML came into play: evading your security to gain access to your system. Of course, if the hacker is using an ML application to perform the attack, detecting the invasion may prove difficult. That's why you need something a little better than traditional techniques to detect and mitigate incursions.

One of the more important methods of keeping your system safe from evasion attacks is to keep testing the applications you wish to protect. You may employ safe coding techniques, but the only way to ensure those techniques remain viable is to test the application out using specially crafted code that fully fakes an attack, without actually launching an attack. Of course, not everyone has time to create a test harness to perform this level of testing and may not have the skills, even if the time is available. That's why you want to check resources such as the following:

- **The Adversarial Robustness Toolbox**: This is an IBM product that comes with full documentation and complete source code that can be downloaded from GitHub (`https://github.com/Trusted-AI/adversarial-robustness-toolbox`). It provides support for checking your application against evasion, model poisoning, membership inference, and model stealing attacks.

- **CleverHans**: This product (named after a really smart horse) comes in two parts. The blog at `http://www.cleverhans.io/` provides great advice on how to harden your application, while the downloadable source at `https://github.com/cleverhans-lab/cleverhans` provides you with a testing application written in Python. You may need to install Jax, PyTorch, or TensorFlow 2 to use this product.

- **SecML**: A library you can use to enhance the robustness of your ML application against various attacks – most importantly, evasion attacks. The documentation appears at `https://secml.readthedocs.io/en/v0.15/`, while the Python downloadable source can be found at `https://gitlab.com/secml/secml`. Note that this package requires that you install additional support, such as NVIDIA GPU libraries, to gain the full benefit.

- **TensorFlow**: Commonly used to implement DL tasks, it's also the only common-use library that currently provides guidance on how to avoid, detect, and mitigate attacks. You can find a tutorial on these techniques at `https://www.tensorflow.org/tutorials/generative/adversarial_fgsm`.

Because evasion attacks are the most common form of attack and the consequences are so far-ranging in scope, you want to have some sort of process in place when working on new and existing application development. Here are some steps you can take to help mitigate the threat:

1. Create a threat assessment summary for the type of application you want to create and the kinds of data you need to protect.

2. Include adversarial examples as part of the training pipeline so that your application learns to recognize them.

3. Perform extensive testing on your application to ensure that it can combat both traditional evasion tactics and new tactics that hackers create using ML or DL technology.

4. Train everyone associated with the application to recognize threats and report them. In addition, everyone should know how to avoid infection, such as not opening the attachment coming from a party they don't know.

5. Use tools, such as the Adversarial ML Threat Matrix discussed earlier in this chapter, to constantly update your threat assessment after the application goes into production.

6. Update the application as often as possible to deal with new adversarial threats.

7. Keep current on the tools supplied to anyone (including developers, DBAs, data scientists, and researchers) working with the ML applications so that they can create effective strategies against evasion attacks.

When it comes to biometric-based evasion attacks, the best defense is to use the latest technology for detection. As stated in the article entitled *Liveness in biometrics: spoofing attacks and detection* at `https://www.thalesgroup.com/en/markets/digital-identity-and-security/government/inspired/liveness-detection`, the techniques demonstrated in the movie *Minority Report* don't work anymore. Most of these advances rely on DL techniques to perform the required analysis, which makes you wonder whether the DL technology can be spoofed in other ways (to allow an obvious fake biometric to pass for real).

## Detecting and mitigating a model poisoning attack

Model poisoning focuses on some type of change that will cause the model to perform unexpectedly. In most cases, you can't automate the process of detecting the model poisoning because once the model is poisoned, it tells you that everything is fine. Instead, you must rely on data monitoring and your observation of the results. Here are some rules of thumb you can employ:

- **Verify how the data is sampled**: It isn't reasonable to obtain all of the user input data used to train your model from a few IP addresses. You should also look for suspicious data patterns or the use of only a few kinds of records for training. In addition, it's essential not to overweight false positives or false negatives that users provide as input. Defenses against this sort of attack vector include limiting the number of inputs that a particular user or IP address can provide and using decaying weights based on the number of reports provided.

- **Compare a newly trained classifier to a known good classifier**: You can send some of the incoming data to the new classifier and the rest to the known good classifier to observe differences in results. You could also rely on a **dark launch** (where the classifier is tested on a small group of users without being announced in any way) to test the newly trained classifier in a production environment or use **backtesting** techniques (where you test the newly trained classifier using historical, real-world data to determine whether you obtain a reliable result).

- **Create a golden test dataset**: Your classifier should be able to pass a specific test using a baseline dataset. This golden test dataset should contain a mix of real and constructed data that demonstrates all of the features that the model should possess, including the ability to ward off attacks of various sorts. The purposeful inclusion of attacks in the dataset sets this kind of testing apart from backtesting, where you use actual historical data that may not contain any attack data at all.

- **Avoid creating a direct loop between input and weighting**: Using a direct loop allows hackers to control your model without much effort. For example, by providing negative feedback, a hacker could hope to trigger some sort of penalization against an opponent. You should authenticate any sort of input in some manner and combine authentication with other verification before accepting it.

- **Never assume that the supposed source of input is providing the input**: There are all sorts of methods of spoofing input sources. In some cases, input not provided by the source will penalize the source in some manner, so the application that performs the penalization is acting on behalf of a hacker, rather than maintaining input purity.

By adding these rules of thumb to any processes you use to create, upgrade, update, and test models, you can avoid many of the attack vectors that hackers currently use to quietly infiltrate your setup. Of course, hackers constantly improve their strategies, so your best line of defense is to stay informed by purposely setting time aside each day for security update reading.

## Detecting and mitigating a membership inference attack

One of the most important issues to consider when mounting a defense against membership inference attacks is that any strategy must consider two domains:

- Data, network, and application security

- Privacy

Creating an effective security strategy can't override any privacy considerations that the organization may need to meet. In addition, security and privacy concerns can't reduce the effectiveness and accuracy of the application results beyond a reasonable amount. Defense becomes a matter of striking the correct balance between all of the various concerns.

You will find that there are many different defenses currently thought of as effective against generalized membership inference attacks, but two stand out from the crowd:

- **Adversarial training**: This is where adversarial examples are added to the training dataset. This approach tends to avoid overfitting, the technique used to infer membership, in a manner that helps avoid the multi-step **projected gradient descent** (**PGD**) attack method. However, this approach can increase privacy leakage under certain conditions, such as the kind of adversarial examples used for training.

- **Provable defense**: This relies on computing an upper bound of the loss value in the adversarial setting, quantifying the robust error bound for the defended model. In other words, it defines the fraction of input examples that can be perturbed under the predefined perturbation constraint. Unfortunately, this approach can reduce the accuracy of the model.

There is a school of thought currently that by masking the confidence level of a model's output, it becomes impossible to perform a membership inference attack. Unfortunately, this is no longer the case. The article entitled *Label-Only Membership Inference Attacks* at `https://arxiv.org/pdf/2007.14321v2.pdf` describes how to perform a label-only attack. The article doesn't provide any code examples at the time of writing, but, likely, the site will eventually provide such code. The point is that merely hiding the confidence level won't protect the model or underlying data. One of the suggestions from the paper's authors is to employ methods that reduce or eliminate the overfitting of the model. They strongly support the use of transfer learning as part of a strategy to keep hackers at bay. In addition, the research shows that these regularization methods work best:

- Strong L2 regularization
- Training with differential privacy
- Dropout

To obtain a better idea of just how effective various forms of regularization can be in deterring a membership inference attack, review *On the Effectiveness of Regularization Against Membership Inference Attacks* at `https://arxiv.org/pdf/2006.05336v1.pdf`. In this particular case, the authors recommend the use of distillation, spatial dropout, and data augmentation with random clipping as the best forms of regularization to use with a minimal loss of accuracy. The takeaway from most research is that some form of dropout regularization will reduce vulnerability to membership inference attacks and seems to be the approach that everyone can agree upon.

There is some discussion of the effectiveness of differential privacy when looking at membership inference attacks because using differential privacy training can reduce the utility of an application. However, if you're working with sensitive data, such as health records, using differential privacy training seems to be a given.

It's also important to note that combining strategies can significantly improve a model's resistance to membership inference attacks, but combination selections must proceed with care and receive full testing. For example, combining early stopping with random cropping hurts accuracy considerably (to the point of making some models useless), but the effect on the accuracy of combining early stopping with distillation is minimal. The reason for this difference is that distillation speeds up training, while random cropping slows training down. You can read more about the effects of combined techniques in section 5, *Combining Multiple Mechanisms*, in the *On the Effectiveness of Regularization Against Membership Inference Attacks* whitepaper cited earlier in this section.

## Detecting and mitigating a Trojan attack

Earlier in this chapter, you discovered that Trojan attacks come in all sorts of forms, some traditional, some AI-specific. If you're working with a marketplace model, there is always the potential for downloading a Trojan. It's important to know who you're working with and what measures they have in place for detecting and mitigating Trojans. However, a good method for keeping Trojan attacks off your network is to use disconnected test systems to check out an API every time a new version becomes available. For the most part, you want to think about the potential threats first, and then act. Assuming that any source of information is safe is a truly bad idea, especially when hackers are working hard to make their nefarious payloads as stealthy as possible.

Researchers are currently working on several methods to identify and mitigate AI-based Trojan attacks. One of these systems, **STRong Intentional Perturbation** (**STRIP**), detects Trojan attacks by looking for variances in predictions based on inputs. When an attacker relies on triggers or perturbed inputs to force a model to work differently than intended, detecting those inputs can be difficult because how the trigger or perturbation is implemented is a secret known only to the hacker. The whitepaper, *STRIP: A Defence Against Trojan Attacks on Deep Neural Networks*, at `https://arxiv.org/pdf/1902.06531.pdf`, provides insights into how to implement a detection and mitigation system that will reduce the work required to deal with such attacks.

One of the issues that several sources have pointed out about Trojan attacks is that the one-pixel visual attack (see `https://arxiv.org/abs/1710.08864` for details) isn't practical in the real world, even though they can be quite effective in lab environments. Imperfections in cameras and other sensors would make such an attack impracticable. To ensure that the DNN receives the trigger, the trigger would need to be bigger and more noticeable to humans. Because the machinery controlled by the DNN often works autonomously, the fact that humans can see the trigger is less important unless the system has some sort of anomaly detection built in to alert a human to a possible trigger. Oddly enough, the issue of making triggers noticeable enough to overcome physical problems in sensors or variances in the software itself provides defense and mitigation. Simply ensuring that your maintenance staff checks the environment in which a DNN will operate, and locating the triggers that a human can detect is a good starting defense against problems.

It's also important to consider the proper vetting of datasets and training strategies for any type of ML application. In many cases, it's possible to locate the Trojan examples in a dataset before the dataset is used to train the model. However, when an application relies on third-party sources and vetting the data (or even checking the model closely) may not be possible, you need some other method of detecting that the model you create has a Trojan in it. In addition, some forms of data vetting and model checking won't work with certain kinds of Trojan attacks, such as an all-in-all attack where Trojan records appear in every class of the dataset, making comparison checks less likely to expose the Trojan elements.

The whitepaper entitled *Detecting AI Trojans Using Meta Neural Analysis* (`https://arxiv.org/pdf/1910.03137.pdf`) points out techniques for recognizing inputs that have a trigger associated with them. In this case, the strategy creates a special ML model that relies on **Meta Neural Trojan Detection** (**MNTD**) to test the target model without making any assumptions about the types of attack vectors it might contain. To avoid the problem of an attacker coming up with methods to fool the MNTD, the system performs fine-tuning during runtime. This means that the system is modifying itself as it runs and that the randomness of the change will likely keep the hacker from modifying the attack strategy.

## Detecting and mitigating backdoor (neural) attacks

There is some overlap between the AI-specific Trojan attacks and backdoor (neural) attacks. Mainly, they both rely on triggers of some sort. However, the triggers for a Trojan often modify a specific input, such as placing a sticker on a stop sign, while the triggers for a backdoor are generalized to the trigger itself, rather than its association with any particular kind of input. For example, a backdoor trigger might rely on a red sticker with a yellow square on it. No matter where this sticker appears, be it on a stop sign or an advertisement for a product, the trigger still works. Some of the whitepapers you read will combine the two terms and add some additional new terms for good measure, so it remains to be seen as to which terms and definitions eventually stick around long enough to become common usage. Of course, there is always the triggerless approach to consider in this case, which doesn't rely on specialized inputs but is triggered instead by a behavior or event in the neural network itself.

A backdoor attack always has some means of providing the incorrect output (generally favoring a hacker's requirements), whether it uses a trigger or not. The whitepaper entitled *Neural Cleanse: Identifying and Mitigating Backdoor Attacks in Neural Networks* at `https://people.cs.uchicago.edu/~ravenben/publications/pdf/backdoor-sp19.pdf` is more along the lines of a true backdoor treatment because it considers what is happening deep within the neural network, rather than simply focusing on particular kinds of inputs. In this case, the researchers have come up with a potential method for detecting and reverse engineering triggers hidden deep within the neural network. This particular whitepaper emphasizes the role of third parties in the infection process, in that a third party trained the model using a tainted dataset or added the backdoor after the fact after uploading it to a marketplace or service.

Unfortunately, there aren't many resources at this time for detecting and mitigating a triggerless backdoor except to perform specific tests to check for variances in output, given the same input. For example, if the trigger relies on the use of a dropout layer, some portion of the inputs that rely on the affected neuron will fail to produce the required output. The best strategy for detecting such a backdoor is to monitor the outputs for unexplained randomness or unexplained failures.

## Summary

This chapter began by defining adversarial ML, which is always the result of some entity purposely attacking the software to elicit a specific result. Consequently, unlike other kinds of damage, the data may not have any damage at all, or the damage may be so subtle as to defy easy recognition. The first step in recognizing that there is a problem is to determine why an attack would take place – to get into the hacker's mind and understand the underlying reason for the attack.

A second step in keeping hackers from attacking your software is to understand the security issues that face the ML system, which defies a one size fits all solution. A hospital doesn't quite face the same security issues that a financial institution does (certainly they face different legal requirements). Consequently, analyzing the needs of your particular organization and then putting security measures in place that keep a hacker at bay is essential. One of the most potent ways to keep hackers out is to employ ensemble ML in a manner that makes poisoning data used to train your model significantly harder.

The third step is to know that a determined hacker will break into your system, which means that building a high-access wall only goes so far in keeping your ML application safe. Detection is an essential part of the security process and it's ongoing. Fortunately, you can make the ML security functionality used to protect your system do most of the work. Even so, a human needs to monitor the process and look at the analysis of data flows to ensure that a hacker isn't fooling the system in some way.

The fourth step is mitigation. Realizing that adversarial ML by a determined hacker will succeed in some cases and defy efforts of immediate detection by even the most astute administrator is a critical goal. Once the system has been breached, it's essential to do something about it. Often, that means restoring data, reinstalling the application, possibly retraining the model, and then putting updated security processes in place based on lessons learned.

*Chapter 4, Considering the Threat Environment*, takes another step by considering the threat environment for both businesses and consumers as a whole. Many organizations only address business environment threats, and then only weakly. *Chapter 4, Considering the Threat Environment*, provides a good understanding of why paying attention to the entire environment in depth is critical if you want to keep your ML application and its data safe.

## Further reading

The following resources will provide you with some additional reading that you may find useful in understanding the materials in this chapter:

- Gain a better understanding of how membership inference attacks work: *Machine learning: What are membership inference attacks?*: `https://bdtechtalks.com/2021/04/23/machine-learning-membership-inference-attacks/`

- Discover more about the creation of shadow models: *Membership Inference Attacks Against Machine Learning Models*: `https://www.researchgate.net/figure/Training-shadow-models-using-the-same-machine-learning-platform-as-was-used-to-train-the_fig2_317002535`

# Part 2 – Creating a Secure System Using ML

This part challenges the assumption that only data scientists work with ML applications by reviewing the topic of ML security from a number of perspectives, especially from that of researchers.

This section includes the following chapters:

- *Chapter 4, Considering the Threat Environment*
- *Chapter 5, Keeping Your Network Clean*
- *Chapter 6, Detecting and Analyzing Anomalies*
- *Chapter 7, Dealing with Malware*
- *Chapter 8, Locating Potential Fraud*
- *Chapter 9, Defending against Hackers*

# 4

# Considering
# the Threat Environment

*Chapter 2* considered threats to your data, while *Chapter 3* considered threats to your application and model. This chapter considers the threats to your environment as a whole and divides environments into two parts: business and consumer. Business threats focus on the ability to earn money, serve clients, provide a useful infrastructure, and address business requirements (such as accounting and meeting legal needs). Consumer threats focus on communication between individuals, entertainment, buying products, and addressing personal needs (such as interacting with government entities or making appointments with your doctor).

(And, yes, you need to worry about the consumer element because your users will always incorporate consumer elements into the business environment.)

The previous chapters examined parts of the whole to make the threats easier to see and understand. This chapter is an introduction to the whole picture, of how things work together to create a particular kind of threat. You've seen the trees; now it's time to see the forest. With these issues in mind, this chapter discusses the following topics:

- Defining an environment
- Understanding business threats
- Considering social threats
- Employing **Machine Learning** (**ML**) in security in the real world

# Technical requirements

This chapter requires that you have access to either Google Colab or Jupyter Notebook to work with the example code. The *Requirements to use this book* section of *Chapter 1, Defining Machine Learning Security*, provides additional details on how to set up and configure your programming environment.

The *Accessing GitHub using OAuth-type authentication* section requires that you have a GitHub account, which you can create at `https://github.com/join`. When testing the code, use a test site, test data, and test APIs to avoid damaging production setups and to improve the reliability of the testing process.

Using the downloadable source code is always highly recommended. You can find the downloadable source on the Packt GitHub site at `https://github.com/PacktPublishing/Machine-Learning-Security-Principles` or my website at `http://www.johnmuellerbooks.com/source-code/`.

# Defining an environment

An **environment** is the sum of the interaction an object has with the world—whether it's an application running on a network, with the network or the internet as its environment, a robot running an assembly line, with the building housing the assembly line as its environment, or a human working in an office with the real world as an environment is immaterial. An environment defines the surroundings in which an entity operates and therefore interacts with other entities. Each environment is unique but contains common elements that make it possible to secure the environment. An ML environment includes the following elements, which are used as the basis for discussion as the chapter progresses:

- Data of any type and from any source

- An application model

- Ancillary code, such as libraries

- Interfaces to third-party code such as services

- An **Application Programming Interface** (**API**)

- Third-party applications that interact directly (such as applications that augment an organization's offerings) or indirectly (such as the shopping site that users surreptitiously use during work hours) with the environment

- Users (those who use the application, but don't control it)

- Managers (those who define organizational, environmental, or application policies)

- Developers (those who create any application code, including data scientists, computer scientists, researchers, database administrators, and so on)

- Security professionals (those who control application access)

Many of the tactics currently available to secure applications (including biometric and physical security) are equally applicable to any environment but you develop and interact with them in different ways. Any environment can benefit from authentication and filtering, but it's hardly likely that you'll find biometric authentication used to access a consumer product site, such as Amazon.com. On the other hand, a site devoted to governmental research will likely include several layers of authentication, including biometric authentication and guards at the door. This chapter doesn't include a discussion of physical security in the form of locked down server rooms and guards at the door, but it does cover a considerable range of application-specific security types, such as implementing passwords, locking down resources, looking for odd data patterns, and removing potentially malicious data.

## Understanding business threats

Business software solutions have become more complex over the years and so have the security threats facing them. Many businesses run a hybrid setup today where part of the business software resides locally on a network (some of which forms a private cloud-based configuration) and the other part is hosted online as one of the "as a service" options, such as **Platform as a Service** (**PaaS**). Consequently, security often comes in layers for businesses.

Traditional security is a starting point for the local part of the infrastructure and service-level security is part of the cloud-based component. The *Cloud Adoption Statistics for 2021* article at `https://hostingtribunal.com/blog/cloud-adoption-statistics/` is enlightening because it shows that, even if you consider only the cloud component of an organization, 69 percent rely on a hybrid solution for their cloud presence, and that some organizations leverage up to 5 different hosting solutions. It's unlikely that your ML application will be able to rely on a single data source or reside on a single setup when your organization is large enough. Consequently, you need to be ready to work with security professionals to secure your application to keep the data safe. Oddly enough, communicating your needs to security professionals who are used to dealing with monolithic applications and microservices is difficult unless you speak their lingo.

Unfortunately, a starting point isn't a solution. For example, when hosting your cloud-based solutions on **Amazon Web Services** (**AWS**), you have a choice of 26 different security-related services (as of the time of writing, you can be sure there will be more soon). Most of them are oriented toward protecting the cloud part of your software, so you still need other layers for the local part of your solution. Amazon does provide help for organizations using its services.

The security picture may seem overwhelming unless you begin to break it down into manageable pieces and review those pieces without necessarily looking at a particular solution until you need to get into the details of configuration and code writing. For example, it's important to know at the outset that you need to encrypt your data (even the open source material because you don't want anyone modifying it after you begin to use it); especially when that data is in a place where a hacker can reach it. However, you don't necessarily need to think about using AWS encrypted **Simple Storage Service** (**S3**) buckets until you choose to implement a part of your solution on AWS (see `https://docs.aws.amazon.com/AmazonS3/latest/userguide/bucket-encryption.html` if you'd like to see what's involved). Becoming mired in details at the outset is problematic because you begin to see individual trees and miss the forest entirely.

## Protecting consumer sites

The majority of attacks against businesses that host consumer sites are through websites or APIs. A website offers the flexibility a hacker needs to experiment and look for potential holes in coverage. It's possible for a hacker to attack through an API, for example, but an API allows better tracking of the incoming request so that the hacker's activities are easier to expose. However, an online store is unlikely to notice the hacker providing requests as little changes in an order through the website interface (as an example) while looking for an exploitable hole.

There are a considerable number of threats against consumer sites, many of which you can handle using traditional methods, but are better handled using ML solutions. For example, as shown in the *Manipulating filtered data* and *Creating an email filter* sections source code examples later, you can use ML applications to monitor the inflow of data for unusual data patterns or unwanted data. Backend ML applications can detect and remove bad data from the database using a technique similar to that shown in the *Starting with a simple removal* section. *Figure 4.1* shows the most common attacks against ML applications through a website and provides some ideas on how to protect against them:

| Attack Type | Major Consideration | Possible Remedy | Resources | Whitepaper/ Example |
| --- | --- | --- | --- | --- |
| **Cross-site scripting (XSS)** | A script is injected as input to a web form, API, or another endpoint, which ends up in a database. The script then executes each time a page is presented, the API is queried for specific data, or appears as part of an analysis performed by an ML application. | Train ML classifiers to detect the scripts or script changes found in the database and then act upon them. The technique in the *Starting with a simple removal* section is also helpful. | `https://www.researchgate.net/publica-tion/322704986_Detecting_Cross-Site_Scripting_Attacks_Using_Machine_Learning` | `https://thesai.org/Downloads/Volume11No5/Paper_85-Ensemble_Methods_to_Detect_XSS_Attacks.pdf` |

| Attack Type | Major Consideration | Possible Remedy | Resources | Whitepaper/ Example |
|---|---|---|---|---|
| **SQL injection** | Carefully crafted incorrect data is injected through a web form or an API to corrupt database data, steal data, or cause other damaging effects. | Begin by sanitizing the data using techniques such as the one shown in the *Manipulating filtered data* section. Then use an ensemble of classifiers (such as those shown in *Chapter 3, Mitigating Inference Risk by Avoiding Adversarial Machine Learning Attacks,* in the *Using ensemble learning* section) to detect and act upon the attack. | `https:// scholar- works.sjsu. edu/cgi/ viewcontent. cgi?arti- cle=1727&- context=etd_ projects` | `https:// scholarworks. sjsu.edu/cgi/ viewcontent. cgi?arti- cle=1649& contex- t=etd_projects` and `https:// portswig- ger.net/ daily-swig/ machine-learn- ing-of- fers-fresh-ap- proach-to-tack- ling-sql-in- jection-vul- nerabilities` |
| **Command injection** | This is a superset of XSS and SQL injection. An application creates a link between itself and a server that a hacker detects. The user issues commands as normal and receives the expected responses. A hacker also uses the same link to issue commands and the results of these commands are sent to the hacker, instead of the user. Part of the problem with this particular attack is that it's hard to detect because activity between the user and the server remains normal. | Normal filtering techniques can prove ineffective because of the stealthy nature of this attack. However, constant signature updates do make filters a little more effective. The best solution currently available is specially designed ML applications such as **Code-Injection Detection With Deep Learning** (**CODDLE**) and **Applications Management and Digital Operations Services** (**AMDOS**). | `https:// ieeexplore. ieee.org/ stamp/stamp. jsp?arnum- ber=8835902` and `https:// www.ibm.com/ downloads/ cas/DW5DGM8K` | `https://www. researchgate. net/ publication/ 335801314_ CODDLE_COde- injection_ Detection_ with_Deep_ LEarning` |

| Attack Type | Major Consideration | Possible Remedy | Resources | Whitepaper/ Example |
|---|---|---|---|---|
| **File-path traversal** | A hacker gains access to sensitive data (such as data used to train a model) by using specially configured paths. These paths either rely on the inherent weaknesses of relative paths that use the `../../` notation or known absolute paths. Once the hacker gains access to the directory, it's possible to look at the `config` file settings, corrupt data, or perform other malicious permanent storage modifications. | Ensure that every resource always has the required protection (see the *Understanding the kinds of application security* section in this chapter) and that any user access relies on the principle of least privilege. It's also possible to rely on special filtering of input data and API requests using a technique similar to that shown in the *Manipulating filtered data* section and pattern detection using the technique in the *Creating an email filter* section. | `https://www.geeksfor-geeks.org/path-travers-al-attack-pre-vention/` and `https://www.trendmi-cro.com/en_us/research/20/j/contentprovid-er-path-traver-al-flaw-on-esc-app-reveals-info.html` | `https://jisajour-nal.sprin-geropen.com/arti-cles/10.1186/s13174-019-0115-x` |
| **Distributed Denial-of-Service (DDoS)** | Packets of useless data and commands are sent from a group of systems under the hacker's control to overwhelm the victim's system and cause it to fail. Given that ML applications often require large amounts of network bandwidth, this class of application is inordinately affected by a DDoS attack. | Most methods today rely on detecting the attack and dealing with it on the victim's system. One proposed solution is to detect the attack from the source (such as the hacker's control machine or the various bots) using ML techniques and then cut off those attackers from the inputs to the victim. | `https://ieeex-plore.ieee.org/docu-ment/7013133` and `https://www.mdpi.com/2504-3900/63/1/51/pdf` | `http://palms.princeton.edu/system/files/Machine_Learn-ing_Based_DDoS_Attack_Detec-tion_From_Source_Side_in_Cloud_camera_ready.pdf` |

Figure 4.1 – Common threats against consumer sites

You can also find other kinds of attacks that aren't ML-specific but could have an effect on ML activities by affecting the environment in which the application executes. These attacks include the following:

- **Session hijacking**: This form of attack is also known as cookie poisoning. The hacker intercepts and modifies a cookie using a **man-in-the-middle** (**MITM**) attack to steal data, corrupt security measures, or both.

- **Scraping**: An activity that some view as a legitimate process for obtaining the huge quantities of data needed for certain types of ML analysis while others see the damage that the process can cause. However, for many websites, scraping is a serious threat that steals private data, obtains information for membership inference attacks, performs reverse engineering of the site for the purpose of illegitimate replication, discovers application operations, and performs other malicious acts as discussed in *Chapter 2* and *Chapter 3*. Consequently, you also see other articles that discuss the opposite side of the coin, such as *Web scraping protection: How to protect your website against crawler and scraper bots* at `https://datadome.co/bot-management-protection/scraper-crawler-bots-how-to-protect-your-website-against-intensive-scraping/`. The fact is that this kind of attack is more of a human nature issue and deals with the mistruth of perspective as described in the *Defining the human element* section in *Chapter 1, Defining Machine Learning Security*.

- **Carding**: Hackers gain access to lists of credit or gift cards in some manner, usually on the dark web. They usually start by making small purchases with each card to determine whether the card is still active. Once the hacker knows the card is legitimate, it's used to make a huge purchase of some type. Vendors are currently fighting back by using ML-based services to detect this kind of attack. Because this is such a huge topic, you will find it covered in detail in *Chapter 8*.

The point of all these threat sources is that consumer sites are security sieves. There are so many holes that it may seem impossible to plug them all, but with diligence, administrators of all stripes and developers can work together to plug most holes and detect intrusions left by others. The essential element in all these possible intrusions is to know what threats are currently in use and then guard against them using techniques such as those found in *Figure 4.1*.

## Understanding malware

The term **malware** refers to any software installed on a host system that causes harm to the host or the client systems attached to it through any means, including stealing anything such as data or company secrets. The software might damage the systems physically, steal data, corrupt data, encrypt the data for ransom demands, or perform a very wide range of other malicious tasks. By damaging the system, the malware can also put people's lives at risk, such as in a medical facility. Mind you, sometimes giants play this game, such as the cyberwar brewing between the US and Russia; see `https://www.nytimes.com/2019/06/15/us/politics/trump-cyber-russia-grid.html` and `https://www.bloomberg.com/news/features/2022-01-26/what-happens-when-russian-hackers-cyberattack-the-u-s-electric-power-grid`.

ML is adept at detecting, preventing, and fixing certain types of malware attacks, but you must write the application in a very flexible way. Fortunately, you can now find websites that can assist with both static and dynamic identification of various kinds of malware on local systems and, to a lesser extent, online sites, such as VirusTotal (`https://support.virustotal.com/hc/en-us/categories/360000160117-About-us`), so it may be less necessary to build low-level skills to disassemble and identify various malware executables and better to develop research skills to locate the information that others have found quickly. You can read more about this kind of attack in *Chapter 7*.

## Understanding network attacks

Most people think about **Denial of Service** (**DoS**) or DDoS attacks when hearing about network attacks. In fact, you see DDoS listed in *Figure 4.1*, discussed later in this chapter, and explored fully in *Chapter 5*. However, network attacks are often subtle, as discussed in the *Eyeing the small stuff* section in this chapter. Hackers might only sniff data from your network in a manner that's nearly impossible to detect (see `http://blog.johnmuellerbooks.com/2011/06/07/sniffing-telnet-using-wireshark/` for one of thousands of exploits). In fact, network attacks are extremely prevalent and hit the big players too. The fact is that if your network isn't safe, customers will hear about it (usually pasted in big letters at the beginning of an article about security), and your business will suffer.

## Eyeing the small stuff

Before proceeding further with the big issues that most people think about, it's important to look at the small things as well. It's easy to miss the small stuff in an organization. For example, the **Internet of Things** (**IoT**) makes an appearance almost everywhere today in devices that most people take for granted. These devices just seem to work, so it's easy to forget about them. However, these devices are connected to the internet and they are becoming more complex as people want them to do more. A thermostat may not seem like a big thing, but consider the fact that you can now find thermostats with an ML application controlling them, as described in *Swiss Researchers Create Machine Learning Thermostat* at `https://www.rtinsights.com/empa-machine-learning/`.

Of course, the first question is what a thermostat, even a really smart thermostat, has to do with ML application security. It turns out that some researchers have found a way to hack a thermostat, as described in the article *#DefCon: Thermostat Control Hacked to Host Ransomware* at `https://www.infosecurity-magazine.com/news/defcon-thermostat-control-hacked/`. Now, imagine what would happen if the thermostat somehow connected to someone's network, perhaps for the purpose of recording statistics, you see that a thermostat really can be a security threat.

In the thermostat scenario, a hacker gains access to the thermometer, adjusts the software to emit corrupted data to the logs, and waits for someone to read the log data. Once on a host system, the corrupted data allows the hacker to gain access to the host system, perhaps using a Trojan. At this point, the hacker can use the access to perform identity theft or hold the system for ransom. To prevent such an attack, you must layer defenses, employing the following:

- Standard forms of security such as passwords (see the *Understanding the kinds of application security* section in this chapter) on the thermostat and host system

- Ensembles to detect errant data input streams (see the *Using ensemble learning* section in *Chapter 3*) as part of the input to the ML application

- Trojan detection (see the *Understanding trojan attacks* section in *Chapter 3*) within the ML application

One of the biggest small holes in security today falls into the category of **Supervisory Control and Data Acquisition** (**SCADA**) systems. They run a great many things, including essentials such as pacemakers and water supply systems, not to mention electrical plants. The article *Understanding the Relative Insecurity of SCADA Systems* at `http://blog.johnmuellerbooks.com/2011/11/28/understanding-the-relative-insecurity-of-scada-systems/` seems outdated, but unfortunately, no one has bothered to secure these small systems, as described in the article *Biggest threats to ICS/SCADA systems* at `https://resources.infosecinstitute.com/topic/biggest-threats-to-ics-scada-systems/` (among many others). Your ML application may connect to these systems and these small holes are a real threat that you need to consider fixing.

## Dealing with web APIs

You see a lot of coverage about APIs in this book because modern businesses can't do without them. *Chapter 6* provides a special focus on the use of anomalies to create holes in APIs that hackers use to gain particular kinds of access. In addition, you see them mentioned as part of *Figure 4.1*. However, this section discusses a prevention and mitigation strategy called confidential computing, which is the use of a specially configured CPU to keep data encrypted until actual processing.

Desktop systems have a **Trusted Platform Module** (**TPM**) that the operating system can use to make working with specially designed applications considerably more secure (see `https://docs.microsoft.com/windows/security/information-protection/tpm/trusted-platform-module-overview` as an example). The TPM makes this latest version of Windows considerably harder to infect with malware, as described in the *Tom's Guide* article at `https://www.tomsguide.com/news/what-is-a-tpm-and-heres-why-you-need-it-for-windows-11` (which is one of many reasons that this book spends less time on local systems and more time on the cloud, IoT, and networks). Fortunately, confidential computing doesn't necessarily require a TPM because cloud providers also make it available as a service:

- **AWS**: `https://aws.amazon.com/ec2/nitro/nitro-enclaves/` (the site doesn't specifically call it confidential computing, but articles such as the one at `https://www.forbes.com/sites/janakirammsv/2020/10/30/aws-nitro-enclaves-bring-confidential-computing-to-amazon-ec2/?sh=45d98f771c8e` make it apparent that it is)

- **Azure**: `https://azure.microsoft.com/en-us/solutions/confidential-compute/`

- **Google**: `https://trustedcomputinggroup.org/resource/trusted-platform-module-tpm-summary/`

- **Intel**: `https://www.intel.com/content/www/us/en/security/confidential-computing.html`

The Confidential Computing Consortium (`https://confidentialcomputing.io/`) was formed by companies such as Alibaba, Arm, Baidu, IBM, Intel, Google Cloud, Microsoft, and Red Hat to make data protection in the cloud easier. Of course, there is no free lunch. While your data is a lot more secure using confidential computing, the services cost you considerably more and you need to also think about the performance loss in using them.

## Dealing with the hype cycle

When considering which security strategies to employ, you need to consider the hype cycle as described in *Hype Cycle for Security Operations, 2020* at `https://www.gartner.com/en/documents/3986721`. What this article tells you is that security follows a cycle:

- **Innovation**: Some event or new technology triggers new security strategies.

- **Inflated expectations**: Everyone gets sold a technology that is never going to work as anticipated.

- **Disillusionment**: People drop a perfectly good technology because it failed as a marketable item, rather than as a good technology. Being able to make a profit on technology within a specific timeframe is what keeps the technology alive through infusions of investment capital.

- **Enlightenment**: The early adopters begin to experience realistic expectations for the technology.

- **Productivity**: The technology is now in common use.

At least you now understand the security cycle to some extent. It's used for every new innovation in technology, not just ML. Unfortunately, developing and fully implementing a new security strategy can take up to ten years, during which time hackers raise some real havoc if they have come up with a zero-day exploit. To ensure that your ML application remains secure, you need to invest in the security technologies and strategies that have made it beyond the disillusionment phase as a minimum.

It may seem as if you could keep business and social concerns separate, but people are social beings, so social threats will creep into your business-related security setting as well. The next section describes social threats from a business perspective. Yes, many of these issues also affect people's personal lives, but the focus is on how these social threats will affect your business.

## Considering social threats

Social threats affect individuals the most. A social threat is something that entices the user to perform a risky behavior or compromises the individual in some way. Here are a few ideas to consider:

- **Social media**: A user is enticed to do things such as discuss company policies or strategies in the interest of being social.

- **Ads**: Someone presents an ad that discusses some new swanky product, but the ad ends up compromising the individual in some way, such as providing access to a social media account, a shopping site, or even your local network. ML makes it possible to create convincing ads based on actual buyer shopping habits.

- **Utilities**: A special tool allows the individual to do something interesting, such as changing the color of their Facebook site. You find utilities all over the place because people naturally want to fiddle with whatever it is that they think requires an update or change. A utility can plant a Trojan on the individual's machine or grab the individual's information.

- **Videos**: Have you seen enough cat videos yet? Well, try this video of someone surfing on a shark in Australia instead! Individuals become unresponsive while watching videos, giving a hacker the opportunity to steal them blind without them even noticing.

- **Followers**: Some interesting person needs more followers. ML makes it possible to infer who might be so interesting that the individual needs to follow them. Click the link and you can become one of those people who followed them into hacker heaven.

- **Terror**: Deep learning makes it possible to create a fake of anything using any media. So, someone sends the individual a link with the individual running down the middle of the main street naked. A link is supplied so that the outraged individual can complain and the hacker gains access.

- **Social engineering**: Hackers use ML to create a social engineering attack based on individual interests, associates, work environment, and the like. The hacker can pose as a salesperson, a colleague from another company, or whatever else it takes to gain the person's trust.

- **Blackmail**: Someone gains access to sensitive information that the individual thought was secure. The blackmail doesn't ever end. The individual will continue giving up information, resources, or whatever else the blackmailer requires until there is nothing left.

No matter how the entry is gained, the idea that a hacker has compromised your personal data or that a website is stalking you is frightening. However, social threats affect businesses as well. People run businesses and when an employee encounters a new threat, the threat applies to the organization too. ML actually makes the hacker's job a lot easier, if it wasn't incredibly easy already. *Chapters 7 through 9* discuss how ML has a big part to play in making social threats considerably more effective.

In some cases, a hacker will use social threats to gain information about an organization or individuals to gain a foothold in the organization itself. For example, profiling makes it possible for a hacker to perform social engineering attacks with greater ease. Tracking a user's activity also provides useful information. If the hacker really wants to create problems, identity theft makes it possible for the hacker to pose as the user of the organization from a remote location. In short, social threats are just as important to the security environment as business threats are, but in a different way.

## Spam

Spam is a major avenue of attack for most hackers (see https://www.howtogeek.com/412316/how-email-bombing-uses-spam-to-hide-an-attack/ for some examples). A hacker could use a spam attack to hide those error messages from your network, intrusion messages from an account, or just about anything else. Spam can also include subtleties such as capturing unintended clicks when a user clicks one after accidentally opening it. Even the best spam filters provide 99.9 percent coverage, which means that if you get 100 emails in a day, one piece of spam is likely to make it through each 10-day period (and most spam checkers just aren't that good). Hackers constantly modify the approach used to create spam to keep any new techniques for detecting spam off balance. The *Developing a simple spam filter example* section of this chapter shows one ML method for detecting spam, but even it isn't perfect. Consequently, it's likely that users will encounter spam and that the spam will eventually provide a vector for social engineering, phishing (see https://www.kaspersky.com/resource-center/threats/spam-phishing), or other attacks against the user's machine, the network, and your ML applications. *Chapters 8* and *9* detail how you can effectively guard against the fraud and hacker aspects of spam.

## Identity theft

Some people see identity theft as a user issue. It's true that the user will spend a great deal of time and money overcoming the effects of identity theft. However, depending on how the identity theft is perpetrated, the effects could be more significant to your business than the loss of money from purchases that no one will pay for after the merchandise is delivered. Even though identity is normally associated with credit or other personal issues, it can also affect your ML application in the following ways:

- The data in your database is corrupted when the identity thief poses as a legitimate user. It's hard to tell the real user's data from that of the identity thief.

- A hacker gains entry to your network using a stolen identity.

- Services are misdirected to the holder of the stolen identity, rather than the legitimate user.

- Analysis of social or other identity-based statistics becomes impossible. For example, which person in which area of town do you use for a profile?

- Top employees can lose security credentials or be compromised in other ways, causing harm to your business by making them unavailable for various tasks.

This is just the tip of the iceberg. Society has always depended to some degree on being able to positively identify individuals. However, that dependency is growing as more technology is added and a positive ID becomes essential.

## Unwanted tracking

Many users don't want businesses or other entities to know every detail of their lives, even if the business is legitimate. The popularity of articles such as *Here's how to avoid unwanted tracking online* at `https://www.techradar.com/news/avoiding-unwanted-tracking-online` and *4 Ways to Protect Your Phone's Data From Unwanted Tracking* at `https://preyproject.com/blog/en/4-ways-to-protect-phone-data-unwanted-tracking/` indicate that the desire for privacy is real. However, when a hacker begins tracking a person, things can get really interesting because, now, the loss of privacy affects more than just the user. For example, a hacker can use tracking to begin a social engineering attack or profile an organization for other kinds of attacks.

## Remote storage data loss or corruption

Employees typically store some amount of business data on their local hard drive (assuming their device has one). If you have remote access to that hard drive, then you can move the data or at least back it up. However, if the employee stores data on a remote server to which you lack access, the data now becomes a problem. You can't back the data up and a hacker could compromise the data, including any company secrets that the user left in plain sight. Even if the data isn't compromised, the fact that the user has it in an undisclosed location means that any corruption will also go unnoticed, which can ultimately affect your ML application in a number of ways (see *Chapter 2, Mitigating Risk at Training by Validating and Maintaining Datasets*). The two best ways to mitigate this threat are through employee monitoring and training.

## Account takeover

According to a number of online sources, users typically have 150 or more personal online accounts, each of which requires a password. However, users are unlikely to create a unique password for all of those accounts. For one thing, few users could memorize all of those passwords, and making each of those passwords strong is nearly impossible. While you may think that users would rely on a password manager (password wallet), the *Password Manager* survey results at `https://www.passwordmanager.com/password-manager-trust-survey/` point out that 65 percent of users don't trust them at all and that 48 percent won't use one. Interestingly enough, only 10 percent of users see **Multi-Factor Authentication** (**MFA**) as a viable alternative to using a password manager. What many users do is create an acceptably strong password and then use the same password everywhere. Consequently, when a hacker takes over a user's account, the hacker also gains insights into the user and possibly finds methods to discover the user's entire list of passwords, including the password for your ML application.

One of the best ways to detect this sort of attack is through behavioral analysis, as described in *Eliminating Account Takeovers with Machine Learning and Behavioural Analysis* at `https://www.brighttalk.com/webcast/17009/326415/eliminating-account-takeovers-with-machine-learning-and-behavioural-analysis` (you need to sign up for the free account). However, behavioral analysis can be time-consuming and requires intimate knowledge of the user.

So far, you have discovered both business and social threats, gained some ideas on how to detect them, and obtained a few tips on either preventing or mitigating them. All this material assumes that you have a stable environment and that the hardening you perform on your systems and applications remains effective. Unfortunately, nothing is stable and hackers have a habit of overcoming obstacles. The next section discusses how to make your setup flexible enough to adapt.

# Employing ML in security in the real world

The real world is ever-changing and quite messy. You may think that there is a straightforward simple solution to a problem, but it's unlikely that the solution to any given security problem is either straightforward or simple. What you often end up with is a layering of solutions that match the requirements of your environment. Consequently, you might find that an ML application designed to detect threats is part of a solution, the flexible part that learns and makes a successful attack less likely. However, you likely need to rely on traditional security and service-based security as well. It's also important to keep user training in mind and not neglect those small things.

The reality of ML is that it's a tool like any other tool and not somehow a magic wand that will remove all of your security problems. If *Chapter 3* shows you anything, it demonstrates that ML security exploits exist in great quantities and that users are often the worst enemies of ML-based solutions. However, with layering, it becomes possible to protect a network in a number of ways, including relying on ensembles to combine the best models for your particular environment. Two of the more common security-specific approaches to protecting a network are as follows:

- Ensuring user authentication (the validation that the user's identity is real) and authorization (giving the user the correct rights) go as planned
- Filtering out potentially hazardous data before the user even gets to see it

There are a number of ways to perform either of these tasks, but this chapter focuses on simple ML examples. The reason you want to use an ML application to perform these tasks is that the application has the potential to learn about new threats without waiting for signature updates or reprogramming. As the ML application becomes more aware of the techniques that a hacker employs to get through, the use of reinforcement learning can augment the training originally provided to the neural network and keep the hacker at bay (at least for a while).

## Understanding the kinds of application security

Banish any thought that there is just one type of security. Application security comes in many forms and each form works best in a particular scenario. *Chapter 5* considers the issue of keeping your network clean, which means using some type of security, but security must extend to the environment as a whole. Security comes down to a matter of control, but the biggest problem is determining what sort of control to use. Useful control must consider both the needs of the individual and the requirements of the organization. Making security measures too onerous will make adherence to policies less likely. Security that isn't robust enough leaves an organization open to attack. Consequently, you see mixes of security measures in the following forms in most environments:

- **Role-based**: Depends on the role that a user is performing at any given time, so that the same user may have more privileges in some situations than in others. For example, the user may have more privileges when accessing a resource locally than when accessing the same resource off-site from a mobile application. This is a flexible form of security, but also the most confusing for users. It works well for critical resources that contain sensitive information.

- **Attribute-based**: Used as an alternative to role-based security where the characteristics of a resource determine who can access it or what actions are acceptable. The focus is on the specifics of the resource, rather than on the role of the user.

- **Resource-based**: Depends on the resource that the user wants to access, with consistent access in all situations. This form of security is useful for less critical resources that users may need to access continually, so consistency is more important than other considerations.

- **Group-based**: Defines security measures based on the needs of a group, such as a workgroup or a department. Every individual in the group has the same access. This form of security is most useful for teams or people who perform the same task on common resources. It tends to reduce training costs. The criticality of the resource is dependent on the trust potential of the group as a whole.

- **Identity-based**: Focuses on the needs of an individual to provide access to somewhat critical resources. Because it provides equal access to the resource at all times no matter what role the user performs at the time, this form of security could potentially lead to leaks.

Many ML applications currently lack any of these forms of security, making them wide open to attack by any user who can gain access to them. Locking down an application means taking the following steps:

1. Requesting authorization
2. Authenticating the individual
3. Monitoring and logging their access
4. Verifying that each action is allowed by the application security profile

Following these steps will help you begin the process of ensuring that your ML application remains safe. Keeping track of what is and isn't effective is important because these steps will require augmentation depending on the particulars of your application.

## Considering the realities of the machine

An ML application can't think, isn't creative, and has extremely limited flexibility. This fact is often brought home to anyone who tries to process textual data to prevent the use of derogatory terms or to prevent the resulting corpus from becoming unfair in some way. The terms in this section are offensive, but I used discretion to try to avoid even more derogatory terms found on the internet. Obviously, I didn't want to include these terms in the book.

A modern form of derogatory comment takes the form of people's names, such as calling someone a Karen, Stacy, Becky, Kyle, Troy, Chad, or any of a number of other names. If you're interested, the definition at `https://www.dictionary.com/e/slang/karen/` provides some insights into the use of the term Karen.

Obtaining useful results from ML applications means removing the derogatory terms, all of them, from the data. Yet, the sentence, "Karen gave the salesperson a hard time about the price marked on the item." is impossible for the ML application to detect, so it remains in place. If left in place, a large enough selection of data with unfortunate terms poses a security risk because it can skew the results of an analysis or cause an ML application to act in a disastrous manner.

Some terms aren't even offensive depending on where they're used. If you use the name Wally in the US, it's just someone's name. However, the same name in some other English-speaking countries could mean that the person is stupid or foolish, which is something that you definitely want to remove from your data (see `https://www.phrases.org.uk/bulletin_board/46/messages/636.html` for details). That's why the technique in the *Developing a simple spam filter example* section might prove so helpful. It will at least move suspect data out of the dataset so a human can interpret it when a machine can't.

## Adding human intervention

Humans differ from each other considerably, which is a good thing because being different has helped the human race survive over the years. However, when considering security, being different isn't always a good thing because the security expert who plans a security strategy has no idea of how other humans in an organization will react to it. Humans can wreck any security strategy, sometimes without much thought, and usually, without ill intent. Simply entering data into a form in a manner never envisioned by the form's designer can create a problem. Failing to follow procedures or getting bored can cause all sorts of failures and ML applications aren't exempt from their effects. Users sometimes play games of what will happen if they do something unexpected, possibly hoping to see a software Easter egg.

When creating any security solution, it pays to employ all the stakeholders in an organization in some manner, especially the users who interact with the application and its attendant security on a daily basis. If a user can break your security, it's not the user's fault; it's how the security is implemented that is to blame. In fact, users who break your security are actually providing a service because if they can break your security, a hacker surely will, and it's unlikely that a hacker will tell you about it.

## Developing a simple authentication example

Online ML examples never incorporate any sort of access detection because the author is focusing on showing a programming technique. When creating a production application or an experimental application that uses sensitive data, you need some way to determine the identity of any entities accessing your ML application using authentication. When you authenticate a user, you only determine the user's identity and nothing else. Before the user can do anything, you must also authorize the user's activities.

You can find the code for the following examples in the `MLSec; 04; Authentication and Authorization.ipynb` file of the downloadable source code.

### Working with basic and digest authentication

There are many ways to accomplish authentication and the techniques used are defined by the following points:

- The kind of access
- The type of server
- The server security setup
- The application security setup

Here is an easy local application-level security access technique:

```
import getpass

user = getpass.getuser()
pwd = getpass.getpass("User Name : %s" % user)

if not pwd == 'secret':
    print('Illegal Entry!')
else:
    print('Welcome In!')
```

The code obtains the user's name and then asks for a password for that name. Of course, this kind of access only works for a local application. You wouldn't use it for a web-based application. This version is also simplified because you wouldn't store the passwords in plain text within the application itself. The password would appear in an encrypted database as a hash value and you'd turn whatever the user types into a hash using the technique shown in the *Relying on traditional methods example* section of *Chapter 2*. After you've hashed the user's password, you'd locate the username in the external database, obtain the hash from the database, and compare the user's hashed password to the hash found in the database.

Online authentication can also follow a simple strategy. Here's an example of this sort of access:

```
import requests
from requests.auth import HTTPDigestAuth

resource = 'http://localhost:8888/files/MLSec/Chapter04/
TestAccess.txt'
```

```
authenticate = HTTPDigestAuth('user', 'pass')
response = requests.get(resource, auth = authenticate)
print(response)
```

In this case, you use a basic technique to verify access to a particular resource, this one on the local machine through `localhost`. You build an authentication object consisting of the username and password, and then use it to obtain access to a resource. A response code of 200 indicates success. Most sites use a response code of 401 for a failed authentication, but some sites, such as GitHub, use a 404 response code instead.

Note that this example uses `HTTPDigestAuth`, which encrypts the username and password before sending it over the network. It's not the most secure method because it's vulnerable to a MITM attack but much better than using `HTTPBasicAuth` for a public API, because basic authentication sends everything in Base64 encoded text. Some security professionals recommend basic authentication for private networks where you can use SSL security, as described at `https://mark-kirby.co.uk/2013/how-to-authenticate-apis-http-basic-vs-http-digest/`. The request library also supports **Open Authentication** (**OAuth**) (see `https://pypi.org/project/requests-oauthlib/` for details), Kerberos (see `https://github.com/requests/requests-kerberos` for details), and **Windows NT Lan Manager** (**NTLM**) (see `https://github.com/requests/requests-ntlm` for details) methodologies.

### Accessing GitHub using OAuth-type authentication

Let's look at the specific example of accessing GitHub, which relies on an OAuth-type access strategy:

To use this example, you must first create an API access token by signing in to your GitHub account and then accessing the `https://github.com/settings/tokens` page.

After you click **Generate New Token**, you see a **New Personal Access Token** page where you provide a token name and decide what access rights the token should provide. For this example, all you really need is **repo**, **package**, and **user access**.

When you are finished with the configuration, click **Generate New Token**. Make sure you copy the token down immediately because you won't be able to access it later. (If you make a mistake, you can always delete the old token and create a new one.)

This simple example shows what you need to do to obtain a list of repositories for a given account. However, that's not really the point of the example. What you're really looking at is the authentication technique used to access specific resources and the use of GitHub isn't that pertinent—it could be any API (any API securing a protected resource). Use these steps to create the example:

1.  Import the required libraries:

    ```
    import requests
    import json
    ```

2.  Obtain the sign-in information. Note that you must replace `Your User Name` with your actual username and `The Token You Generated` with the token you created earlier:

```
resource = 'https://api.github.com/user/repos'
username = 'Your User Name'
token = 'The Token You Generated'
```

3.  Create the reusable session object:

```
session = requests.Session()
session.auth = (username, token)
```

4.  Request the list of repos for this user:

```
repos = json.loads(session.get(resource).text)
```

5.  Output the repo names:

```
for repo in repos:
    print(repo['name'])
```

This code is really an extension of the examples in the previous section. Note that you must supply your username and token (not your GitHub password) before running this example or you'll see an error. In this case, the code creates a GitHub session, then uses it to obtain a list of repositories owned or accessible by the user from `https://api.github.com/user/repos`. The example loads the repository information, which includes everything about the repository, not just the name, in JSON format. It then prints a list of names. The names you get depend on the repositories you have set up or shared with other GitHub users. The session object will also allow you to perform tasks such as creating new repositories. The tasks you can perform are limited by the token you generate. You can find extensive documentation about the GitHub REST API at `https://docs.github.com/en/rest/overview`.

This example demonstrates something else, authorization. Once you authenticate the user, you authorize certain actions by that user, such as by using the `get(resource)` session call. When generating the GitHub token, you define the actions that the token will allow. One user might be authorized to do everything, while another user might only be able to list the repository content and download files.

## Developing a simple spam filter example

Most people associate spam with email and text messages, and you do see ML applications keeping spam away from people all the time. However, for an ML application, spam is any data in any form that you don't want the application to see. Spam is the sort of information that will cause biased or unusable results because, unlike a human, an ML application doesn't know to ignore the spam. In most cases, spam is an annoyance rather than a purposeful attempt to attack your model. You can find the code for the following examples in the `MLSec_ 04_ Remove Unwanted Text.ipynb` file of the downloadable source code.

### Starting with a simple removal

When creating a secure input stream for your ML application, you need to think about layers of protection because a hacker is certainly going to pile on layers of attacks. Even if you've limited access to your application and its data sources, and provided an ensemble to predictively remove any data source that is most definitely bad, hackers can still try to get data through seemingly useful datasets. Consider the simple text file shown here (also found in `TestAccess.txt`):

```
You've gained access to this file.
This is a bad line.
This is another bad line.
This line is good.
And, this line is just sort of OK.
This is yet another bad line for good measure.
You don't want this bad line either.
Finally, this line is great!
```

Imagine that every line that has the word bad in it really is bad. Perhaps the data includes a script or unwanted values. In fact, perhaps the data just isn't useful. It's not necessarily bad, but if you include it in your analysis, the result is biased or perhaps skewed in some way. In short, the line with bad in it is some type of limited spam. It's not selling you a home in outer whatsit, but it's not helping your application either. When this sort of issue occurs, you can remove the bad lines and keep the good lines using code similar to that shown in the following steps:

1. Import the required libraries. When you perform these imports, the **Integrated Development Environment** (IDE) will tell you that it has downloaded stopwords needed for the example:

```
import numpy as np
import os
import nltk
nltk.download('stopwords')
from nltk.corpus import stopwords
nltk.download('punkt')
nltk.download('wordnet')
from collections import Counter
from sklearn.naive_bayes import MultinomialNB
from skleyer.metrics import confusion_matrix
```

2.  Create a function that accepts a filename and a target to remove unwanted lines. This function opens the file and keeps processing it line by line until there are no more lines:

```
def Remove_Lines(filename, target_word):
    useful_lines = []
    with open(filename) as entries:
        while True:
            line = entries.readline()
            if not line:
                break
            if not target_word.upper() in line.upper():
                useful_lines += [line.rstrip()]
    return useful_lines
```

3.  Define the file and target data to search, then create a list of good entries in the dataset and print them out:

```
filename = 'TestAccess.txt'
target = 'bad'

good_data = Remove_Lines(filename, target)
for entry in good_data:
    print(entry)
```

There is nothing magic about this code—you've used something like it before to process other text files. The difference is that you're now using a file-processing technique to add security to your data. Notice that you must set both the current word and the target word to uppercase (or lowercase as you like) to ensure the comparison works correctly. Here's the output from this example:

```
You've gained access to this file.
This line is good.
And, this line is just sort of OK.
Finally, this line is great!
```

Notice that all of the lines with the word bad in them are now gone.

### *Manipulating filtered data*

Most people who work with data understand the need to manipulate it in various ways to make the data better suited for analysis. For example, when performing text analysis, one of the first steps is to remove the stop words because they don't add anything useful to the dataset. Some of these same techniques can help you find patterns in input data so that it becomes harder for a hacker to sneak something in even after you remove the bad elements. For example, you might find odd repetitions of words, number sets, or other data that might normally appear infrequently, if at all, in a dataset that will alert you to potential hacker activity. The following steps show how to create a simple filter that helps you see unusual data or patterns. This code relies on the same libraries you imported in the previous section:

1.  Define a function to remove small words such as "to," "my," and "so" from the text:

```
def Remove_Stop_Words(data):
    stop_words = set(stopwords.words('english'))
    new_lines = []
    for line in data:
        words = line.split()
        filtered = [word for word in words
                    if word.lower() not in stop_words]
        new_lines += [' '.join(filtered)]
    return new_lines
```

2.  Define a function that will list each word individually, along with the count for that word:

```
def Create_Dictionary(data):
    all_words = []
    for line in data:
        words = line.split()
        all_words += words

    dictionary = Counter(all_words)
    return dictionary
```

3.  Define a function that creates a matrix showing word usage:

```
def Extract_Features(data, dictionary):
    features_matrix = np.zeros(
        (len(data),len(dictionary)))
    lineID = 0
```

```
    for line in data:
        words = line.split()
        for word in words:
          wordID = 0
          for i,d in enumerate(dictionary):
            if d == word:
              wordID = i
              features_matrix[lineID, wordID] += 1
        lineID += 1
    return features_matrix
```

4. Create a filtered list of text strings from the original text that has the stop words removed:

```
filtered = Remove_Stop_Words(good_data)
print(filtered)
```

5. Create a dictionary of words from the filtered list:

```
word_dict = Create_Dictionary(filtered)
print(word_dict)
```

6. Create a matrix showing which words are used and when in each dataset row:

```
word_matrix = Extract_Features(filtered, word_dict)
print(word_matrix)
```

Each of the functions in this example shows a progression:

1. Remove the stop words from each line in the dataset that was created from the original file.
2. Create a dictionary of important words based on the filtered dataset.
3. Define a matrix that shows each line of the dataset as rows and the words within that row as columns. A value of 1 indicates that the word appears in the specified row.

There are some interesting bits of code in the example. For example, Remove_Stop_Words() relies on a list comprehension to perform the actual processing. You could also use a for loop if desired. You must also use join() to join the individual words back together and place them in a list to perform additional processing. The output looks like this:

```
["You've gained access file.", 'line good.', 'And, line
  sort OK.', 'Finally, line great!']
```

A dictionary is essential for many types of processing. `Create_Dictionary()` makes use of the `Counter()` function found in the `collections` library to make short work of creating the dictionary in a form that will make defining the matrix easy. Here's the output from this step:

```
Counter({'line': 3, "You've": 1, 'gained': 1, 'access': 1,
  'file.': 1, 'good.': 1, 'And,': 1, 'sort': 1, 'OK.': 1,
  'Finally,': 1, 'great!': 1})
```

The output doesn't appear in any particular order and it's not necessary that it does. Each unique word in the dataset appears as an individual dictionary key. The values show the number of times that the word appears. Consequently, you could use this output to perform tasks such as determining word frequency. In this case, the example simply creates a matrix to show where the words appear within the dataset. There are possibly shorter ways to perform this task, but the example uses a straightforward approach that processes each word in turn and finds its position in the matrix by enumerating the dictionary. Here's the output from this step:

```
[[1. 1. 1. 1. 0. 0. 0. 0. 0. 0. 0.]
 [0. 0. 0. 0. 1. 1. 0. 0. 0. 0. 0.]
 [0. 0. 0. 0. 1. 0. 1. 1. 1. 0. 0.]
 [0. 0. 0. 0. 1. 0. 0. 0. 0. 1. 1.]]
```

If you look at the first row, the first four entries have a 1 in them for `You've`, `gained`, `access`, and `file`. None of these words appear in the other rows, so the entries are 0 in the other rows. However, `line` does appear in three of the rows, so there is a 1 for that entry in each of the rows. The next section takes these techniques and shows how to apply them to multiple files in an email dataset.

### Creating an email filter

Emails can contain a great deal of useless or harmful information. At one time, email filters worked similarly to the example in the previous section (a simple filter). However, trying to keep track of all of the words that hackers use to get past the filter became impossible. Even though the simple filtering technique is still useful for certain needs, email filtering requires something better—an approach that is flexible enough to change with the techniques that hackers use to attempt to get past the filter. One such approach is to use an ML application to discover which emails are useful and which are spam.

The example in this section performs a simple analysis of the useful (ham) versus spam orientation of each email in the Ling-Spam email corpus described at `http://www2.aueb.gr/users/ion/docs/ir_memory_based_antispam_filtering.pdf` and available for download at `http://www.aueb.gr/users/ion/data/lingspam_public.tar.gz`. The original dataset is relatively complex and somewhat unwieldy, so the example actually uses a subset of the messages split into two folders: `Email_Train` and `Email_Test`. To save some time and processing, the example relies on the content of the `\lingspam_public\lingspam_public\lemm_stop\` folder, which provides the messages with the stop words already processed and the words normalized using lemmatization (see the *Choosing between stemming and lemmatization* section for details). The messages in the Email_Train folder come from the part1, part2, and part3 folders (867 messages in total with 144 spam messages), while the messages in the Email_Test folder come from the part4 folder (289 messages in total with 48 spam messages). You can tell which messages contain spam because they start with the letters spmsg (for spam message).

---

**Recognizing the benefits of targeted training and testing data**

Even though this example uses a generic database, it's always better to use your organization's email to train and test any model you create. Doing so will greatly decrease the number of false positives and negatives in the production environment because the data will reflect what the users actually receive. For example, an engineering firm specializing in fluid dynamics can expect to receive a lot more emails about valves than a financial firm will. This same principle holds true for all sorts of other filtering needs. The data from your organization will always provide better results than generic data will. Of course, you need to make sure that any organizational data you use meets privacy requirements and is properly sanitized before you use it, as described in *Chapter 13.*

---

Each of the text files contains three lines. The first line is the email subject, the second line is blank, and the third line contains the message. In processing the emails, you look at just the third line with regard to content and know that you can label the training messages as spam if the filename begins with spmsg or ham when the filename begins with something else. With this in mind, the following code shows a spam filter you can create using techniques similar to those used in the previous section but using multiple files in this case. This code relies on the same libraries you imported in the *Starting with a simple removal* section (make sure you use the 1.0.*x* version, originally version 0.23.*x*, of scikit-learn, as described in *Chapter 1,* for this part of the chapter or you may encounter errors):

1. Set the paths for the training and testing messages:

```
train_path = "Email_Train"
train_emails = \
    [os.path.join(train_path,f) for f
    in os.listdir(train_path)]

test_path = "Email_Test"
test_emails = \
    [os.path.join(test_path,f) for f
    in os.listdir(test_path)]
```

2. Create a dictionary function to build the required dictionary. Then, remove non-word items that include numbers, special characters, end-of-line characters, and so on:

```
def Create_Mail_Dictionary(emails):
    cvec = CountVectorizer(
        stop_words='english',
        token_pattern=r'\b[a-zA-Z]{2,}\b',
        max_features=2000)
    corpus = [open(email).read() for email in emails]
    cvec.fit(corpus)
    return cvec


train_cvec = Create_Mail_Dictionary(train_emails)
```

3. Create a features matrix function. Instead of lines and words, this code uses documents and words for the matrix:

```
def Extract_Mail_Features(emails, cvec):
    corpus = [open(email).read() for email in emails]
    return cvec.transform(corpus)


train_feat = Extract_Mail_Features(train_emails,
    train_cvec)
test_feat = Extract_Mail_Features(test_emails,
    train_cvec)
```

4. Create labels showing which messages are ham (0) and spam (1):

```
train_labels = np.zeros(867)
train_labels[723:867] = 1

test_labels = np.zeros(289)
test_labels[241:289] = 1
```

5. Train the Multinomial Naïve Bayes model:

```
MNB = MultinomialNB()
MNB.fit(train_feat, train_labels)
```

6.  Predict which of the messages in the test group are ham or spam and output the correctness of the prediction as a confusion matrix:

```
result = MNB.predict(test_feat)
print(confusion_matrix(test_labels, result))
```

7.  Display the confusion matrix in a nicely plotted form:

```
matrix = plot_confusion_matrix(MNB,
                               X=test_feat,
                               y_true=test_labels,
                               cmap=plt.cm.Blues)
plt.title('Confusion matrix for spam classifier')
plt.show(matrix)
plt.show()
```

The listing shows that simple techniques often provide the basis for more complex processing. The `Create_Mail_Dictionary()` and `Extract_Mail_Features()` functions provide the ability to work with multiple files and to provide additional data cleaning. Notice that this example uses a more efficient method of creating the dictionary using `scikit-learn CountVectorizer()`. The concept and the result are the same as what you see in the previous section, but this approach is shorter and more efficient. The `Extract_Mail_Features()` function is also made shorter by using list comprehensions in addition to calling the `cvec.transform()` function on the resulting corpus. Again, the output is the same and the process is the same under the covers, but you're using a more efficient approach.

The Multinomial Naïve Bayes model will vary in its ability to correctly predict ham or spam messages after you fit it to the training data. In this case, the result shows that there are 241 ham messages and 48 spam messages in the test dataset. A larger test dataset is likely to show a less impressive result, but according to *Machine learning for email spam filtering: review, approaches and open research problems* at `https://www.sciencedirect.com/science/article/pii/S2405844018353404`, some companies, such as Google, have achieved rates as high as 99.9 percent. In this case, however, the companies use advanced ML strategies, rather than the more basic Multinomial Naïve Bayes model. In addition, the strategies rely on ensembles of learners as suggested in the *Using ensemble learning* section of *Chapter 3*.

## Choosing between stemming and lemmatization

There are two common techniques for normalizing words within documents: stemming and lemmatization. Each has its uses. **Stemming** simply removes the prefixes and suffixes of words to normalize the root word. For example, player, plays, and playing would all be stemmed from the root word play. This technique is mostly used for word analysis, such as determining how often particular words appear in one or more documents. **Lemmatization** processes the words in context, so that the words running, runs, and ran all appear as the root word run. You use this technique most often for text analysis, such as determining the relationships of words in a spam message versus a usable (ham) message. Here is an example of stemming:

```
from nltk.stem import LancasterStemmer
from nltk.tokenize import word_tokenize

LS = LancasterStemmer()
print(LS.stem("player"))
print(LS.stem("plays"))
print(LS.stem("playing"))

tokens = word_tokenize("Gary played the player piano while
playing cards.")
stemmed = [LS.stem(word) for word in tokens]
print(" ".join(stemmed))
```

The example imports the required libraries, creates an instance of LancasterStemmer() and then uses the instance to stem three words with the same root. It then does the same thing for a sentence containing the three words. The output shows that context isn't taken into account and it's possible to end up with some non-words:

```
play
play
play
gary play the play piano whil play card .
```

Lemmatization takes a different route, as shown in this example (note that you may have to add the nltk.download('omw-1.4') statement after the import statement if you see an error message after running this code):

```
from nltk.stem import WordNetLemmatizer

WNL = WordNetLemmatizer()
print(WNL.lemmatize("player", pos="v"))
print(WNL.lemmatize("plays", pos="v"))
print(WNL.lemmatize("playing", pos="v"))

tokens = word_tokenize("Gary played the player piano while
playing cards.")
```

```
lemmatized = [WNL.lemmatize(word, pos="v") for word in tokens]
print(" ".join(lemmatized))
```

Notice the pos argument in the lemmatize() calls. This argument provides the context for performing the task and can be any of these values: adjective (a), satellite adjective (s), adverb (r), noun (n), and verb (v). In choosing verbs, the example provides this output, which you can contrast with stemming:

```
player
play
play
Gary play the player piano while play card .
```

The point is that you must choose carefully between stemming and lemmatization when creating filters for your ML application. Choosing the right process will result in significantly better results in most cases.

## Summary

This chapter helped you understand both business and social threats to your ML application, what to look for, how to mitigate attacks when they occur, and how to keep them from happening in the first place. The goal is to provide a flexible setup that makes the hacker work so hard that going somewhere else becomes attractive. Never assume that the hacker can't break your security. In fact, presenting any sort of challenge will keep a hacker interested until your security does break, so always assume that any security threat can gain access if wanted.

Layering is an essential part of any security solution. Using layers adds complexity, which is a double-edged sword. On the one hand, it makes the hacker's job harder by putting up barriers that change over time, as administrators learn and correct misconceptions about how security should appear. On the other hand, as anyone who does reliability studies will tell you, more parts mean more things to break, which reduces the reliability of the setup being protected. Consequently, more layers are good, but more layers than you actually need only makes your system unreliable.

Thinking about complexity, the next chapter will zoom in on the network itself. Most hackers are after your network, not an individual machine. Given that users generally have at least two systems they use to access ML applications, infecting just one machine likely isn't enough to provide the hacker with a carte blanche to enter your application. Keeping your network clean is a requirement if you want to keep your ML application safe and you need to consider both the local network and the network in the cloud.

# Further reading

The following links provide you with some additional reading that you may find useful to further understand the materials in this chapter:

- This link helps you discover more about the ML component of a SageMaker application: *Building secure machine learning environments with Amazon SageMaker*:

  `https://aws.amazon.com/blogs/machine-learning/building-secure-machine-learning-environments-with-amazon-sagemaker/`

- Learn more about cookie poisoning: `https://www.f5.com/services/resources/glossary/cookie-poisoning`

- See how to perform scraping appropriately: *How to scrape websites without getting blocked*: `https://www.scrapehero.com/how-to-prevent-getting-blacklisted-while-scraping/`

- Discover how to use ML techniques to perform scraping: `https://towardsdatascience.com/web-scraping-for-machine-learning-5fffb7047f70`

- Discover how to use ML to detect scraping efforts:

  `https://kth.diva-portal.org/smash/get/diva2:1117695/FULLTEXT01.pdf`

- Learn some additional detail on the carding attack type: *How to Use AI and Machine Learning in Fraud Detection*:

  `https://spd.group/machine-learning/fraud-detection-with-machine-learning/`

- Discover how malware can cause physical network damage: *Emotet Malware Causes Physical Damage*: `https://securityboulevard.com/2020/04/emotat-malware-causes-physical-damage/`

- Learn more about how malware can cause bodily harm: `https://www.bbc.com/news/technology-54204356`

- Read about the effect of loss of SCADA control led to a power plant hack in Ukraine: *Everything We Know About Ukraine's Power Plant Hack*: `https://www.wired.com/2016/01/everything-we-know-about-ukraines-power-plant-hack//`

- Provides detailed information about how SCADA and IoT are linked in ways that could cause serious problems: *What are SCADA and IoT?*: `https://www.datashieldprotect.com/blog/what-is-scada-iot`

- Learn more about how confidential computing works: *Confidential Computing* article at

  `https://www.ibm.com/cloud/learn/confidential-computing`

- Discover how a TPM works in detail: `https://trustedcomputinggroup.org/resource/trusted-platform-module-tpm-summary/`

- Discover why Windows 11 requires the use of the TPM 2.0 chip: *What is TPM 2.0 — the chip you need to run Windows 11:* `https://www.laptopmag.com/articles/tpm-chip-faq`

- Learn about specific exploits against older TPM chips that may affect your ML application: *Researchers Detail Two New Attacks on TPM Chips:* `https://www.bleepingcomputer.com/news/security/researchers-detail-two-new-attacks-on-tpmchips/`

- Understand why it takes up to 10 years to develop and implement a new security strategy:

  `https://www.hindawi.com/journals/je/2020/5267564/`

- Provides insights into how to perform behavior analysis: *Using machine learning to understand customer's behavior:* `https://towardsdatascience.com/using-machine-learning-to-understand-customers-behavior-f41b567d3a50`

- Learn more about attribute-based access: `https://www.okta.com/blog/2020/09/attribute-based-access-control-abac/`

- Gain a better understanding of the difference between basic and digest authentication: `https://www.hackingarticles.in/understanding-http-authentication-basic-digest/`

- Understand the difference between the way GitHub apps and OAuth apps work: `https://docs.github.com/en/developers/apps/differences-between-github-apps-and-oauth-apps`

- Find a discussion of the features used for advanced machine learning classifiers used for spam detection: *Deep convolutional forest: a dynamic deep ensemble approach for spam detection in text:* `https://www.ncbi.nlm.nih.gov/pmc/articles/PMC9039275/`

- See an example of how to create a spelling corrector that could be combined with spam detection: *Spelling Recommender With NLTK:* `https://jorgepit-14189.medium.com/spelling-recommender-with-nltk-2fb6fe94a7b3`

- Discover a technique for detecting fake news that can also be used to help with spam detection: *Detecting Fake News with Python and Machine Learning:* `https://data-flair.training/blogs/advanced-python-project-detecting-fake-news/`

# Keeping Your Network Clean

A **network** is the sum of all environments within an organization, even those not directly controlled by the organization. For example, an environment could consist of a database management application that resides partly on local servers and partly on hosted servers in the cloud, so part of the environment is controlled directly by the organization and another part is controlled by a third party. The same holds true for applications that rely on third-party services or access data through third-party APIs. In addition, users often rely on more than one device to perform work, and some of those devices are owned by the user, rather than the organization.

The current environment demands new ways of ensuring control of resources through a combination of traditional and other means that are more flexible and have a broader range than protections in the past. Because hackers often employ **zero-day exploits** nowadays (those that occur immediately after a new threat is exposed, often before an organization knows about it), real-time analysis and mitigation are critical. In fact, a modern network requires some level of predictive protection so that a hacker doesn't find it easy to break in. These sorts of protections are impossible using just traditional techniques; you really need **machine learning** (**ML**) methods to detect, assess, and mitigate the threats.

With these issues in mind, this chapter will discuss these topics:

- Defining current network threats
- Considering traditional protections
- Adding ML to the mix
- Creating real-time defenses
- Developing predictive defenses

# Technical requirements

This chapter requires that you have access to either Google Colab or Jupyter Notebook to work with the example code. The *Requirements to use this book* section of *Chapter 1, Defining Machine Learning Security*, provides additional details on how to set up and configure your programming environment. When testing the code, use a test site, test data, and test APIs to avoid damaging production setups and to improve the reliability of the testing process. Testing over a non-production network is highly recommended but not absolutely necessary. Using the downloadable source is always highly recommended. You can find the downloadable source on the Packt GitHub site at `https://github.com/PacktPublishing/Machine-Learning-Security-Principles` or my website at `http://www.johnmuellerbooks.com/source-code/`.

# Defining current network threats

Network threats go well beyond the application level, and it's unlikely that a single individual would provide support for every protective means that a network will require. For example, developers aren't going to handle physical security – a security company will likely handle it. However, your ML application may interact with the physical security system by monitoring cameras and other sensors. If you think this is a little futuristic, companies such as Bosch (`https://www.boschsecurity.com/xc/en/solutions/video-systems/video-analytics/`) and Nelly's Security (`https://www.nellyssecurity.com/blog/articles/what-is-deep-learning-ai-and-why-is-it-important-for-video-surveillance`) have products available today. An ML application can look for trends, such as an attacker who is casing a business before attempting to break in. The human monitoring the cameras may not see that the same person shows up on various nights, yet never enters the building. This chapter focuses more on the software end of networks, but it's important to keep the hardware element in mind.

## Developing a sense of control over chaos

It's essential to define control in the context of security. Chaos reigns on the internet today because so many people have so many views on how to manage it. Because there is no centralized strategy for managing issues such as security, hackers are able to place **data wedges** (exploits such as code insertion, viruses, and trojans) in various ways that can create a security nightmare for your business's network. However, this chaos also applies to casual use of the internet by users who do their shopping at work when the organization allows such an activity, because the user is using organizational resources to interact with the world. Many of these interactions are currently controlled through traditional means, such as honeypots, to detect hacker activities. A **honeypot** (essentially, a fake monitored system) can help detect intrusions that piggyback on legitimate requests.

Businesses rely on huge numbers of interconnected machines that share expensive resources to promote data sharing and keep costs low in the form of networks. These interconnected machines work both separately and together at the behest of their operators to perform tasks of all sorts. The term *network* applies to the entirety of connected systems, whether local or cloud-based, whether part of the organization or part of a third party, and whether directly or indirectly connected. When you consider a network in these terms, the vastness of the configuration can become overwhelming, yet you must consider the network at this level or suffer the consequences wrought by hackers, outsiders, insiders, well-meaning users, and inept managers. Keeping a network clean means constantly monitoring conditions for any form of threat that would damage infrastructure, mangle data, or prevent users from accomplishing something useful. Unfortunately, networks are inherently unsafe because someone, somewhere, is almost certainly going to leave the door open to some form of attack. Consequently, knowing these attack types is essential.

Traditional techniques are effective to an extent when managed well and updated constantly to address new threats. However, the task of managing and updating security, plus checking data and keeping logs, is daunting and impossible to perform manually. ML strategies help an administrator keep up to date by performing many tasks automatically so that they can focus on tasks that require special attention. In fact, ML-based tools can often help the administrator discover those tasks that are most important to complete.

However, it's in the proactive nature of a real-time defense that networks can gain the biggest advantage against adversaries. Real-time defenses depend on ML applications that can detect specific patterns of attack, adjust neural networks to address new types of attack, and generally anticipate certain types of human behavior. Predictive ML methodologies take protection one step further by making it possible to create a defense before the attack comes. Of course, the efficacy of predictive measures relies on tracking trends, and these measures can't address black swan events (see the *Understanding the Black Swan theory* section in *Chapter 3, Mitigating Inference Risk by Avoiding Adversarial Machine Learning Attacks*, for details). Now that you have a sense of what control means with regard to security, it's time to look at the categories of network control in the ML domain.

## Implementing access control

As someone who works with ML code for any reason, you likely know that it's important to ensure that you have firm control over the data you download, create, or interact with in any other way. However, hackers don't just work with data files. They also use any other sort of access available. For example, Microsoft used to make Telnet (a notoriously high-security risk) a default part of Windows. Telnet Server has not been available on its server products since 2016. However, you can still install Telnet Client as an optional product on Windows Server 2019 (`https://www.rootusers.com/how-to-enable-telnet-client-in-windows-server-2019/`), Windows 10 (`https://social.technet.microsoft.com/wiki/contents/articles/38433.windows-10-enabling-telnet-client.aspx`), and even the supposedly more secure Windows Server 2022 and Windows 11 (`https://petri.com/enable-telnet-client-in-windows-11-and-server-2022/`).

Generally speaking, a network administrator will know not to install Telnet because there are other, more secure, methods of achieving the same thing. The most common replacement is **Secure Shell (SSH)**. You can read about the differences between Telnet and SSH at `https://www.tutorialspoint.com/difference-between-ssh-and-telnet`. Some Windows administrators now use a PowerShell equivalent, as explained at `https://www.techtutsonline.com/powershell-alternative-telnet-command/`, but this solution only works on Windows. If you don't want to work at the command line, then you might consider using PuTTY instead (`https://www.makeuseof.com/tag/windows-10-ssh-vs-putty/`). The GUI interface shown in *Figure 5.1* makes working with network connections a lot easier, especially if you're not used to working with them.

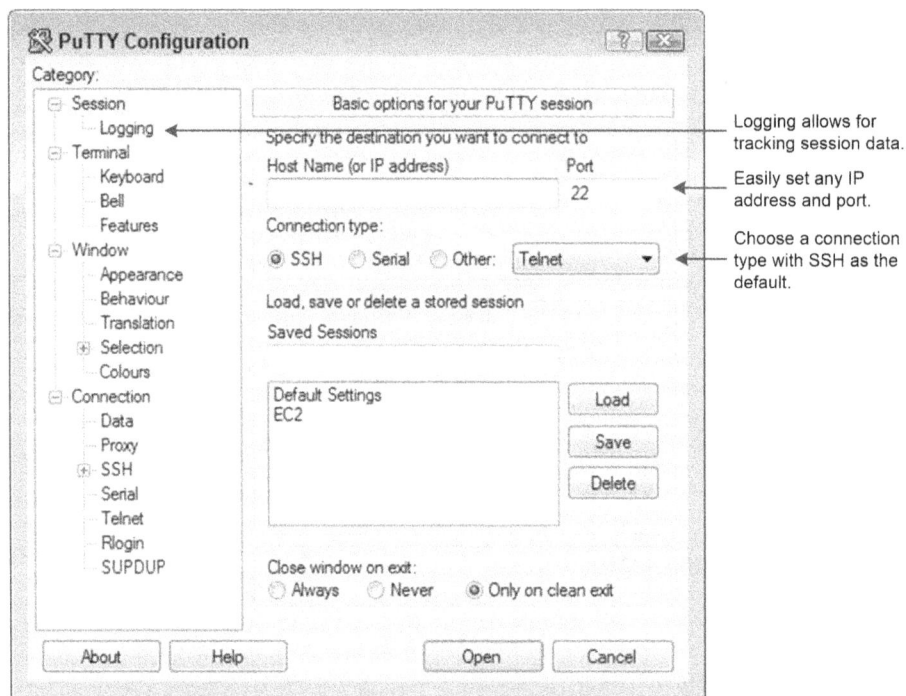

Figure 5.1 – ML developers need not fight with a command line
interface when testing security-related ML applications

What the network administrator won't know about is some product vulnerabilities, such as the **Remote Code Execution (RCE)** problem with Python (`https://www.zdnet.com/article/python-programming-language-hurries-out-update-to-tackle-remote-code-vulnerability/`). If you use Python to create your ML model, then you also need to keep track of required updates or you could inadvertently create a security hole for your organization while trying to protect it. NumPy (`https://numpy.org/`) has a similar RCE flaw (`https://www.cybersecurity-help.cz/vdb/SB2019012101`). (Other Python libraries also have various vulnerabilities, including RCE issues.)

*Figure 3.1* and *Figure 3.2* show the use of ensembles of learners to control access to specific applications. It turns out that you can also use this tactic for controlling and monitoring access to your network, as described in *AI, machine learning and your access network* at `https://www.networkworld.com/article/3256013/ai-machine-learning-and-your-access-network.html`. What this article points out is that using ML tools can focus administrator attention on actual problems after the ML tool has already scoured the incoming data.

## Ensuring authentication

Authentication must work at several levels to ensure the integrity of a network and its constituent parts. When someone is authenticated, the system has ensured that the person's identity is known. The *Developing a simple authentication example* section of *Chapter 4, Considering the Threat Environment*, discusses the role of authentication in application security, and the same need is evident at the network level. The example code in the *Accessing GitHub using OAuth-type authentication* section of *Chapter 4* demonstrates authentication and another requirement, authorization, which provides actual access to a resource by a known individual. Networks tend to focus on authentication, after which a user must go through an additional hurdle of authenticating to a specific application before gaining access to a resource. The problem with multiple layers of security is that the user quickly tires of it all and finds ways around it, which is where adaptive authentication comes into play, as described in *Adaptive Authentication and Machine Learning* at `https://towardsdatascience.com/adaptive-authentication-and-machine-learning-1b460ae53d84`. The short definition of **adaptive authentication** is the deployment and configuration of **multi-factor authentication** (**MFA**) such that it becomes possible to select the correct kind of authentication, depending on a user's risk profile and tendencies. When writing an ML application to deal with network-level authentication, the developer can take a significant number of factors into consideration, such as the following:

- Device type and/or name
- Location
- Network type
- Operating system
- User risk profile
- User tendencies

All of this happens in the background without the user's knowledge. What the user sees is an MFA setup that provides the access needed for both the network and common applications. Adaptive authentication is far more flexible and less likely to encounter problems than most monolithic solutions in place today.

## Detecting intrusions

ML is seeing major use in **Network-Based Intrusion Detection Systems** (**NIDSs**), an automated technique to determine whether a particular access attempt is an attack or benign. The use of ML allows the NIDS to react quickly and also adapt its behavior to new threats faster than most humans can. The main point of contention is how best to implement NIDS. Articles such as *Deep learning methods in network intrusion detection: A survey and an objective comparison* (`https://www.sciencedirect.com/science/article/abs/pii/S1084804520302411`) express a preference for deep feedforward networks over autoencoders and **deep belief networks** (**DBNs**) (see `https://www.analyticsvidhya.com/blog/2022/03/an-overview-of-deep-belief-network-dbn-in-deep-learning/` for details). However, not everyone agrees that using this approach will work well. In fact, there are some complex studies available to show the effectiveness of various NIDS strategies, such as *Network intrusion detection system: A systematic study of machine learning and deep learning approaches* at `https://onlinelibrary.wiley.com/doi/full/10.1002/ett.4150`.

NIDSs necessarily rely on a combination of hardware and software. Because the NIDS is detecting a particular event, rather than interacting directly with the network, it often appears in a separate server in a supervisory capacity. *Figure 5.2* shows a typical NIDS setup, with a generalized ML configuration.

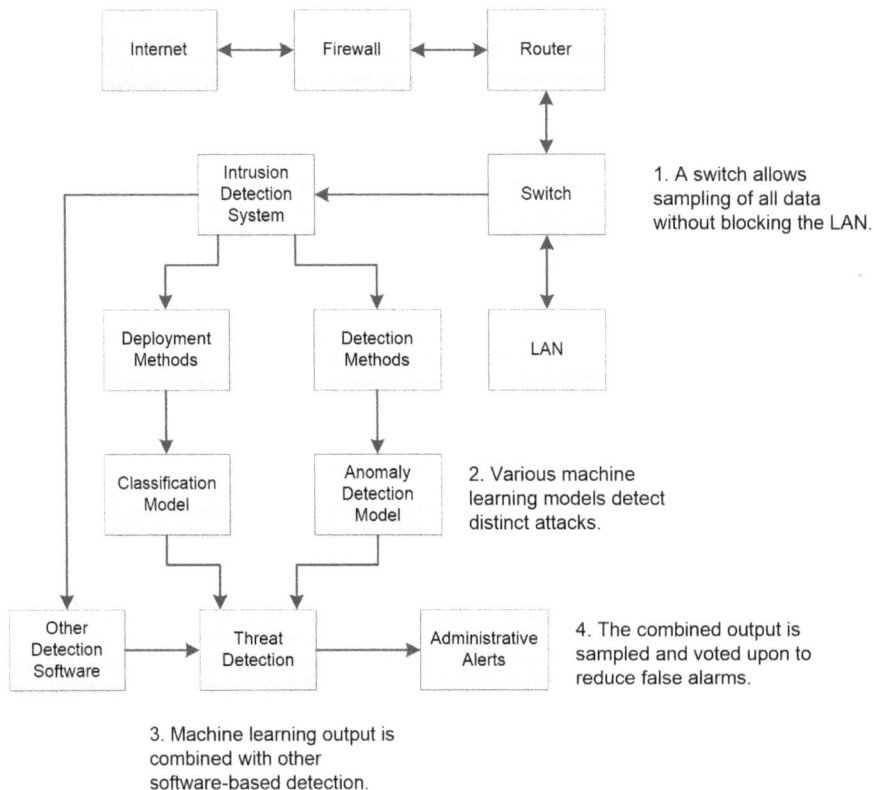

Figure 5.2 – Developing a generalized NIDS solution with the NIDS in a supervisory role

An essential part of this design is that you use multiple independent models (in contrast to an ensemble, where models work together), each of which is designed to detect a particular threat. Most of these models will output a classification regarding threat potential (*Chapter 7, Dealing with Malware*, discusses malware detection) or detect suspicious activity as represented by anomalies (*Chapter 6, Detecting and Analyzing Anomalies*, discusses how to detect anomalies). The reason that you need multiple models is that no one model will provide both classification and anomaly detection, and no one model will even cover a single detection method completely. The process will likely include output from other types of non-ML software and the output of all of the detection methods centralized in a threat detection model. This module provides a vote upon the various detection models and final output to the administrator in the form of an alert when enough components agree there is a threat.

## Defining localized attacks

The previous sections have focused on attacks from the outside or protection from outside influences. Networks are attacked at the local level in a wide range of ways, some of which aren't even purposeful but rather due to user error. In many cases, the use of network analytics can detect issues such as aberrant behavior or various kinds of probing. However, network analytics are ill-equipped to detect changes in user behavior. Because an administrator can't follow every user around looking for potential problems, an ML approach is helpful. One such technique appears in *Deep Learning and Machine Learning Techniques for Change Detection in Behavior Monitoring* at `http://ceur-ws.org/Vol-2559/paper3.pdf`. In this case, the paper is talking about monitoring the elderly or those with special needs, but the same techniques can be employed in network scenarios, as described in *Network Anomaly Detection and User Behavior Analysis using Machine Learning* at `https://www.ijcaonline.org/archives/volume175/number13/vadgaonkar-2020-ijca-920635.pdf`. The former techniques detect physical behavioral changes (which can drastically affect your network), while the latter techniques detect usage behavioral changes, both of which are required for complete user coverage.

However, tracking down user issues is only part of the localized attack detection requirement. It's also necessary to detect attacks against the ML models on your network, especially those involved in NIDS solutions (as described earlier in the chapter). Unfortunately, not a lot of research has been done in this area, so the best approach is to ensure you track any local input into the NIDS that could disrupt its functionality from hacker attacks, in which users actually facilitate hacker entry into the network.

## Understanding botnets

**Botnets** are a kind **Distributed Denial of Service** (**DDoS**) attack, but the goal is to steal information, spread spam, infect systems, or cause damage to infrastructure in most cases, rather than simply bring a network down. The problem with botnets is that some of them simply won't die. Law enforcement takes the botnet down, but it comes back again like an unwanted weed. You can read about some of these exploits in *Botnet Detection with ML* at `https://medium.com/@EbubekirBbr/botnet-detection-with-ml-afd4fa563d31`. What is truly terrifying is that one such botnet consisted of 2 million zombie computers.

The essential element in botnet detection is that the botnet will produce some sort of anomaly that the system can detect and monitor, using an approach similar to that shown in *Figure 5.2*. The anomalies are detectable by looking for techniques such as the following:

- **TCP syn (synchronization) scanning**: The hacker tries to create a handshake with every port on the server

- **DNS monitoring**: The hacker tries to modify DNS records

- **Botnet attack and propagation models**: The hacker relies on multiple botnet variations that replicate themselves

A legitimate user won't engage in any of these activities, so looking for them provides you with an anomaly to verify. *The Machine Learning Based Botnet Detection* white paper at `http://cs229.stanford.edu/proj2006/NivargiBhaowalLee-MachineLearningBasedBotnetDetection.pdf` provides insights into the effectiveness of specific algorithms in detecting botnet attacks.

There are some interesting examples of detecting botnets using ML **classifiers** (special applications designed to categorize entries based on learned characteristics), similar to the technique shown in the *Developing a simple spam filter example* section of *Chapter 4*. One such site is *Build botnet detectors using machine learning algorithms in Python* at `https://hub.packtpub.com/build-botnet-detectors-using-machine-learning-algorithms-in-python-tutorial/`. In this case, you work with the approximately 2 GB CTU-13 dataset found at `https://mcfp.felk.cvut.cz/publicDatasets/CTU-13-Dataset/` and described at `https://mcfp.weebly.com/the-ctu-13-dataset-a-labeled-dataset-with-botnet-normal-and-background-traffic.html`. The dataset explodes in size to around 74 GB when you extract it. It's best to download the dataset separately and extract the data because doing so within the example code can sometimes make it appear that the example has frozen. You may also want to temporarily disable virus detection on your system (after disconnecting from the internet). Otherwise, your antivirus application will dutifully remove all of the example files before you can use them for testing. The second example also relies on a Twitter dataset that's apparently no longer available, but it's possible to modify the code to use the dataset found at `https://www.kaggle.com/davidmartngutirrez/twitter-bots-accounts`.

Now that you have some idea of what the threats are, it's time to look at some protections. Protections currently come in four levels: traditional, ML, real-time, and predictive. This next section provides an overview of traditional protects that work well in ML application environments.

## Considering traditional protections

Understanding the threats to your network is a good first step because knowing about the threat is the first step in avoiding it. However, now it's time to do something about the threats. Anything that protects your network directly because of some type of detection practice is part of an **Intrusion Detection System (IDS)**. It doesn't matter whether the protection is a firewall, virus scanner, or other software that checks data in some manner, an actual security element designed to fool the attacker in some manner, or (as described later) an ML application. All of this protection reports an intrusion after detecting it, making it an IDS. Of course, you often find the term IDS cloaked in some sort of mystical way (depending on the organization/author), but really, they're straightforward. As described in the previous section, attacks come in waves and at different levels. Consequently, you need multiple layers of security (**defense in depth**) to address them. Each layer is part of your IDS.

## Working with honeypots

A honeypot is a security mechanism that purposely attracts hackers, as described at `https://www.imperva.com/learn/application-security/honeypot-honeynet/`. The idea is to create a fake network that includes security holes that a hacker can use to gain access. In order to be successful, the fake network must look real enough so that the hacker is fooled into believing it actually is real. However, the honeypot can be completely disconnected from anything else in the organization so that the hacker gains nothing of value in exchange for the effort of breaking into the system. For a security specialist, a honeypot provides an opportunity to discover how hackers perform tasks while maintaining the safety of the real network. One of the more interesting pieces on this sort of effort is *Honeypots: Free psy-ops weapons that can protect your network before defences fail* at `https://www.theregister.com/2017/02/08/honeypots_feature_and_how_to_guide/`.

From an ML perspective, the most useful honeypot is a high-interaction honeypot – one that looks completely real in every respect. In some cases, these honeypots are actually part of the functioning network and provide a means of detecting an intrusion. The most common use for AI today is in creating a honeypot that provides increased intrusion detection capability, as described in *AI-powered honeypots: Machine learning may help improve intrusion detection* at `https://portswigger.net/daily-swig/ai-powered-honeypots-machine-learning-may-help-improve-intrusion-detection`. Unfortunately, hackers have also been busy, as explained in *Automatic Identification of Honeypot Server Using Machine Learning Techniques* at `https://www.hindawi.com/journals/scn/2019/2627608/`. The consensus is that knowledgeable hackers now have tools to detect low-interaction and medium-interaction honeypots, but that high-interaction honeypots are still viable.

To create an ML-based honeypot, it's essential to know where to place the honeypot application and how to create the required agent. An article entitled *A Smart Agent Design for Cyber Security Based on Honeypot and Machine Learning*, at `https://www.hindawi.com/journals/scn/2020/8865474/`, provides some ideas on how to perform this task. In most cases, you're depending mostly on linear regression classification that is trained using hacker profile data to determine whether an access attempt is normal or an attack. Some sources also combine this type of agent with another agent that detects whether the actor in an interaction is a legitimate user or a hacker, based on activity – how the actor interacts with the system.

Honeypots aren't necessarily limited to detection. You can use them in a number of other ways. For example, you could possibly use a honeypot to provide a competitor with fake data or even a tainted version of a real ML model used within your organization. Because of the nature of ML models, detecting a reasonable fake would be incredibly difficult. For that matter, you could simply let the hacker steal the model and implant some sort of phone home code so you know the hacker's location (or, at least, the location of one of the hacker's zombies). Oddly enough, there is currently a patent application for such technology (see `https://www.freepatentsonline.com/y2020/0186567.html` for details).

## Using data-centric security

**Data-centric security** focuses on protecting data, rather than infrastructure. When a user opens an application, the application has privileges required to access precisely the data it needs and no more. The user has no rights to the data at all because the user isn't accessing the data; the application is. Any attempt by the user to access the data directly would result in a denial by the system and requisite notification to the administrator of the attempt. Even if the user were to somehow manage to evade the security, a secondary level of data encryption would thwart any attempt to actually use the data. This method of securing a network has these advantages:

- The data remains encrypted except when the application is actually using it
- It's easier to monitor data access because you track one application, rather than multiple users
- The actual location of the data is masked
- Creating policies governing data usage becomes more straightforward

Organizations normally combine data-centric security with other security measures for sensitive data. A number of vendors now provide support for data-centric security, as described at `https://www.g2.com/categories/data-centric-security`. These vendors are moving to ML strategies because using ML makes solutions more flexible and adaptive. In fact, if you read *The Worldwide Data-centric Security Industry is Expected to Reach $9.8 Billion by 2026 at a CAGR of 23.1% from 2020* at `https://www.globenewswire.com/en/news-release/2021/05/07/2225398/28124/en/The-Worldwide-Data-centric-Security-Industry-is-Expected-to-Reach-9-8-Billion-by-2026-at-a-CAGR-of-23-1-from-2020.html`, you'll find that data-centric security is an emerging technology that's attracting a lot of attention. According to the Forbes article at `https://www.forbes.com/sites/forbestechcouncil/2020/02/14/12-tips-to-help-shift-your-business-to-data-centric-cybersecurity/?sh=2098f174555d`, the two main contributions of ML to data-centric security are as follows:

- Implementing analytics-based security controls
- Monitoring data flows

When creating an ML application to implement data-centric security, you create one model that performs anomaly detection on access logs. All access to the data should follow easily recognized patterns, and anything that falls outside those patterns is suspect. The second model would categorize data flow. If only *applications* are supposed to access the data, then a *user* accessing the data would represent a threat. However, the categorization must go further. When an application is restricted to local access only, seeing it access the data from a remote location is cause for concern.

Part of data-centric security measures is to improve the quality of the data itself. For example, if the data isn't clean or of the right type, the outliers contained within it could be viewed as a potential security issue, rather than simply a failure to clean the data correctly. The article *Big Data To Good Data: Andrew Ng Urges ML Community To Be More Data-Centric And Less Model-Centric* (`https://analyticsindiamag.com/big-data-to-good-data-andrew-ng-urges-ml-community-to-be-more-data-centric-and-less-model-centric/`) is enlightening because it points to a need to create better results. By ensuring that data is more correct by reviewing it for missingness, consistency, and other issues, it's easier for a model to detect a botnet or malware, as opposed to real data, because there are fewer anomalies to deal with. The article makes clear that code is important, but data quality is far more important because it takes up 80 percent of a data scientist's time. Consequently, in addition to creating models that detect anomalies and classify access, ensuring data remains clean so that the models can do their jobs is essential.

Data-centric security also relies partly on the same techniques used for privacy programming, using products such as PyGrid (`https://github.com/OpenMined/PyGrid`) and PySyft (`https://github.com/OpenMined/PySyft`). *Chapter 13* looks into the matter of ensuring that data remains private through the use of federated training techniques. As a data scientist creates data-centric security models, the need to train on sensitive or encrypted data is important. Using the same measures that developers rely on to keep data private will also ensure the efficacy of the security model.

## Locating subtle intrusion indicators

If you only use NIDS to protect your network, then you're actually leaving it wide open to attack. The **Internet of Things** (**IoT**) is quickly changing the security landscape because IoT makes subtle, backdoor attacks possible. Upon viewing *Multi-level host-based intrusion detection system for Internet of things* at `https://link.springer.com/article/10.1186/s13677-020-00206-6`, you'll find that IoT devices are largely unsecured now and lack any sort of intrusion detection. However, a NIDS won't work in this case. What you need instead is **Host-Based Intrusion Detection Systems** (**HIDSs**). Of course, you may wonder how the IoT can present any sort of threat. Consider this process:

1. A hacker gains entry to a smart device connected to the internet for monitoring purposes.
2. The hacker changes data in an unobtrusive manner on the device, such that the data will produce an unexpected result when processed by analytics software.
3. The user or host service accesses the device from a desktop system, tablet, or other device attached to the network.

4. The data modifications produce an unexpected result.

5. The network is now potentially open to attack due to the result produced by the analytics software.

Because the smart device continues to operate as expected, no one suspects that it has become a time bomb. The hacker modified the data, not the device, so interacting with the device wouldn't show any difference. The attack only becomes apparent when performing an analysis of the data. Besides direct data manipulation, IoT devices represent these sorts of threats to your business and network:

- An attacker monitors users of interest to see whether they will tell family members about potentially sensitive information. There is an app available to hackers to make this possible with very little effort (see `https://www.siliconrepublic.com/enterprise/`

- `amazon-alexa-google-home-smart-speaker-research` for details). IoT devices, such as smart speakers, are sensitive enough to hear a heartbeat (`https://www.ncbi.nlm.nih.gov/pmc/articles/PMC7943557/`), so whispering won't prevent the divulging of sensitive information to hackers who are listening.

- The use of a group of IoT devices can create a DDOS attack, such as the Mirai attack, where the botnet turned IoT devices running on ARC processors into a group of zombies (`https://www.csoonline.com/article/3258748/the-mirai-botnet-explained-how-teen-scammers-and-cctv-cameras-almost-brought-down-the-internet.html`).

- The attacker gains access to a home network with access to your business network through an IoT device.

- In order to reduce employee effectiveness, the attacker bricks company-issued IoT devices using a botnet, such as the BrickerBot malware (`https://www.trendmicro.com/vinfo/us/security/news/internet-of-things/brickerbot-malware-permanently-bricks-iot-devices`).

One way in which to combat subtle intrusions such as IoT devices is to ensure all data sources receive proper checks. As noted in the *Developing a simple spam filter example* section of *Chapter 4, Considering the Threat Environment*, you need to parse the incoming data looking for anomalies of any sort, including data that falls out of range or simply doesn't follow a predictable pattern. For example, a thermometer indicating 5 hours of intense furnace use on a 90-degree day signals that there is either something wrong with the thermostat and it should be replaced, or that someone is tampering with it. ML techniques can take data from multiple sources, such as outside temperature monitors and thermostats, and combine it to detect threats that would otherwise go unnoticed.

It's also possible to mitigate potential tampering using what is known as a trace – a little piece of software in each device that produces trace data. A trace is essentially data that monitors device activity with regard to network communication. The *Obtaining Data for Network Traffic Testing* section talks about using Wireshark to obtain network traffic. Wireshark can also be used to create trace data, as described at `https://2nwiki.2n.cz/pages/viewpage.action?pageId=52265299`.

When working with IoT devices, it's necessary to create a trace point, which is a small piece of code that collects information in instrumented kernels, such as Linux or Android devices. When a tracer, such as LTTng (`https://lttng.org/`), hits the trace point, the trace point provides the device state in the **Common Trace Format** (**CTF**) (`https://diamon.org/ctf/`). When an IoT device isn't instrumented, it's often possible to add tracing ability through software such as barectf (`https://github.com/efficios/barectf`). To make binary data compatible with ML code, you can use the Babeltrace API (`https://babeltrace.org/`). *Figure 5.3* shows what a typical trace scenario might look like.

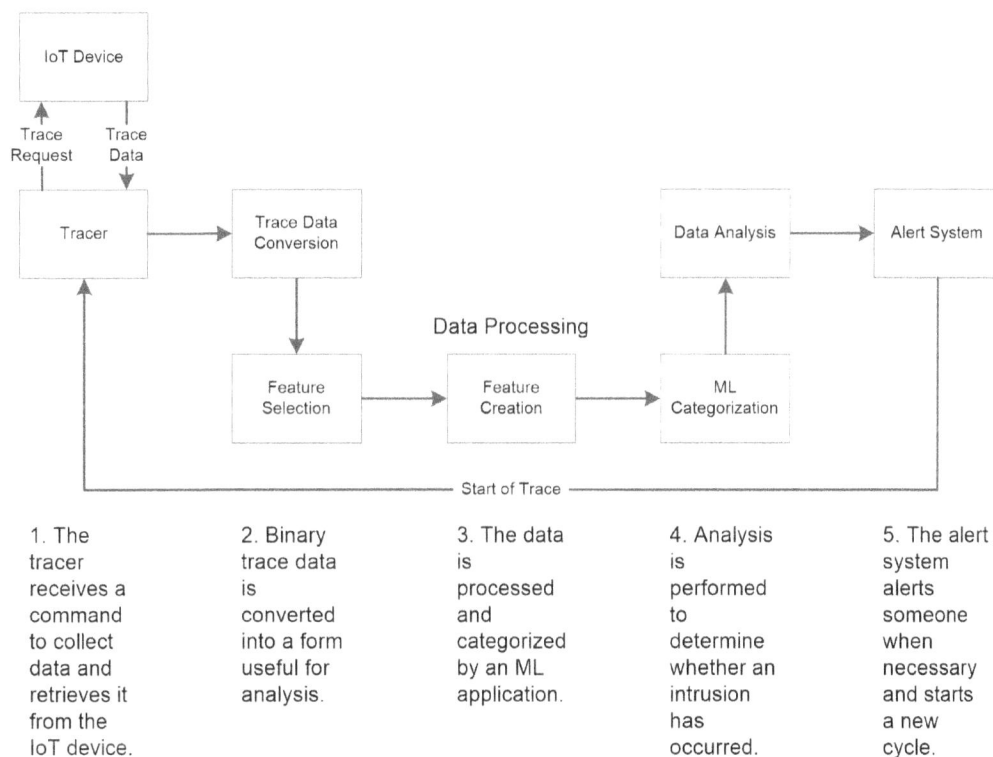

1. The tracer receives a command to collect data and retrieves it from the IoT device.

2. Binary trace data is converted into a form useful for analysis.

3. The data is processed and categorized by an ML application.

4. Analysis is performed to determine whether an intrusion has occurred.

5. The alert system alerts someone when necessary and starts a new cycle.

Figure 5.3 – Using trace data to monitor IoT devices directly and look for intrusions

Note that this approach doesn't depend on a particular kind of IoT device. It works with any IoT device that contains enough intelligence to communicate with the outside world. Interestingly enough, even newer Wi-Fi garage door openers can provide an opening for hackers (`https://smarthomestarter.com/can-garage-door-openers-be-hacked-smart-garages-included/`). Although this particular article is about home systems, the techniques work just fine for businesses too.

## Using alternative identity strategies

People lose passwords, create passwords that are too easy to guess, and generally don't use passwords correctly. So, your traditional protection might be easier to hack than you think because of the human factor. Biometric security has become quite common and is used in a number of ways, as discussed in the *Biometrics: definition, use cases and latest news* article at `https://www.thalesgroup.com/en/markets/digital-identity-and-security/government/inspired/biometrics`. The use of biometrics would seem to be perfect because a person can't easily lose their fingerprint without also losing the associated finger. In addition, many biometrics, such as DNA, would seem to be hard, if not impossible, to duplicate. So, using biometrics should also reduce fraud by making it hard for one person to impersonate another.

Unfortunately, it's easy to find articles that discuss techniques hackers use to overcome some forms of biometric security, such as fingerprints, voiceprints, iris scans, palm prints, and facial impressions. However, researchers keep working on new approaches that will be harder to overcome. Of course, you could always force everyone to provide a drop of blood for DNA testing (see `https://www.ibia.org/biometrics-and-identity/biometric-technologies/dna` for some current uses of DNA for biometrics). One such alternative is finger vein biometrics, as discussed in the *Finger-vein biometric identification using convolutional neural network* article at `https://www.researchgate.net/figure/Error-rates-of-the-CNN-models-tested-in-cross-validation-process_fig4_299593157`. Since finger veins are inside the body, it's harder to overcome the biometric technology involved.

The point is that if a hacker is determined enough, not even biometrics will prevent fraud or other uses of a person's identity without permission. Creating security measures that are harder to overcome keeps honest people honest and prevents a determined hacker from succeeding.

## Obtaining data for network traffic testing

Unlike other ML tasks, finding network traffic data can prove difficult, partly because network traffic consists of so many kinds of data. It's possible to find a few sources online, such as the network traffic datasets at `https://sites.google.com/a/udayton.edu/fye001/simple-page/network-traffic-classification`, which discusses two datasets. The first is the Curtin University dataset that simulates standard network traffic and includes the **SYN (synchronize)**, **RST (reset)**, **FIN (finalize)**, and **ACK (acknowledge)** sequences. (See the *TCP Flags* article at `https://www.geeksforgeeks.org/tcp-flags/` for more information about how these TCP flags work.) The second is the DoS dataset that simulates a DOS attack. The entirety of both datasets is 40 GB, so it's not something you'd download. In fact, the site tells you where to send a hard drive.

Unfortunately, getting a canned dataset in this case won't provide you with a model for your network traffic. At best, you'll come to understand the network traffic for another organization. Some white papers, such as *Evaluation of Supervised Machine Learning for Classifying Video Traffic* at `https://core.ac.uk/download/pdf/51093126.pdf`, suggest using a product such as Wireshark to obtain data. Wireshark has an established reputation and people have used it for a great number of tasks, as illustrated in the blog post at `http://blog.johnmuellerbooks.com/2011/06/07/sniffing-telnet-using-wireshark/`. The best part about Wireshark is that you can choose precisely what you want to track and then save the data to a text file. The resulting data reflects your actual network traffic. You can also generate specially crafted network traffic using a data generator that you build to reflect your actual network traffic, as described in the *Building a data generator* section of this chapter.

When creating an ML application to perform specific regression, classification, or clustering tasks on your network, you need to consider the data used to train the model carefully. Otherwise, the model might not detect the type of traffic you want to monitor accurately. Fortunately, you can perform a great many security tasks without necessarily relying on packet-level methods. For example, as shown in the *Developing a simple spam filter example* section of *Chapter 4*, you can look for email spam without going to the packet level. So, it's also important to consider the level at which you choose to monitor network traffic adequately.

Now that you have an idea of how traditional security techniques with augmentation can improve protections for both ML applications and associated data, it's time to look at how you can use ML itself to make the traditional techniques more flexible. Traditional techniques can be brittle and easily broken by a hacker because they're easily diagnosed, according to the traits that they present. ML can use algorithms to analyze and anticipate changes that hackers will make to avoid traditional protections.

## Adding ML to the mix

Once you get past the traditional defenses, you can use ML to implement **Network Traffic Analytics (NTA)** as part of an IDS, as shown in *Figure 5.2*. Most ML strategies are based on some sort of anomaly detection. For example, it's popular to use convolutional auto-encoders for network intrusion detection. A few early products still in the research stage, such as nPrintML, discussed in *New Directions in Automated Traffic Analysis* at `https://pschmitt.net/`, have also made an appearance. Here are just a few of the ways in which you can use ML to augment traditional security layers:

- Perform regression analysis to determine whether certain packets are somehow flawed compared to normal packets from a given source. In other words, you're not dealing with absolutes but, rather, determining what is normal from a particular sender. Anything outside the normal pattern is suspect.

- Rely on classification to detect whether incoming data matches particular suspect patterns. Unlike signature matching, this form of analysis relies on training a neural network to recognize classes of data that it hasn't seen before. Consequently, even if an attacker changes a signature, the model can still likely recognize the data class.

- Use clustering to detect attack patterns and as part of forensic analysis in real time. For example, suddenly seeing groups of requests from a particular set of IP addresses that all have the same characteristics is a type of suspect pattern.

As part of creating new layers for your IDS, you also need to consider the people who are part of that implementation strategy. For example, an updated **security plan** (the document that discusses how to deal with security issues so that people know what to do when a security event occurs) will describe how to look at the reports generated by the ML application and use them to determine when a potential threat is real. In addition, the people who are managing the network will need input on just what to do with the threat because it might not match threats they've seen in the past.

ML can be used in a variety of ways that many administrators haven't considered possible. For example, you can add ML to applications to detect unusual usage patterns or to a cloud environment to detect unusual API call patterns. Endpoint security is an area in which ML can excel, but only when the application knows what to look at and you maintain good records of existing trends. Each endpoint type is unique, so strategies that work on a workstation may not work as well on a server, and not at all in your cloud environment. Because the incoming data for each endpoint is also different, you need some means of preparing the data for comparison purposes, which is the real benefit of using ML to protect from a coordinated attack using unusual vectors (such as gaining access to a network through an IoT device).

## Developing an updated security plan

If you plan to employ ML as part of your security strategy, then you need to update the security plan. For example, users need to know that the ML application exists, how to access it, and what to do about the information it generates. This might seem like a straightforward requirement, but many security plans don't receive updates and are therefore useless in the event of an attack. The new security plan should have an eye toward ML techniques such as the following:

- Determining what sort of data to collect before, during, and after an attack

- Tracking user activities for analysis as part of detecting attack vectors

- Creating and testing models specifically designed for security needs and then providing instructions for deploying them during an attack

Immediately after the update, you need to provide user training on it and go through various scenarios. Just having something in writing won't prepare the people who have to react during an emergency. Remember that they're not going to be thinking as well as they could; they'll be excited, frantic even. Running through the security plan when everyone is calm and thinking correctly will help ensure its success.

## Determining which features to track

**Features** are essentially specific kinds of data that you want to track to create a dataset for your ML model. For example, if you're protecting an API, then tracking which IP addresses make specific API calls and when these calls are made are potential features. A hacker will present a different API calling pattern than a benign user because the hacker is searching for an entry point into your system. Often, that means the hacker uses less frequently accessed API calls, on the assumption that these API calls could contain bugs not found by users who rely on common API calls. Of course, the hacker will try to disguise this activity by making other calls, which adds noise to the dataset that the model must remove to see the true pattern. You have various options to format these features as data, depending on the model you want to create:

- Make each API call a separate feature in a two-dimensional table that lists the IP address making the call. Each API call could appear across the top of the table and the IP addresses could make up the rows (similar to one-hot encoding). This setup would work best with regression.

- Create a three-column, two-dimensional table that has the API call, IP address, and time the call is made as separate columns. This setup could work with either regression or classification.

- Define a three-dimensional table with one dimension being the API call, a second dimension being the IP address, and the third dimension being the times that the calls are made. This setup would work well with clustering.

- Provide a two-dimensional table, with the API calls as the rows, the IP addresses as the columns, and the number of calls made as the data. A bubble chart would work best in this case, with the ML model using size (showing the number of calls) and bubble color (perhaps based on hacker activity probability) to show patterns.

When adding ML to the security layers of your IDS, it's essential to think outside the box and to look at endpoints as a significant place for the installation of potential protections. For example, when working with an application, users usually resort to certain usage patterns. These workflows are based on the tasks that users perform, and there isn't a good reason for the user to deviate from them. However, there is no precise step-by-step way to define a workflow because each user will also express some level of uniqueness in their approach to a task. ML can learn a user's methodologies using unsupervised learning techniques and then use what the model learns to predict the next step in a process. When the user begins to deviate from the normal process employed specifically by them, it's possible that they aren't actually controlling the application at all. These features of application usage rely on data, such as keystroke analysis, to keep a network clean.

Note that users are the most difficult source of potential network attacks for traditional security measures to detect and mitigate. The detection of user security issues is a perfect way to use ML. Even though users are unpredictable and there is no source of labeled data for a dataset, user behaviors can point to potential issues. It's possible to detect user behavior problems in these ways:

- Track user behaviors such as login time, the time between breaks, and other factors using regression

- Employ known user factors, such as meeting times, to classify users by peer group (such as a workgroup or users who exercise during lunch)

- Use clustering techniques to detect users who have unusual habits or aren't part of known groups (the outliers)

Analysis can go further than predictive measures. Customizing the detection of certain kinds of data input can greatly decrease your risk. Most off-the-shelf software works well in general cases. However, some attacks are specific to your organization, so you can provide added levels of security to detect them. An email server could classify certain types of messages as ransomware, malware, or spyware based on previous patterns of attack against your organization that the off-the-shelf products missed. Such an application could track features such as the message source, specific subjects, some types of content, or the types of attachments provided. *Figure 5.4* shows a variety of attack vectors, implementation methods (as it concerns an ML solution), the ML model that is likely to work best, and an exploit site that demonstrates the attack.

| Attack type | Implementation method | Best model type | Example exploit site |
|---|---|---|---|
| **Authentication bypass** | Any of a number of methods that could include direct page requests, parameter modification, session ID prediction, HTML form SQL injection, and gray box testing | Regression | `https://owasp.org/www-project-web-security-testing-guide/latest/4-Web_Application_Security_Testing/04-Authentication_Testing/04-Testing_for_Bypassing_Authentication_Schema` |
| **Cross-Site Scripting** (XSS) | The use of scripts as part of a data injection strategy | Classification | `https://owasp.org/www-community/attacks/xss/` (specifically the sections on alternative syntax) |
| DDoS | An automated attack that relies on a number of zombie machines to overwhelm network defenses | Clustering | `https://www.imperva.com/learn/ddos/ddos-attacks/` |
| **Remote Code Execution** (RCE) | A method that relies on automation to locate vulnerabilities in software, in an effort to gain administrative privileges | Classification | `https://ozguralp.medium.com/simple-remote-code-execution-vulnerability-examples-for-beginners-985867878311` |

| Attack type | Implementation method | Best model type | Example exploit site |
|---|---|---|---|
| **Server Side Request Forgery (SSRF)** | Specially crafted web requests to web applications or to server resources that focus on infrastructure vulnerabilities that are normally hidden by a firewall | Regression | `https://www.acunetix.com/blog/articles/server-side-request-forgery-vulnerability/` |
| **SQL Injection (SQLi)** | Any number of methods that insert unexpected data or SQL scripts into SQL requests | Classification | `https://portswigger.net/web-security/sql-injection` |
| **XML External Entity (XXE)** | Request anomalies – a type of XML malformation that affects the receiver's ability to process XML data | Regression | `https://portswigger.net/web-security/xxe` |

Figure 5.4 – Common endpoint attacks and strategies

It's also possible to use ML to detect issues with various organizational processes. Regression techniques can help track the usual pattern of processes even in a large organization and predict what should happen next, even if the next step isn't necessarily related to the previous step. For example, if just one truck leaves with goods on a night other than the usual night, humans may not notice the change in pattern, but a regression model would. You could also use classification to detect these unusual changes in the pattern as potential fraud. When working with multiple organizational units that each perform similar processes, comparing one unit with another at a detailed level could show outliers in what would become clustered data.

Understanding the massive number of threats that your network faces and how ML methods can help reduce them are only part of the picture. The techniques described in the previous two major sections are static and reactive in nature. A hacker is already making an attack when you employ them, and now you must put your finger in the dike (so to speak) to keep the flood of hacker activity under control. What is really needed are real-time defenses that react immediately, at the same time the hacker is making an attack, as described in the next section.

# Creating real-time defenses

The previous section discussed how to use ML to augment your existing security, but it didn't mention when the solution will kick in. A problem exists for most network administrators and the developers who support them in that most strategies are either static or reactive. A real-time defense would be proactive and dynamic because hackers aren't going to wait until a network administrator can marshal forces formidable enough to keep them at bay. Networks can become overwhelmed by a lack of adequate real-time protection. What is really needed is real-time **Detection, Analysis, and Mitigation (DAM)**.

Mary Mapes Dodge published the novel *Hans Brinker, or The Silver Skates* in 1865. The novel is the story about Hans and the silver skates he wants to win, but it also contains an interesting little side story about a Dutch boy who plugs a hole in a dam with his finger and saves his people (see `https://marleenswritings.wordpress.com/2015/02/16/the-story-of-hans-brinker/` for details). The problem is that this is wishful thinking of the same sort employed by network administrators because a finger simply won't cut it. The article *Why the Little Dutch Boy Never Put his Finger in the Dike* at `https://www.dutchgenealogy.nl/why-the-little-dutch-boy-never-put-his-finger-in-the-dike/` provides a dose of reality. So, when creating DAM for your network to keep the waters of outside influences at bay, you really must consider the reality that constant maintenance and vigilance are essential or the whole thing is bound to fall down.

One of the most exciting elements of ML is that it's finally possible to create DAM for your network – one that is fully proactive and dynamic. The ML applications you create won't get tired, won't make mistakes, and will monitor the network constantly for those stresses that a finger simply can't address. However, as presented in the previous four chapters, hackers can and do overwhelm ML applications, so constant human supervision is also needed. The ML application may help detect an intrusion pattern, but it takes a human to interpret that pattern and address it when the ML application simply can't do the job.

## Using supervised learning example

Most of the ML models you create for security needs rely on supervised learning because this approach provides a better result, with known issues that you can track in real time. This is the approach that works best for issues such as determining when a hacker wants to break into your API or evade security through trial and error. It's also the method most commonly used to detect malware, fake data, or other types of exploits. The example in this section is somewhat generic but does demonstrate how you can collect features, such as the number of API calls made by a specific IP address, and then use this information to create a model that can then detect unwanted activity. The following sections show how to perform this task using a decision tree. You can also find this code in the `MLSec; 05; Real Time Defenses.ipynb` file of the downloadable source.

### *Getting an overview*

The example is somewhat contrived in this case because a production system would use an active data stream to acquire data, and the dataset itself would be much larger. *Figure 5.5* shows an example of how a production system might acquire and analyze data using the techniques found in this example. However, it's also important that the example is understandable, so the example dataset is static and relatively small to make it easy to see how various elements work.

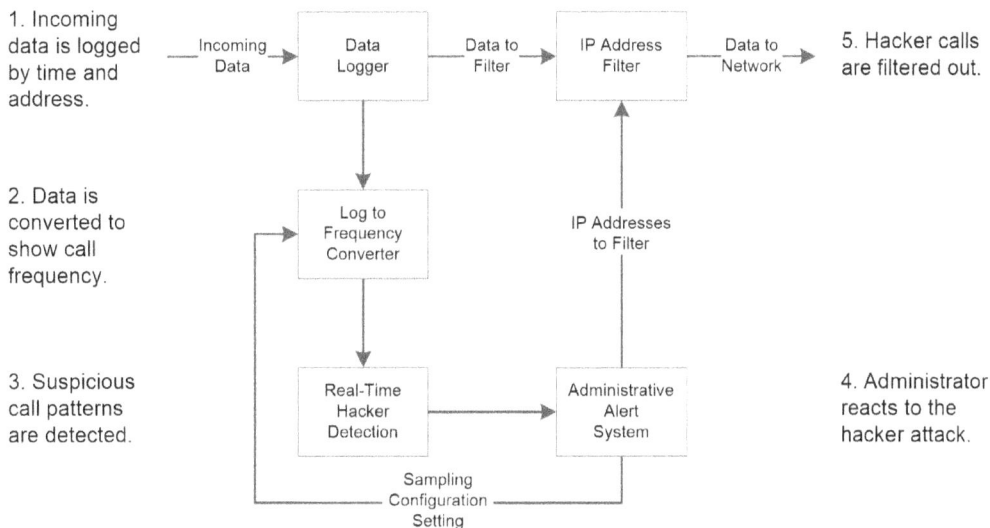

1. Incoming data is logged by time and address.

Incoming Data → Data Logger → Data to Filter → IP Address Filter → Data to Network

5. Hacker calls are filtered out.

2. Data is converted to show call frequency.

Log to Frequency Converter

IP Addresses to Filter

3. Suspicious call patterns are detected.

Real-Time Hacker Detection → Administrative Alert System

4. Administrator reacts to the hacker attack.

Sampling Configuration Setting

Figure 5.5 – Obtaining and analyzing API call data

The goal of this setup is to not only disrupt the flow of data as little as possible but also ensure that any hacker activity is detected and immediately stopped. The data logger simply creates continuous output files on disk. In the example, the output files are in .csv format, with just the time, IP address, and API call made by everyone who is using the system. It isn't unusual to have such setups anyway, so most developers won't have to implement anything new to obtain the data; they'll just have to know where to find it and ensure that it contains the information needed for the next step.

The next step is to read and convert the data in the logs at specific intervals, perhaps every 2 minutes. The conversion process takes the log data and counts how many times each IP address is making each API call. The sections that follow show how this works, but the transformation process needs to be as fast and simple as possible.

Creating a model based on call patterns comes next. A human will need to go through the frequency logs and label IP addresses to decide whether they're benign or a hacker, based on the call patterns. During real-time use, when the model detects a hacker, it sends the information to the administrator, who can then filter out the IP address and change the timespan to a shorter interval used to detect new intrusions. In the meantime, the benign data continues to flow into the network.

### Building a data generator

In a perfect situation, you have access to data from your network to use in building and testing a model to detect an API attack or other security issue (because this example focuses on an API attack, this is where the discussion will focus from this point on). However, you might not always have access to the required labeled data. In this case, you can construct and rely upon a data generator to create a simulated dataset with the characteristics you want. Start by importing the packages and classes needed for this example as a whole, as shown in the following code block:

```
from datetime import time, date, datetime, timedelta
import csv
import random
from collections import Counter
import pandas as pd
import numpy as np
from sklearn.ensemble import RandomForestClassifier
from sklearn.metrics import accuracy_score
```

Note that memory usage is kept down by not importing everything from `sklearn`.

### Creating the CreateAPITraffic() function

You need a function to actually generate the API traffic. The following steps outline what goes into this function (it appears as a single block of code in the online source:.

1.  Define a function that creates the actual data, based on what the logs for your network look such as. This example uses a relatively simple setup that includes the time, IP address of the caller, and the API call made, as shown here. The code starts by defining arguments you can use to modify the behavior of the data generation process. It's a good idea to provide default values that are likely to prove useful so that you're not dealing with missing argument problems when working with the function:

```
def CreateAPITraffic(
    values = 5000,
    benignIP = ['172:144:0:22', '172:144:0:23',
        '172:144:0:24', '172:144:0:25',
        '172:144:0:26', '172:144:0:27'],
    hackerIP = ['175:144:22:2', '175:144:22:3',
        '175:144:22:4', '175:144:22:5',
        '175:144:22:6', '175:144:22:7'],
```

```
apiEntries = ['Rarely', 'Sometimes', 'Regularly'],
bias = .8,
outlier = 50):
```

2.  Define the variables needed to perform tasks within the function. You use data to hold the actual log entries for return to the caller. The `currTime` and `updateTime` variables help create the log's time entries. The `selectedIP` variable holds one of the IP addresses, provided as part of the `benignIP` or `hackerIP` argument, and this is the IP address added to the current log entry. The `threshold` determines the split between benign and hacker log entries. The `hackerCount` and `benignCount` variables specify how many of each entry type appears in the log:

```
data = []
currTime = time(0, 0, 0)
updateTime = timedelta(seconds = 1)
selectedIP = ""
threshold = (len(apiEntries) * 2) - \
    (len(apiEntries) * 2 * bias)
hackerCount = 0
benignCount = 0
```

3.  A loop for generating entries comes next. This code begins by defining the time element of an individual log entry:

```
for x in range(values):
    currTime = (datetime.combine(date.today(),
        currTime) + updateTime).time()
```

4.  Selecting an API entry comes next. The code is written to accommodate any number of API entries, which is an important feature of any data generation function:

```
apiChoice = random.choice(apiEntries)
```

5.  The hardest part of the function is determining which IP address to use for the data entry. The `CreateAPITraffic()` function uses a combination of approaches to make the determination, based on the assumption that the hacker will select less commonly used API calls to attack because these calls are more likely to contain bugs, which is where `threshold` comes into play. However, it's also important to include a certain amount of noise in the form of outliers as part of the dataset. This example uses `hackerCount` as a means of determining when to create an outlier:

    ```
    choiceIndex = apiEntries.index(apiChoice) + 1
    randSelect = choiceIndex * \
        random.randint(1, len(apiEntries)) * bias
    if hackerCount % outlier == 0:
        selectedIP = random.choice(hackerIP)
    else:
        if randSelect >= threshold:
            selectedIP = random.choice(benignIP)
        else:
            selectedIP = random.choice(hackerIP)
    ```

6.  It's time to put everything together as a log entry. Each entry is appended to data in turn, as shown here. In addition, the code also tracks whether the entry is a hacker or benign:

    ```
    data.append([currTime.strftime("%H:%M:%S"),
        selectedIP, apiChoice])
    if selectedIP in hackerIP:
        hackerCount += 1
    else:
        benignCount += 1
    ```

7.  The final step is to return `threshold`, the benign log count, the hacker log count, and the actual log to the caller.

    ```
    return (threshold, benignCount, hackerCount, data)
    ```

When creating your own data generation function, you need to modify the conditions under which the log entries are created to reflect the real-world logs that your network generates automatically. The need for randomization and inclusion of noise in the form of outliers is essential. Just how much randomization and noise you include depends on the kinds of attacks that you're trying to prepare your model to meet.

## Creating the SaveDataToCSV() function

Part of developing a test log to use is to provide some method of saving the data to disk in case you want to use it again, without regenerating it each time. The `SaveDataToCSV()` function serves this purpose, as shown here.

```
def SaveDataToCSV(data = [], fields = [],
    filename = "test.csv"):
    with open(filename, 'w', newline='') as file:
        write = csv.writer(file, delimiter=',')
        write.writerow(fields)
        write.writerows(data)
```

This code consists of a loop that takes the log data as an input and writes one line at a time from the log to a `.csv` data file on disk. Note that this function makes no assumptions about the structure of the data that it writes. Again, the point is to provide significant flexibility in the use of the function.

Note that the `fields` input argument is used to write a heading to the `.csv` file so that it's possible to know which columns the file includes. It's likely that any real-world server logs you use will also include this information, so adding this particular data is important to ensure that any analysis code you create works properly.

## Defining the particulars of the training dataset

To make the generation process as flexible as possible, it's helpful to provide variables that you can easily modify to see their effect on the output. In this case, the example specifies API calls and IP addresses that differ from the defaults created earlier:

```
callNames = ['Rarely',
             'Sometimes1', 'Sometimes2',
             'Regularly1', 'Regularly2', 'Regularly3',
             'Often1', 'Often2', 'Often3', 'Often4',
             'Often5', 'Often6', 'Often7', 'Often8']
benignIPs = ['172:144:0:22', '172:144:0:23',
             '172:144:0:24', '172:144:0:25',
             '172:144:0:26', '172:144:0:27',
             '172:144:0:28', '172:144:0:29',
             '172:144:0:30', '172:144:0:31',
             '172:144:0:32', '172:144:0:33',
             '172:144:0:34', '172:144:0:35',
             '172:144:0:36', '172:144:0:37']
```

Note that the example is using the default `hackerIP` values, but you could modify the example to include more or fewer hackers, as desired.

### Generating the CallData.csv file

Now that everything is in place, you need to actually generate and save the data file using the following code:

```
random.seed(52)
threshold, benignCount, hackerCount, data = \
    CreateAPITraffic(values=10000,
        benignIP=benignIPs,
        apiEntries=callNames)
print(f"There are {benignCount} benign entries " \
        f"and {hackerCount} hacker entries " \
        f"with a threshold of {threshold}.")
fields = ['Time', 'IP_Address', 'API_Call']
SaveDataToCSV(data, fields, "CallData.csv")
```

The call to `random.seed(52)` ensures that you obtain the same output every time during the code testing process. When you finally start using the code to generate real data, you comment out this call so that the data can appear random (this is why they call it a pseudo-random number generator).

The call to `CreateAPITraffic()` comes next, with the output being unpacked into local variables. The `threshold`, `benignCount`, and `hackerCount` variables provide output that tells you about the functioning of the data generation process. When using the random seed of 52, you should see this output from the function:

```
There are 9320 benign entries and 680 hacker entries with a
threshold of 5.599999999999998.
```

The final step is to call `SaveDataToCSV()` to save the data to disk. At this point, you've completed the data generation process.

### *Converting the log into a frequency data table*

Log entries don't provide information in the right form for analysis. What you really need to know is which IP addresses made what calls and how often. In other words, you need an aggregation of the log entries so that your model can use the calling pattern as a means to detect whether a caller is a hacker or a regular user. This process takes place in a number of steps that include reading the data into the application and performing any manipulations required to put the data into a form that the classifier can understand. The following sections show how to perform this task.

### Creating the ReadDataFromCSV() function

The first step is to create a function that can read the .csv file from disk. More importantly, this function automates the process of labeling data as either benign or from a hacker. This particular part of the function is exclusive to this part of the example and serves to demonstrate that you can try really hard to make every function generic, but you may not always succeed. The following steps show how this function works:

1.  Define the function and read in the data file from disk:

    ```
    def ReadDataFromCSV(filename="test.csv"):
        logData = pd.read_csv(filename)
    ```

2.  Obtain a listing of the unique API calls found in the file:

    ```
    calls = np.unique(np.array(logData['API_Call']))\
    ```

3.  Aggregate the data, using IP_Address as the means to determine how to group the entries and API_Call as the means to determine which column to use for aggregation:

    ```
    aggData = logData.groupby(
        'IP_Address')['API_Call'].agg(list)
    ```

4.  Create a DataFrame to hold the data to analyze later. Begin labeling the data based on its IP address (which is most definitely contrived, but it does add automation to the example instead of forcing you to label the entries by hand). A value of 0 denotes a benign entry, while a value of 1 denotes a hacker entry. Note the use of ipEntry.sort() to place all alike IP entries together:

    ```
    analysisEntries = {}
    analysisData = pd.DataFrame(columns=calls)
    for ipIndex, ipEntry in zip(aggData.index, aggData):
        ipEntry.sort()
        if ipIndex[0:3] == '172':
            values = [0]
        else:
            values = [1]
    ```

5.  Create columns for the DataFrame based on the API calls:

    ```
    keys = ['Benign']
    for callType in calls:
        keys.append(callType)
        values.append(ipEntry.count(callType))
    ```

6. Define each row of the DataFrame, using the number of calls from the IP address in question as the values for each column:

```
analysisEntries[ipIndex] = pd.Series(values,
    index=keys)
```

7. Create the DataFrame and return it to the caller:

```
analysisData = pd.DataFrame(analysisEntries)
return (analysisData, calls)
```

At this point, you're ready to read the data from disk.

## Reading the data from disk

Now that you have a function for reading the data, you can perform the actual act of reading it from disk. The following code shows how:

```
analysisData, calls = ReadDataFromCSV("CallData.csv")
print(analysisData)
```

It's important to look at the data that you've created to ensure it contains the kinds of entries you expected with the pattern you expected. *Figure 5.6* shows an example of the output of this example (the actual output is much longer).

|            | 172:144:0:22 | 172:144:0:23 | 172:144:0:24 | 172:144:0:25 \ |
|------------|--------------|--------------|--------------|--------------|
| Benign     | 0            | 0            | 0            | 0            |
| Often1     | 48           | 49           | 38           | 50           |
| Often2     | 23           | 31           | 60           | 48           |
| Often3     | 38           | 41           | 47           | 50           |
| Often4     | 43           | 43           | 38           | 48           |
| Often5     | 43           | 40           | 55           | 43           |
| Often6     | 47           | 41           | 54           | 31           |
| Often7     | 55           | 55           | 44           | 49           |
| Often8     | 57           | 48           | 55           | 57           |
| Rarely     | 33           | 22           | 28           | 24           |
| Regularly1 | 40           | 51           | 33           | 40           |
| Regularly2 | 46           | 47           | 35           | 43           |
| Regularly3 | 51           | 38           | 51           | 39           |
| Sometimes1 | 29           | 32           | 30           | 42           |
| Sometimes2 | 42           | 54           | 39           | 40           |

Figure 5.6 – This figure shows the output of the data creation process

Note that each API call appears as a row, while the IP addresses appear as columns. Later, you will find that you have to manipulate this data so that it works with the classifier. The Benign column uses 0 to indicate a benign entry and 1 to indicate a hacker entry.

## Manipulating the data

The classifier needs two inputs as a minimum for a supervised model. The first is the data itself. The second is the labels used to indicate whether the data is of one category (benign) or another category (hacker). This means taking what you need from the DataFrame and placing it into variables that are traditionally labeled X for the data and y for the labels. Note that the X is capitalized, which signifies a matrix, while y is lowercase, which signifies a vector. Here is the code used to manipulate the data:

```
X = np.array(analysisData[1:len(calls)+1]).T
print(X)
y = analysisData[0:1]
print(y)
y = y.values.ravel()
print(y)
```

The X variable is all of the data from the `DataFrame` that isn't a label. This isn't always the case in ML examples, but it is the case here because you need all the aggregated data shown in *Figure 5.6*. Note the T at the end of the `np.array(analysisData[1:len(calls)+1])` call. This addition performs a transform on the data so that rows and columns are switched. Compare the output shown in *Figure 5.7* with the output shown in *Figure 5.6*, and you can see that they are indeed switched. In addition, this is now a two-dimensional array (a matrix).

```
[[48 23 38 43 43 47 55 57 33 40 46 51 29 42]
 [49 31 41 43 40 41 55 48 22 51 47 38 32 54]
 [38 60 47 38 55 54 44 55 28 33 35 51 30 39]
 [50 48 50 48 43 31 49 57 24 40 43 39 42 40]
 [59 39 48 40 38 40 34 43 24 46 42 45 44 33]
 [52 45 55 41 38 54 50 39 30 39 35 33 35 48]
 [45 45 47 42 47 48 49 39 31 41 38 44 41 49]
 [41 41 38 45 52 60 29 44 28 45 44 43 31 29]
 [40 36 47 41 40 48 41 52 31 28 32 55 29 37]
 [40 57 58 39 39 42 48 42 29 44 45 47 28 38]
 [36 37 49 37 56 34 52 45 25 55 50 39 44 31]
 [55 39 43 50 37 47 39 43 26 38 39 38 32 42]
 [52 36 43 38 46 57 35 37 29 27 38 27 38 40]
 [50 43 47 43 42 47 41 54 22 35 39 44 33 43]
 [40 48 47 37 49 43 37 47 25 37 40 38 30 36]
 [47 44 52 40 56 51 41 45 24 37 46 39 26 38]
 [ 0  0  0  0  0  0  0  1 54  4  6 10 23 11]
 [ 0  1  0  0  0  0  0  0 46 12  7  9 23 15]
 [ 0  1  0  0  0  0  0  0 41  9  6  5 22 18]
 [ 0  0  0  0  0  0  0  0 52 12 10  4 34 15]
 [ 0  0  0  0  0  0  0  0 54  9  4  9 25 14]
 [ 0  1  1  1  0  0  0  0 43  5  7 12 28 16]]
```

Figure 5.7 – The X data is turned into a transformed data matrix

The y variable consists of the labels in the Benign row. However, when you print y out, you can see that it retains the labeling from the DataFrame, as shown in *Figure 5.8*.

```
        172:144:0:22  172:144:0:23  172:144:0:24  172:144:0:25  172:144:0:26  \
Benign             0             0             0             0             0

        172:144:0:27  172:144:0:28  172:144:0:29  172:144:0:30  172:144:0:31  \
Benign             0             0             0             0             0

        ...  172:144:0:34  172:144:0:35  172:144:0:36  172:144:0:37  \
Benign  ...             0             0             0             0

        175:144:22:2  175:144:22:3  175:144:22:4  175:144:22:5  175:144:22:6  \
Benign             1             1             1             1             1

        175:144:22:7
Benign             1
```

Figure 5.8 – The y variable requires additional manipulation

To change the data into the correct form, the code calls y.values.ravel(). The values property strips all of the DataFrame information, while the ravel() call flattens the resulting vector. *Figure 5.9* shows the result.

```
[1 rows x 22 columns]
[0 0 0 0 0 0 0 0 0 0 0 0 0 0 0 0 0 1 1 1 1 1]
```

Figure 5.9 – The y variable is now a vector suitable for input to the classifier

The reason that *Figure 5.7* and *Figure 5.9* are so important is that many input errors that occur when working with the classifier have to do with data being in the wrong format. In many cases, the error message doesn't provide a good idea of what the problem is, and trying to figure out precisely why the classifier is complaining can be difficult. In a few cases, the classifier may only provide a warning, rather than an error, so it's even more difficult to locate the problem.

### Creating the detection model

All of the data generation and preparation took a long time in this example, but it's an even longer process in the real world. This example hasn't considered issues such as cleaning data, dealing with missing data, or verifying that data is in the correct range and of the correct type. This is actually the short version of the process, but now it's time to finally build the model and see how well it does in detecting hackers.

## Selecting the correct algorithm

This example relies on `RandomForestClassifier`. There are no perfect classifiers (or any other algorithm, for that matter). The reason for using a random forest classifier in this case is that this particular algorithm works well for security needs. You can use the random forest classifier to better understand how the ML model makes a particular decision, which is essential for tuning a model for security needs. A random forest classifier also has these advantages:

- There is less of a chance of bias providing you with incorrect output because there are multiple trees, and each tree is trained on a different subset of data. This algorithm relies on the power of the crowd to reduce the potential for a bad decision.

- This is a very stable algorithm in that new data that shows different trends from the training data may affect one or two trees but is unlikely to affect all of the trees.

- You can use this algorithm for situations where you have both categorical and numerical features. Security situations often require the use of categorical data to better define the environment, so that something such as an API call can be a hacker in one situation but not another.

- This is also the algorithm to rely on when you can't guarantee that real-world data is scaled well. It also handles a certain amount of missingness. Security data may not always be as complete as you want it to be. In fact, it's in the hacker's interest to make the data incomplete.

Although the random forest algorithm may appear perfect, appearances can be deceiving. You also need to be aware of the disadvantages of using this algorithm:

- The algorithm is quite complex, so training can require more time.

- Tuning the model can prove a lot more difficult, especially with the stealthy manner in which hackers operate. You may find yourself tweaking the model for quite a while to get a good result.

- The resulting model can use a lot more computational resources than other solutions. Given that security solutions may rely on immense databases, you really do need some great computing horsepower to use this algorithm (despite the apparent alacrity of the example due to the small size of the dataset).

These advantages and disadvantages should give you a better idea of what to expect when working with the random forest algorithm in the real world. The critical point here is that you can't skimp on hardware and expect a good result.

## Performing the classification

The actual classification process is all about fitting a model to the data that you've created. **Fitting** is the process of creating a curve that differentiates between benign requests on one side and hackers on the other. When a model is **overfitted**, it follows the data points too closely and is unable to make good predictions on new data. On the other hand, when a model is **underfitted**, it means that the curve doesn't follow the data points well enough to make an accurate prediction, even with the original data. The RandomForestClassifier used in this example is less susceptible to either overfitting or underfitting than many algorithms are. With this in mind, here are the steps needed to perform classification:

1. Create the classifier and fit it to the data generated and manipulated in the previous sections. In this case, the example uses all of the classifier's default settings except for the data. The output of this step simply says RandomForestClassifier(), which tells you that the creation process was successful:

```
clf=RandomForestClassifier()
clf.fit(X,y)
```

2. Generate test data to test the model's performance. This is the same approach that was used to create the training data, except this data is generated using a different seed value and settings to ensure uniqueness:

```
random.seed(19)
threshold, benignCount, hackerCount, data = \
    CreateAPITraffic(benignIP=benignIPs,
        apiEntries=callNames, bias=.95, outlier=15)
print(f"There are {benignCount} benign entries " \
    f"and {hackerCount} hacker entries " \
        f"with a threshold of {threshold}.")
fields = ['Time', 'IP_Address', 'API_Call']
SaveDataToCSV(data, fields, "TestData.csv")
```

Something to note about this step is that the bias and outlier settings are designed to produce the effect of a more realistic attack by making the attacker stealthier. The output of this step is as follows:

```
There are 4975 benign entries and 25 hacker entries with
a threshold of 1.4000000000000021.
```

3. Perform the actual classification. This step includes manipulating the data using the same approach as the training data:

```
testData, testCalls = ReadDataFromCSV("TestData.csv")
X_test = np.array(testData[1:len(calls)+1]).T
y_test = testData[0:1].values.ravel()
y_pred = clf.predict(X_test)
print('Accuracy: %.3f' % accuracy_score(y_test, y_pred))
```

This process ends with a measurement of the accuracy of the model, which is the most important part of the process. Data scientists often use other measures to verify the usefulness of a model and to check which factors contributed most to the model's accuracy. You can also create graphs to show data distributions (https://machinelearningmastery.com/statistical-data-distributions/) and how the decision process is made (https://stackoverflow.com/questions/40155128/plot-trees-for-a-random-forest-in-python-with-scikit-learn and https://mljar.com/blog/visualize-tree-from-random-forest/). In this case, the output says that the model is 95.5 percent accurate (although this number can vary, depending on how the training and testing data is configured). It's possible to improve the accuracy in a number of ways, but the best way would be to provide additional data.

## Using a subprocess in Python example

Real-time defenses often depend on using the correct coding techniques in your ML code. For example, something as innocuous as a subprocess could end up torpedoing your best efforts. A **subprocess** is a task that you perform from the current process as a separate entity. Even though you don't see it used often in Python ML examples, you do run across it from time to time. For example, you might use a subprocess to download a dataset or extract data from a .zip file. It's handy to see how this kind of problem can appear totally innocent and how you might not even notice it until a hacker has exploited it. This first example, which won't run on online environments such as Google Colab because the IDE provides protections against it, shows the wrong way to perform tasks using a subprocess. The reason that this code is included is to provide you with a sort of template to see incorrect coding practices in general. You'll find this example in the MLSec; 05; Real Time Defenses. ipynb file for this chapter:

```
from subprocess import check_output

MyDir = check_output("dir", shell=True)
print(MyDir.decode('ascii'))
```

Even though the code executes as a script at the command line and provides a listing of the current working directory, the use of shell=True creates a potential RCE hole in your code, according to https://docs.python.org/3/library/subprocess.html#security-considerations and https://www.hacksplaining.com/prevention/command-execution. In order to avoid the RCE hole, you can use this form of the code instead (note that this code may not work on your Linux system):

```
from subprocess import check_output

MyDir = check_output(['cmd','/c','dir'])
print(MyDir.decode('ascii'))
```

In this case, you create a new command processor, execute the directory command, and then close the command processor immediately by using the /c command-line switch. Of course, this approach requires some knowledge of the underlying operating system. It's often better to find a workaround that doesn't include calling the command processor directly, such as the solution shown here (which will definitely run on your Linux system):

```
from os import listdir
from os import getcwd

MyDir = listdir(getcwd())
print(MyDir)
```

This version obtains the directory as an easily processed list, in addition to protecting it from an RCE attack. In many respects, it's also a lot easier to use, albeit less flexible, because now you don't have access to the various command-line switches that you would when using the check_output() version. For example, this form of check_output() will obtain only the filenames and not the directories as well: MyDir = check_output(['cmd','/c','dir', '/a-d']). You can also control the ordering of the directory output using the /o command-line switch with the correct sub-switch, such as /os, to sort the directory output by size.

## Working with Flask example

You may want to expose your ML application to the outside world using a REST-type API. Products such as Flask (https://www.fullstackpython.com/flask.html) make this task significantly easier. However, when working with Flask, you must exercise caution because you can introduce XSS errors into your code. For example, look at the following code (also found in the MLSec; 05; Real Time Defenses.ipynb file for this chapter):

```
from flask import Flask, request

app = Flask(__name__)

@app.route("/")
def say_hello():
    your_name = request.args.get('name')
    return "Hello %s" % your_name
```

In this case, you create a Flask app and then define how that app is going to function. You start at the uppermost part of the resulting service (yes, it runs fine in Jupyter Notebook). The `hello()` function obtains a name as a request argument using `request.args.get('name')`. It then returns a string with the name of the website for display. To run this example, you place this call in a separate cell:

```
app.run()
```

When you run the second cell, it won't exit. Instead, you'll see a message such as this one (along with some additional lines that aren't a concern in this example):

```
* Running on http://127.0.0.1:5000/ (Press CTRL+C to quit)
```

Note that the address the server provides may vary, so you need to use the correct URL or the application will fail. When you see the running message, you can open a new tab in your browser and type something such as this as an address: `http://127.0.0.1:5000/?name=John`. The browser page will display `Hello` and then your name. However, if you type the following address instead, `http://127.0.0.1:5000/?name=<script>alert(1)</script>`, you will see an alert dialog. Of course, you could have run any script, not just this one. To stop the server, click the **Stop** button in Jupyter Notebook. Here's a fix for this problem:

```
from flask import Flask, request, escape

app = Flask(__name__)

@app.route("/")
def say_hello():
    your_name = request.args.get('name')
    return "Hello %s" % escape(your_name)
```

The use of `escape(your_name)` means that the server won't actually execute the script. Instead, it escapes the script as text and displays the text on screen. It's important to note that Flask creates a running log for you as you experiment with the test page. This log appears as part of the Jupyter Notebook output, such as this:

```
127.0.0.1 - - [20/Apr/2021 16:56:18] "GET /?name=John HTTP/1.1"
200 -
127.0.0.1 - - [20/Apr/2021 16:56:27] "GET
/?name=%3Cscript%3Ealert(1)%3C/script%3E HTTP/1.1" 200 -
```

From an ML perspective, you could place this log in a file and then use techniques such as the one demonstrated in the *Using supervised learning example* section to perform analysis on it. The log provides everything you need to begin hunting the hacker down in real time, as the hacker is experimenting with the API, rather than waiting until it's too late.

## Asking for human intervention

*Figure 5.2 and Figure 5.5* make it clear that a significant part of this example includes the use of human monitoring to ensure that nothing has gone awry with the supervised learning example. The setup could constantly check the accuracy of predictions that it's making, for example, and send an alert of some type to the administrator when accuracy begins to fall below a certain point. The examples in the *Using a subprocess in Python example* and *Working with Flask example* sections also point to the need for human monitoring of the system. If you take anything at all away from this chapter, it should be that ML augments the ability of humans to detect and mitigate threats – it can't act as a replacement for humans.

Unfortunately, humans are easily overwhelmed. ML can perform constant monitoring in a consistent way and then ask a human for intervention when the monitoring process detects anything unusual. This is the essential component that will make real-time monitoring in your organization possible – the realization that ML can direct a human's attention to the right place at the right time.

This part of the chapter takes you beyond what most organizations do today – that is, addressing threats in real time. However, to have a real edge against hacker attacks, you really need to know what a hacker is likely to do in the future. Predicting the future always comes with certain negative connotations because it really isn't possible to predict anything with complete accuracy. Otherwise, everyone engaged in the stock market would be a millionaire. However, when it comes to security, it's possible for the observant administrator to use ML to predict future attacks with enough accuracy to make the process worthwhile and possibly thwart a hacker before any mischief begins.

## Developing predictive defenses

Being able to predict the future is something that everyone who is involved with security would like to have. The use of ML to help predict things such as network attacks is an ongoing venture, but there aren't any commercial examples of such technology to date, and usable examples are also hard to find. However, it's possible to postulate what a commercial offering might look like and start doing some experimenting of your own, as described in the sections that follow.

## Defining what is available today

What you see most often today are explorations into predictive software based on new **Long Short-Term Memory** (**LSTM**), **Recurrent Neural Network** (**RNN**), and **Multilayer Perceptron** (**MLP**) models, which are described in articles such as *CyberSecurity Attack Prediction: A Deep Learning Approach* at `https://dl.acm.org/doi/fullHtml/10.1145/3433174.3433614` and *A deep learning framework for predicting cyber attacks rates* at `https://link.springer.com/article/10.1186/s13635-019-0090-6`. (You'll have to pay for the privilege of reading the first research paper.) Still, just as ML has become popular in predicting all sorts of other events, researchers will eventually come up with methods of predicting possible attacks.

Another white paper that provides an interesting read, *Conceptualisation of Cyberattack prediction with deep learning* (`https://cybersecurity.springeropen.com/articles/10.1186/s42400-020-00053-7`), also focuses on the need for new algorithms to perform attack prediction, including the use of new ReLU algorithms to provide activation functions in deep learning models. *Figure 1* of this article provides a block diagram of a potential model to use for predictive purposes. The end of the article also provides useful links to datasets you can use while experimenting.

Forensic analysis of events that have already happened using techniques such as those shown in the *Using supervised learning example* section, such as botnet attacks as described in *Botnet Forensic Analysis Using Machine Learning* at `https://www.hindawi.com/journals/scn/2020/9302318/`, are helping to improve the odds of creating good predictive solutions. As you read the article, however, you quickly become aware of the new skills required to actually perform such analysis. In addition, the article demonstrates that the forensic model won't be easy to construct because it consists of ensembles of learners, much such as the ensembles discussed in the *Using ensemble learning* section of *Chapter 3*.

There are also people skills to consider. Not only will the person performing the analysis have to have a good knowledge of ML techniques but also low-level network analysis techniques, as described earlier in this chapter. If the person who is performing the analysis isn't comfortable with using tools such as Wireshark, then the analysis is likely to fail. In addition, botnet forensics only tells you about trends –you still need to create some sort of ML tool to predict future botnet attacks based on current trends.

Even though it may not seem relevant at first, the ability to predict certain classes of worldwide events will also prove helpful in determining when network attacks could possibly peak. For example, *Prediction of Future Terrorist Activities Using Deep Neural Networks* at `https://www.hindawi.com/journals/complexity/2020/1373087/` looks at terrorist activity. Knowing this information could provide clues as to when terrorists will use various kinds of network attacks to bring down the financial sector or simply look for easy ways to finance their various endeavors, by forcing businesses to pay a ransom to get their data back. Keeping track of world events is a viable aid to keeping your network free of various attacks as well.

## Downsides of predicting the future

In many ways, trying to predict whether someone is going to attack your network is like trying to predict the weather. Yes, techniques have improved over time, but ask anyone who has had a picnic ruined by rain and you'll quickly understand that the process is far from bulletproof. Humans have always tried to predict the future using various means, with many of the correct guesses being more a matter of random chance than any skill on the part of the predictor. When trying to predict whether a hacker will attack your network, you need to consider these issues:

- Current trends in hacker attacks

- Historical trends in attacks of the same type

- The probability of a hacker continuing to use similar (albeit updated) attacks

- The likelihood of hackers devising an entirely new attack vector

- Human intuition

The first two items on the list do have a solid mathematical basis found in real-world data, assuming you can find the right data and that it hasn't become biased or manipulated in some way.

The second two items on the list have a statistical basis of the same sort and can predict the outcome of games of chance. Yes, you can create ML models that can provide you with statistics and even choose the options with the highest probability, but in the end, any of the options under review can occur. Winners of the lottery know that sometimes the option with the smallest chance of success actually does occur.

The fifth item on the list is the hard one, and it's the one with a significant effect on your efforts. Interpreting data requires human intuition because a computer has no sense of one attack vector's meaning in a specific situation over another. Human intuition also informs a hacker when it's time to move on to another method because the old one has a dwindling chance of success, based on a new administrator technique. The hacker can simply get a feeling that something is about to happen and switch tactics. There is no way to predict such feelings.

---

### The human element in predictions

There is a strong fear that a combination of ML applications and robotics will replace humans. It's true that ML applications can perform a great many predictable and mundane tasks that humans once performed exclusively. Because an ML application doesn't need to eat and doesn't get tired, it can actually do a better job in many cases. However, computers don't feel, don't have intuition, can't think creatively, and can't do a wealth of things that humans can do. What is more likely to happen with ML is that the application will assist the human security professional, relieving the security professional of performing mundane tasks. In looking at the techniques used to mitigate network attacks, it's important to consider the contributions that the security professional will make, rather than become blinded by the flashing lights of technology.

The problem then is one of unrealistic expectations on the part of those using the prediction techniques. When anyone makes a prediction using any model created for the purpose, even with the addition of human intuition, there is an expectation that the prediction is correct. However, there is only a probability that the prediction is correct, and hackers attacking your network may decide to do something completely different and unexpected. This is how a hacker wins the war over the defenses you build and the predictions you make – by behaving unpredictably.

## Creating a realistic network model

Security requires a realistic, as opposed to a proof-of-concept, model if you truly want to detect attacks in real time and predict future trends. Of course, trying to generate a model that represents the real world is difficult at best. It's possible to obtain perfectly reliable pre-built models for many ML needs and save yourself a lot of time, not to mention guesswork, in creating such a model. Some problem domains are common enough and predictable enough that you can use a pre-built model and feel good about it. Security is no exception to this rule. Using a pre-built model for these needs will likely provide good results:

- Various types of common fraud, such as credit card fraud
- Identity theft of various sorts, including the use of illegally obtained credentials to access your network
- Spoofing techniques, such as those used to gain unlawful access to a system
- Spam mail or malware of various kinds
- Some types of attacks on cloud-based resources that are hosted by a particular vendor

The problem arises when you try to use what someone else has put together for predicting specific types of problem domains. No one can anticipate, much less build, a model that will match the particular characteristics of your organization (even if they attempt to tell you that they can do so). Here are some examples of situations where you need to build your own model to ensure that it works as anticipated:

- Social attacks of various sorts
- Attacks that depend on your organization's structure
- Physical attacks that depend on the layout, structure, and organization of any physical elements of your organization, such as buildings
- Infiltration based on your particular product line or methodologies
- Attacks on your APIs or databases

Building a realistic network model that you can use to predict future activity comes down to examining how the data is trending on your network, rather than on networks in general. This means using statistical analysis to track your particular network and being aware of the threats peculiar to your network. Your network may not be particularly vulnerable to an API attack. Perhaps the hackers interested in you are working in other ways, such as social engineering through various kinds of email entreaties instead.

> **Beware of one-size-fits-all**
>
> A common mistake that leads to security issues is to assume that every network is just like every other network out there. Networks do have similarities, but they also have unique qualities. Hackers poke and prod at a network using every means possible to locate a chink in a network's security armor. However, once they find that chink, they don't ignore it. Rather, they use what is known about the network to create an exploit that will work. This is the reason why you really do need to consider the uniqueness of your network as part of a solution for predicting what will happen in the future, based on what you know now.

## Summary

This chapter has covered a broad range of network topics, which should tell you one thing – keeping your network secure is a team effort that requires the devoted efforts of professionals in several different areas. In order to make the topic a little easier to understand, this chapter broke the requirements down into traditional protections, ML protections, real-time detection, and predictive defenses. Hackers are constantly doing three things to thwart your efforts: finding new ways to break into your network, developing ever-faster techniques, and doing the unexpected to evade your defenses. These hacker methodologies are why you must view network security as a collaboration between humans and various kinds of automation. Without augmentation, humans are hopelessly mired in detail and won't see an attack until it has already finished and the damage is done. Despite this, automation can't possibly deal with a hacker's ability to perform attacks in unexpected ways.

Part of this chapter focused on techniques that work well for security purposes, rather than ML as a whole. For example, you can use any number of algorithms to detect suspicious activity, but the best practice is to start with a random forest classifier for security needs because of the particular advantages it provides. In addition, you use supervised learning to perform detection, again because of the advantages it provides over unsupervised learning in a security context. The point is that there is a difference between ML techniques in general and those used specifically for security needs, so it's important to consider the focus of any security applications you create.

You've also discovered that Python, the language most often used to create ML applications, has security holes that can affect your network as well. Consequently, you need to look out for the issues presented by programming techniques such as subprocesses and the potential for RCE. Even though these security holes always affect Python applications, they're especially important to consider when working with network applications because a hacker can access them with greater ease.

*Chapter 6* studies a topic that's essential to creating and using ML successfully to keep hackers at bay – the detection of **anomalies**, which is the presence of anything unexpected in data, models, or activities, no matter where they occur in a system. An anomaly can be a black swan event, the effects of cosmic radiation, or the subtle intrusion of a hacker, but you won't know until you investigate it.

# 6

# Detecting and Analyzing Anomalies

The short definition of an **anomaly** is something that you don't expect—something strange, out of the ordinary, or simply a deviation from the norm. You don't expect to see values outside a specific numeric range when reviewing data—these values often called **outliers** because they lie outside the expected range. However, anomalies occur in all sorts of ways, many of which don't fall into the category of outliers. For example, the data may simply not meet formatting requirements, or it may appear inconsistently, as with state names that are correct but presented in different ways.

Some people actually enjoy seeking anomalies, finding them amusing or at least interesting. The point is anomalies occur all the time, and they may appear harmless, but they have the potential to affect your business in various ways. The point of this chapter is to help you discover what anomalies are with regard to ML, how to determine what sort of anomaly it is, and how to mitigate its effects when necessary. It's not possible or even required to deal with rare events caused by cosmic radiation, for example. With these issues in mind, this chapter discusses the following topics:

- Defining anomalies
- Detecting data anomalies
- Using anomaly detection effectively in ML
- Considering other mitigation techniques

## Technical requirements

This chapter requires that you have access to either Google Colab or Jupyter Notebook to work with the example code. The *Requirements to use this book* section of *Chapter 1, Defining Machine Learning Security*, provides additional details on how to set up and configure your programming environment.

When testing the code, use a test site, test data, and test APIs to avoid damaging production setups and to improve the reliability of the testing process. Testing on a non-production network is highly recommended but not absolutely necessary. Using the downloadable source code is always highly recommended. You can find the downloadable source on the Packt GitHub site at `https://github.com/PacktPublishing/Machine-Learning-Security-Principles` or my website at `http://www.johnmuellerbooks.com/source-code/`.

# Defining anomalies

In the ML realm, anomalies represent data that lies outside of the expected range. The anomaly may occur accidentally, or someone may have put it there, but an anomaly is usually unexpected and potentially unwanted. Anomalies come in two forms:

- **Outliers**: When the data doesn't fit in with the rest of the data, it's an outlier. An outlier can come in many forms, but the defining characteristic is that it's definitely not wanted because it skews any sort of analysis performed with it in place.

- **Novelties**: Sometimes, the data is outside the normal range, but it actually does fit in with the rest of the data. In this case, the data represents a new example that must be considered as part of any analysis. Otherwise, the analysis will fail to represent the true state of whatever the analysis is supposed to bring to light.

Part of the problem, then, is that both kinds of anomaly lie outside the normal range, but one is wanted and the other isn't. Before any mitigation can occur, the ML application must provide some means of determining whether the data is an outlier or a novelty. In some cases, making a correct decision may actually require that a human review the data to make the determination.

## Specifying the causes and effects of anomaly detection

Before you can know that an anomaly exists, you must detect it. Anomaly detection is a mix of the following methods:

- **Engineering**: Using known rules and laws, which can be done using ML techniques.

- **Science**: Defining a hypothesis and then proving it. Science requires a human to define the hypothesis but can depend on ML to help prove it.

- **Art**: Simply getting a feeling that something is wrong, which is most definitely, exclusively, the realm of humans.

It's important to detect anomalies as soon as possible when they occur. Unlike many areas of security, there aren't any methods of predicting an anomaly will occur because anomalies are, by nature, completely unexpected and unpredicted. Previous chapters discussed purposeful attacks by hackers to change data in ways that cause problems for every security element on your network, but anomalies usually happen in a different manner; they aren't usually planned. When they are planned, they're exceptionally hard to detect because hackers can simply hide the attack within all of the other data. Consequently, this chapter looks at the engineering and science aspects of anomaly detection from the perspective of the unexpected event that could be perpetrated by a hacker but could be caused by a vast number of other sources. You'll have to develop your own sixth sense to feel that an anomaly has occurred (and most people do develop one with experience).

Part of the problem with detecting anomalies is that humans have a tendency toward bias, which means having an inclination not to see the unexpected if it doesn't make enough of an impression. As a real-world example, one person is focused entirely on work, while another repaints the house. The first person comes home from work and doesn't see that the house is painted because bias prevents them from doing so. Their focus is on work; nothing else matters. The same thing happens with anomalies that affect the business environment, especially data. It's possible to look at the data and yet not really see it because the focus is on something else at the time. This is where ML can come to the rescue by providing that extra impression to their human counterparts that something is apparently wrong.

Fortunately, you have other methods of anomaly detection that you can use to prevent break-ins and issues like model stealing. These methods fill in where visual observation and the use of ML tools don't quite fill in all of the gaps. Discovering precisely how you can add these tools to the measures you already have in place is an essential part of discovering anomalies early and mitigating them quickly after discovery.

## Considering anomaly sources

Anomalies occur in different situations, and it isn't always the fault of the data source, which is something that many texts forget to mention. Novelties do occur as part of a change in source data, but outliers aren't so limited. With this in mind, the following list provides some sources to check when you see anomalies in your data:

- **Dataset damage**: Any data source is subject to damage. Users can enter incorrect data, a network outage can corrupt data, hackers can change data, nature can modify data, or the data could simply become outdated to the point that it's not useful any longer. This source of anomalies is always associated with outliers.

- **Concept drift**: The meaning of data can change over time based on physical, political, economic, social, or other external forces, such that the interpretation of the data changes. You often see this kind of anomaly as a novelty, but it can also be an outlier.

- **Environmental change**: Sensors are especially sensitive to this issue and will likely generate anomalies in the form of novelties. However, this sort of anomaly can also appear as an outlier when hackers or other individuals purposely modify the environment to create unusable data.

- **Source data change**: Any data source that you use that isn't under your personal control is subject to unexpected change. Your application will likely detect changes in format or other physical factors. However, it won't detect issues such as policy changes by the organization that creates and manages the data. Data that suddenly uses a different perspective to express ideas or suddenly includes various biases will produce outliers that you need to either remediate or account for in your application.

- **Man-in-the-middle (MITM)**: It's happening more often that the data source hasn't changed, and your code for accessing it hasn't changed, but at some point between sending the information from the source and your receipt of the data, the data gets changed, usually in a subtle manner. The concept of the MITM attack has been around for a long time now and you can find plenty of guides about it, such as *The Ultimate Guide to Man in the Middle (MITM) Attacks and How to Prevent them* at `https://doubleoctopus.com/blog/the-ultimate-guide-to-man-in-the-middle-mitm-attacks-and-how-to-prevent-them/`. This is simply a new twist on a tried and tested method used by hackers to cause problems.

- **User error**: It's not possible to train humans to perform tasks with complete correctness all of the time. If humans always made the same sorts of errors, it might be possible to detect the errors and fix them with relative ease, but humans are also quite creative in the way they make errors, so detection of human errors is hard, and there is no silver bullet solution.

There are many other sources of anomalies. The best advice to consider is that you can't trust any data at any time. All data is subject to anomalies, even if it comes from a well-known source and is supposedly clean. Of course, you can't just wait for anomalies to come along to train your models, so it's also necessary to know about data sources you can use for testing purposes. *Figure 6.1* provides some commonly used sources that you may want to try while training your model. The *Checking data validity* section of this chapter discusses the all-important topic of detecting anomalies in the data stream, which is where you'll likely see them first:

| Dataset | Training Type | Description | Location |
|---------|---------------|-------------|----------|
| Cifar-10 | Image | A set of datasets containing smallish 32x32 images that would prove useful in certain types of anomaly detection training. The datasets come in a variety of sizes, with `Cifar-10` containing 60,000 images in 10 classes and `Cifar-100` containing 60,000 images in 100 classes. | `https://www.cs.toronto.edu/~kriz/cifar.html` |

| ImageNet | Image | An immense dataset at 150 GB that holds 1,281,167 images for training and 50,000 images for validation, organized into 1,000 categories. The dataset is so popular that the people who manage it recently had to update their servers and reorganize the website. All of the images are labeled, making them suitable for supervised learning. | `https://www.image-net.org/update-mar-11-2021.php` |
|---|---|---|---|
| Imagenette and Imagewoof | Image | A set of datasets containing images of various sizes, including 160-pixel and 320-pixel sizes. These datasets aren't to be confused with ImageNet (and the site makes plenty of fun of the whole situation). The `Imagenette` datasets consist of 10 classes using easily recognizable images, while the `Imagewoof` dataset consists of a single class, all dogs that can be incredibly tough to recognize. This series of datasets also include noisy images and purposely changed labels in various proportions to the correct labels. | `https://github.com/fastai/imagenette` |

| | | | |
|---|---|---|---|
| **Modified National Institute of Standards and Technology (MNIST)** | Image | Handwritten digits with 60,000 training and 10,000 testing examples. The digits are already size normalized and centered in a fixed-size image. Many Python packages include this dataset, such as `scikit-learn` (`https://scikit-learn.org/stable/auto_examples/classification/plot_digits_classification.html`). | `http://yann.lecun.com/exdb/mnist/` |
| **Numenta Anomaly Benchmark (NAB)** | Streaming anomaly detection | Contains 58 labeled real-world and artificial time-series data files you can use to test your anomaly detection application in a streamed environment. The data files show anomalous behavior but not on a consistent basis, making it possible to train a model to recognize the anomalous behavior. This dataset isn't officially supported on Windows 10 or (theoretically) 11. | `https://github.com/numenta/NAB` |

| One Billion Word Benchmark | **Natural Language Processing (NLP)** | A truly immense dataset taken from a news commentary site. The dataset consists of 829,250,940 words with 793,471 unique words. The problem with this dataset is that the sentences are shuffled, so the ability to determine context is limited. | `https:// opensource. google/projects/ lm-benchmark` |
|---|---|---|---|
| Penn Treebank | NLP | Contains 887,521 training words, 70,390 validation words, and 78,669 test words with the text preprocessed to make it easy to work with. | `https://deepai. org/dataset/penn- treebank` |
| WikiText-103 | NLP | A much larger version of the `WikiText-2` dataset contains 103,227,021 training words, 217,646 validation words, and 245,569 testing words taken from Wikipedia articles. | `https://blog. einstein.ai/ the-wikitext- long-term- dependency- language- modeling-dataset/` |
| WikiText-2 | NLP | Contains 2,088,628 training words, 217,646 validation words, and 245,569 test words taken from Wikipedia articles. This dataset provides a significantly improved environment over Penn Treebank for training models because preprocessing is kept to a minimum. | `https://blog. einstein.ai/ the-wikitext- long-term- dependency- language- modeling-dataset/` |

Figure 6.1 – Datasets that work well for anomaly training

Even though most of these datasets fall into either the image or NLP categories, you can use them for a variety of purposes. As time progresses, you'll likely find other well-defined, labeled, and vetted datasets suitable for security use in other categories, such as streaming anomaly detection. However, it's hard to find such datasets today. It's not that other datasets are lacking, but they haven't received the scrutiny that these datasets have.

## Understanding when anomalies occur

Anomalies occur all the time, 24 hours a day, 7 days a week. However, determining when anomalies occur can help in determining whether the anomaly represents a threat, whether it's a benign outlier, or whether it's a novelty to include as part of your data. The perception that hackers get up at odd hours of the night to attack your network is simply wrong. The article *Website Hacking Statistics You Should Know in 2021* at `https://patchstack.com/website-hacking-statistics/` points out that it's nearly impossible to define solid statistics anymore, but it does provide you with guidelines on what to expect. For example, the FBI reported a 300 percent increase in the number of cybercrimes during COVID-19. When a hacker attacks your site, network, application, API, or any other part of your infrastructure, the act generates anomalies that you can detect and act upon.

The frequency of occurrence can also tell you a lot. Users may lose a password and try to log into the network five or six times before giving up and finding an administrator, but hackers are far more persistent. To be successful, the hacker has to keep trying, which means that you'll see a lot of login attempts for a particular account.

---

**The benefits of freezing an account after so many password tries**

It almost seems archaic because the technique has been around for so long, since before the internet. However, allowing a user to make a specific number of tries to input the correct name and password has at least three benefits. First, it means that the user is going to have to alert the administrator to the password loss sooner, which may mean looking for any account irregularities sooner before a hacker has a chance to act. The user may not have actually lost their password; a hacker may have accessed their account and changed it for them. Second, it tends to thwart a hacker's use of automation or at least slow it down. If a hacker has to wait 20 minutes (as an example) after every four tries, many of the common forms of automation that hackers rely upon will be far less efficient. Third, if you have a really determined hacker, that pattern of logins will become a lot more noticeable.

---

Thinking through anomalies can also help a lot. An anomaly that occurs during the day when users are logged in has a higher probability of being benign in most cases. Of course, this line of thought assumes that your business isn't running three shifts. You should also consider the ebb and flow of data that occurs as customers check in to determine when products will ship or new products are available for purchase. By looking for patterns based on reasonable activity, you can begin to see the activities of those who are purposely trying to break into your business for some reason, which isn't always apparent until the break-in occurs.

## Using anomaly detection versus supervised learning

Anomaly detection, by definition, uses unsupervised (most often) or possibly semi-supervised (rarely today) learning techniques by definition. It allows the model to robustly handle unknown situations, which is often a requirement when dealing with data from an unknown or alternative source. Supervised learning (the focus of *Chapter 5*, on detecting network hacks) relies on labeled data to correctly train a model to recognize good and bad data. It allows the model to make decisions with fewer false positives and false negatives but can misclassify outliers.

In some cases, a setup designed to detect anomalies and then classify them will start with an unsupervised anomaly detection model. When the model detects an anomaly, it passes it on to the supervised learning model. Since the supervised model is more adept at classifying good versus bad data, it can often detect whether an anomaly is an outlier or simply novel data.

## Using and combining anomaly detection and signature detection

Many forms of **Network Intrusion Detection Systems** (**NIDS**) rely on a combination of anomaly detection and signature detection today. The reason for using both approaches is to create a **defense in depth** (**DiD**) scenario where one detection method or the other is likely to detect any sort of problem. Anomaly detection is flexible and brings with it the ability to handle new threats almost immediately. Signature detection provides a robust solution to known threats because it detects those threats based on a specific signature.

The main advantage of using the combination is that anomaly detection is prone to high false positives in some situations, while signature detection is known to provide very low false positives in precisely the same situations. Anomaly detection is also slower than signature detection in most cases because the data must go through the analysis process before making a decision. However, anomaly detection excels in locating zero-day exploits and is indispensable for advanced hackers who know how to modify the signature of an attack vector enough to fool a signature detection setup.

In order to see an actual threat, the two detection systems must actually receive the data that requires analysis. This means placing such systems in more than one location in the organization:

- **Network interface**: A network interface connects your network to the outside world, to other entities in your organization, or to other segments on the same network. You need protection wherever a boundary exists because hackers are adept at locating these boundaries and exploiting them.
- **Endpoints**: Any endpoint can receive or send anomalous data. Detecting the anomaly before it leaves the endpoint can keep your data safe. Of course, you need different kinds of detection for different kinds of endpoints:
- **Verification**: User systems are especially prone to generating anomalous data, so you need to verify what the user is doing at all times.

- **Monitoring**: Any data generator can create anomalous data accidentally, as an act of nature, as a system failure, or as some other cause, so you need to track what sort of data the generator is creating for you.

- **Filtering**: A disk server is unlikely to generate anomalous data, but it can be damaged by it, so you need to filter the data before the disk server receives it.

- **Web application firewall**: Every web application, whether it's an end-user application or an API called on by other entities, requires constant monitoring. This is a major source of hacker intrusion into your system, so it needs to be locked down as much as possible.

- **Other**: Systems are so interconnected now that even if you have a map of the system, it likely has missing elements, such as **Internet of Things (IoT)** devices and various sensors. Hackers will access your system through any opening you provide, so locking these other entry points down is important.

This list likely isn't complete for your organization. Brainstorming the locations that require detection and ascertaining the kind of detection required is important. Make sure you include items outside the normal purview of developers, such as sensors and IoT devices. The next section moves from defining what an anomaly is to how to detect anomalies using various techniques.

## Detecting data anomalies

Anomaly (and its novelty counterpart) detection is a never-ending, constant requirement because anomalies happen all the time. However, with all this talk of detecting and removing anomalies, you need to consider something else. If you remove the novelties from the dataset (thinking that they are anomalies), then you may not see an important trend. Consequently, detection and research into possible novelties go hand in hand. Of course, the most important place to start is with the data itself, looking for values that don't obviously belong. *Figure 6.2* provides a list of common techniques to detect outliers (the table is definitely incomplete because there are many others):

| Method | Type | Description |
|---|---|---|
| **Cook's distance** | Model-specific | This estimates the variations in regression coefficients after removing each observation one at a time. The main goal of this method is to determine the influence exerted by each data point, with data points having undue influence being outliers or novelties. This technique is explored in the *Relying on Cook's distance* section of the chapter. |

| Interquartile range (IQR) | Univariate | This considers the placement of data points within the first and third quartiles normal. All data points 1.5 or more times outside of these quartiles are considered outliers. This technique is usually displayed graphically using a box plot. It was created by John Tukey, an American scientist best known for creating the **fast Fourier transform (FFT)**. This technique is explored in the *Relying on the interquartile range* section of the chapter. |
|---|---|---|
| **Isolation Forest** | Multivariate | This detects outliers using the anomaly score of the Isolation Forest (a type of random forest). The article at `https://towardsdatascience.com/outlier-detection-with-isolation-forest-3d190448d45e` provides additional insights into this approach. |
| **Mahalanobis distance** | Multivariate | This measures the distances between points in multivariate space. It detects outliers by accounting for the shape of the data observations. The threshold for declaring a data point an outlier normally relies on either **standard deviation (STDEV)** or **mean absolute deviation (MAD)**. You can find out more about this technique at `https://www.statisticshowto.com/mahalanobis-distance/`. |
| **Minimum Covariance Determinant (MCD)** | Multivariate | Based on the Mahalanobis distance, this technique uses the mean and covariance of all data, including the outliers, to determine the difference scores between data points. This approach is often considered far more accurate than the Mahalanobis distance. You can find out more about this approach in the article at `https://onlinelibrary.wiley.com/doi/full/10.1002/wics.1421`. |

| Pareto | Model-specific | This assesses the reliability and approximate convergence of Bayesian models using estimates for the k shape parameter of the generalized Pareto distribution. This method is named after Vilfredo Pareto, the Italian civil engineer, economist, and sociologist. It's the source of the 80/20 rule. The article at `https://www.tandfonline.com/doi/abs/10.1080/00949655.2019.1586903?journalCode=gscs20` provides additional information about this method. |
|---|---|---|
| **Principle component analysis (PCA)** | Multivariate | This restructures the data, removing redundancies and ordering newly obtained components according to the amount of the original variance that they express. Using this approach makes multivariate outliers particularly evident. This technique is explored in the *Relying on principle component analysis* section of the chapter. |
| **Z-score** | Univariate | This describes a data point as a deviation from a central value. To use this approach, you must calculate the z-scores one column at a time. Different sources place different scores as the indicator of an outlier—as low as 1.959 and as high as 3.00. This technique is explored in the *Relying on z-score* section of the chapter. |

Figure 6.2 – Common methods for detecting outliers

The common techniques in *Figure 6.2* are those that you will use most often, and you may not ever need any other method unless your data is structured oddly or contains patterns that don't work well with the kinds of analysis these methods perform. In this case, you need to look for algorithms that meet your specific need rather than try to force a common algorithm to provide an answer that it can't provide to you. For example, the article *5 Anomaly Detection Algorithms in Data Mining (With Comparison)* at `https://www.intellspot.com/anomaly-detection-algorithms/` bases anomaly detection on data mining techniques.

## Checking data validity

One of the things you need to ask with regard to anomalies is whether the data you're using is valid. If there are too many outliers, any results from the data analysis you perform will be skewed, and you'll derive erroneous results. In most cases, the outliers you want to find are those that relate to some sort of data entry error. For example, it's unlikely that someone who is age 2 will somehow have a PhD. However, outliers can also be subtle. For example, a hacker could boost the prices of homes in a certain area by a small amount so that a confederate with homes to sell can get a better asking price.

The point is that data tends to have certain characteristics and fall within certain ranges, so you can use an engineering approach to finding potential outliers. You must then resort to intuition to decide whether the errant values are novelties or actual outliers that you then research and fix.

The real problem is the outlier that deviates from the norm by a wide margin. Depending on the ML algorithms you use, and the method used to determine errors, a carefully placed outlier can cause the model to change the weights it uses for making decisions by a large amount. The K-means algorithm is just one of many that fall into this category. Fortunately, not all algorithms are sensitive to a few outliers with unbelievable values. If you use the K-medoids clustering variant instead, one or two outliers will have much less of an effect. That's because K-medoids clustering uses the actual point in the cluster to represent a data item, rather than the mean point as the center of a cluster. Some algorithms, such as random forest, are supposedly less affected by outliers, according to some people (see `https://heartbeat.fritz.ai/how-to-make-your-machine-learning-models-robust-to-outliers-44d404067d07`) and not so robust according to others (see `https://stats.stackexchange.com/questions/187200/how-are-random-forests-not-sensitive-to-outliers`).

The safe assumption is that outliers create problems and that you should remove them from your data. A number of methods exist to detect outliers. Some of them work with one feature (**univariate**); others focus on multiple features (**multivariate**). The approach you use depends on what you suspect is wrong with the data. The examples in this section all rely on the California housing dataset, which is easily available from scikit-learn and provides well-known values for the comparison of techniques. The `MLSec; 06; Check Data Validity.ipynb` file contains source code and some detailed instructions for performing validity checks on your data. You can learn more about the California housing dataset, which is part of `sklearn.datasets` (`https://scikit-learn.org/stable/modules/generated/sklearn.datasets.fetch_california_housing.html`), at `https://www.kaggle.com/datasets/camnugent/california-housing-prices`.

### Relying on the interquartile range

The univariate approach to locating outliers was originally proposed by John Tukey as **Exploratory Data Analysis (EDA)** IQR. You use a box plot to see the median, 25 percent, and 75 percent values. By subtracting the 25 percent value from the 75 percent value to obtain the IQR and multiplying by 1.5, you obtain the anticipated range of values for a particular dataset feature. The extreme end of the range appears as whiskers on the plot. Any values outside this range are suspected outliers. Use these steps to see how this process works:

1. Import the required libraries:

   ```
   from sklearn.datasets import fetch_california_housing
   import pandas as pd

   %matplotlib inline
   ```

2. Manipulate the dataset so that it appears in the proper form:

```
california = fetch_california_housing(as_frame = True)
X, y = california.data, california.data
X = pd.DataFrame(X, columns=california.feature_names)
print(X)
```

The example prints the resulting dataset so that you can see what it looks like, as shown in *Figure 6.3*. Across the top, you can see the features used for the dataset:

```
       MedInc  HouseAge  AveRooms  AveBedrms  Population  AveOccup  Latitude \
0      8.3252      41.0  6.984127   1.023810       322.0  2.555556     37.88
1      8.3014      21.0  6.238137   0.971880      2401.0  2.109842     37.86
2      7.2574      52.0  8.288136   1.073446       496.0  2.802260     37.85
3      5.6431      52.0  5.817352   1.073059       558.0  2.547945     37.85
4      3.8462      52.0  6.281853   1.081081       565.0  2.181467     37.85
...       ...       ...       ...        ...         ...       ...       ...
20635  1.5603      25.0  5.045455   1.133333       845.0  2.560606     39.48
20636  2.5568      18.0  6.114035   1.315789       356.0  3.122807     39.49
20637  1.7000      17.0  5.205543   1.120092      1007.0  2.325635     39.43
20638  1.8672      18.0  5.329513   1.171920       741.0  2.123209     39.43
20639  2.3886      16.0  5.254717   1.162264      1387.0  2.616981     39.37

       Longitude
0        -122.23
1        -122.22
2        -122.24
3        -122.25
4        -122.25
...          ...
20635    -121.09
20636    -121.21
20637    -121.22
20638    -121.32
20639    -121.24

[20640 rows x 8 columns]
```

Figure 6.3 – The data manipulation creates a tabular view with columns labeled with feature names

3. Create the box plot:

```
X.boxplot('MedInc', return_type='axes')
```

The box plot specifically shows the median income (MedInc) feature values. The output shown in *Figure 6.4* indicates that the full range of values runs from about $9,000 to $80,000. The median value is around $35,000, with a 25 percent value of $25,000 and a 75 percent value of $50,000 (for an IQR of $25,000). However, some values are above $80,000 and are suspected of being outliers:

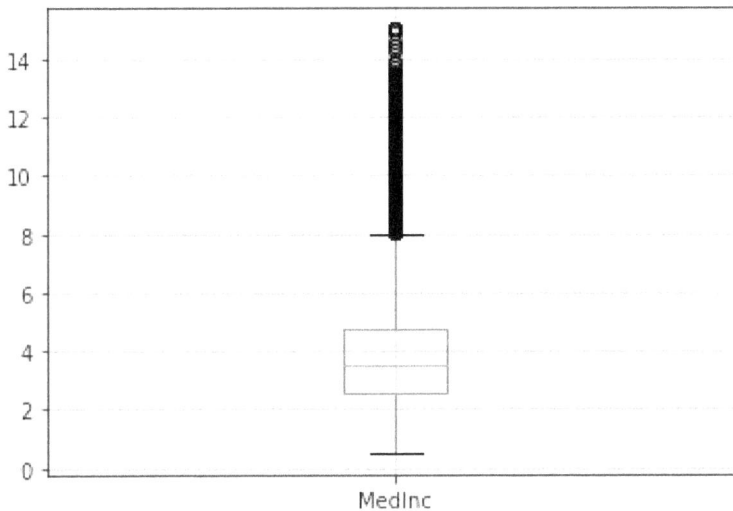

Figure 6.4 – The box plot provides details about the data range
and indicates the presence of potential outliers

The next section relies on the same variable, X, to perform the next outlier detection process, PCA. It shows you the potential outliers in the MedInc feature in another way.

### Relying on principle component analysis

In viewing the output shown in *Figure 6.4*, you can see that there are potential outliers, but it's hard to tell how many outliers, and how they relate to other data in the dataset. The need to understand outliers better is one of the reasons you resort to a multivariate approach such as PCA. The actual math behind PCA can be daunting if you're not strong in statistics, but you don't actually need to understand it well to use PCA successfully.

The number of variables you choose depends on the scenario that you're trying to understand. In this case, the example looks for a correlation between median income and house age. When working through a similar security problem, you might look for a correlation between unsuccessful login attempts and time of day or less-used API calls and network load. It's all about looking for patterns in the form of anomalies that you can use to detect hacker activity.

This example also relies on a scatterplot to show the results of the analysis. A scatterplot will tend to emphasize where outliers appear in the data. The following steps will lead you through the process of working with PCA using the California housing dataset, but the same principles apply to other sources of data, such as API access logs:

1.  Import the libraries needed for this example:

    ```
    from sklearn.decomposition import PCA
    from sklearn.preprocessing import scale
    import matplotlib.pyplot as plt
    ```

2.  Consider the need to scale your example by running the following code:

    ```
    print(X[["MedInc", "HouseAge"]][0:5])
    print(scale(X[["MedInc", "HouseAge"]])[0:5])
    ```

    The output shown in *Figure 6.5* shows how the original data is scaled in the MedInc and HouseAge columns so that it's the same relative size. **Scaling** is an essential part of working with PCA because it makes the comparison of two variables possible. Otherwise, you have an apples-to-oranges comparison that will never provide you with any sort of solid information:

    ```
           MedInc   HouseAge
    0   8.3252       41.0
    1   8.3014       21.0
    2   7.2574       52.0
    3   5.6431       52.0
    4   3.8462       52.0
    [[ 2.34476576   0.98214266]
     [ 2.33223796  -0.60701891]
     [ 1.7826994    1.85618152]
     [ 0.93296751   1.85618152]
     [-0.012881     1.85618152]]
    ```

    Figure 6.5 – Scaling is an essential part of the multivariate analysis process

3.  Use PCA to fit a model to the scaled data:

    ```
    pca = PCA(n_components=2)
    pca.fit(scale(X))
    C = pca.transform(scale(X))
    print(C)
    print("Original Shape: ", X.shape)
    print("Transformed Shape: ", C.shape)
    ```

The pca.transform() method performed dimensionality reduction. You can see the result of the fitting process and dimensionality reduction in *Figure 6.6*:

```
[[ 1.88270434 -0.50336186]
 [ 1.37111955 -0.12140565]
 [ 2.08686762 -0.5011357 ]
 ...
 [ 1.40235696 -1.09642559]
 [ 1.5429429  -1.05940835]
 [ 1.40551621 -0.89672727]]
Original Shape:    (20640, 8)
Transformed Shape:  (20640, 2)
```

Figure 6.6 – Fitting the data and then transforming it provides data you can plot

4.  Plot the data to see the result of the analysis:

```
plt.title('PCA Outlier Detection')
plt.xlabel('Component 1')
plt.ylabel('Component 2')
plt.scatter(C[:,0],C[:,1], s=2**7, edgecolors='white',
    alpha=0.85, cmap='autumn')
plt.grid(which='minor', axis='both')
plt.show()
```

The output in *Figure 6.7* clearly shows the outliers in the upper-right corner of the plot. The *x*-and *y*-axis data appears first in the call to plt.scatter(). The remainder of the arguments affects presentation: s for the marker size, edgecolors for the line around each marker (to make them easier to see), alpha to control the marker transparency, and cmap to control the marker colors:

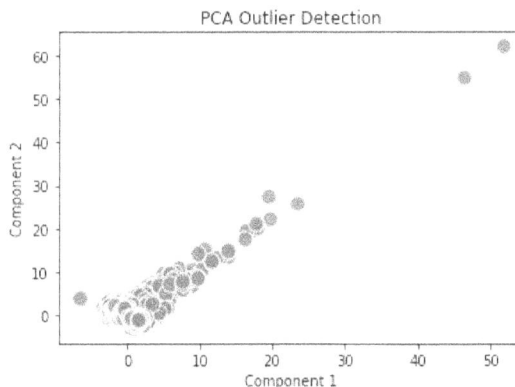

Figure 6.7 – The outliers appear in the upper right corner of the plot

To use this method effectively, you need to try various comparisons to find the correlations that will make the security picture easier for you to discern. In addition, you may find that you need to use more than just two features, as this example has done. Perhaps there is a correlation between invalid logins, IP addresses, and time of day. Until you model the data for your particular network, you really don't know what the anomalies you suspect will tell you about your security picture.

---

**Going overboard with analysis**

It would be quite easy to go overboard with the analysis you want to perform by testing a ridiculous number of combinations and looking for patterns that aren't there. Human intuition is extremely important in determining which combinations to try and figuring out what data actually is an anomaly and which is a novelty. The computer can't perform this particular task for you. PCA only works when the human performing it makes decisions based on previous experience, current trends, network uniqueness, and anticipated or viewed behaviors.

---

### *Relying on Cook's distance*

Cook's distance is all about measuring influence. It detects the amount of influence that a particular observation has on a model. When it comes to anomalies, a very influential observation is likely an outlier or novelty. Using Cook's distance properly will help you see specifically where the outliers reside in the dataset. Removing these outliers can help you create a better baseline of what is normal for the dataset and makes anomaly detection easier.

Note that before you can run this example, you need to have Yellowbrick (https://www.scikit-yb.org/en/latest/) installed. The following command will perform the task for you:

```
conda install -c districtdatalabs yellowbrick
```

However, if you don't have Anaconda installed on your system, you can also add this code to a cell in your notebook to install it:

```
modules = !pip list
installed = False

for item in modules:
    if ('yellowbrick' in item):
        print('Yellowbrick installed: ', item)
        installed = True

if not installed:
    print('Installing Yellowbrick...')
    !pip install yellowbrick
```

In this second case, the code checks for the presence of Yellowbrick on your system and installs it if it isn't installed. This is a handy piece of code to keep around because you can use it to check for any dependency and optionally install it when not present. In addition, this piece of code will work with alternative IDEs, such as Google Colab. Using this check will make your code a little more bulletproof. If Yellowbrick is already installed, you will see an output similar to this showing the version you have installed:

```
Yellowbrick installed:   yellowbrick                    1.5
```

Once you have Yellowbrick installed, you can begin working with the example code using the following steps:

1.  Import the required dependencies:

    ```
    from yellowbrick.regressor import CooksDistance
    ```

2.  Import a new copy of the California housing dataset that's formed in a specific way as show in the following code snippet:

    ```
    california = fetch_california_housing(as_frame = True)
    X = california.data["MedInc"].values.reshape(-1, 1)
    y = range(len(X))
    ```

3.  Printing the result of the import shows how X is formatted for use with Yellowbrick. The y variables are simply values from 0 through the length of the data to act as an index for the output:

    ```
    print(X)
    ```

    *Figure 6.8* shows a sample of the output you should see:

    ```
    [[8.3252]
     [8.3014]
     [7.2574]
     ...
     [1.7   ]
     [1.8672]
     [2.3886]]
    ```

Figure 6.8 – Ensure your data is formatted correctly for the visualizer

4.  Create a `CooksDistance()` visualizer to see the data, use `fit()` to fit the data to it, and then display the result on screen using `show()`:

```
visualizer = CooksDistance()
visualizer.fit(X, y)
visualizer.show()
```

The **stem plot** output you see should look similar to that shown in *Figure 6.9*. Notice that the anomalies are readily apparent and that you know areas that contain records with anomalous data. In addition, the red dashed line tells you the average of the records so that you can see just how far out of range a particular value is:

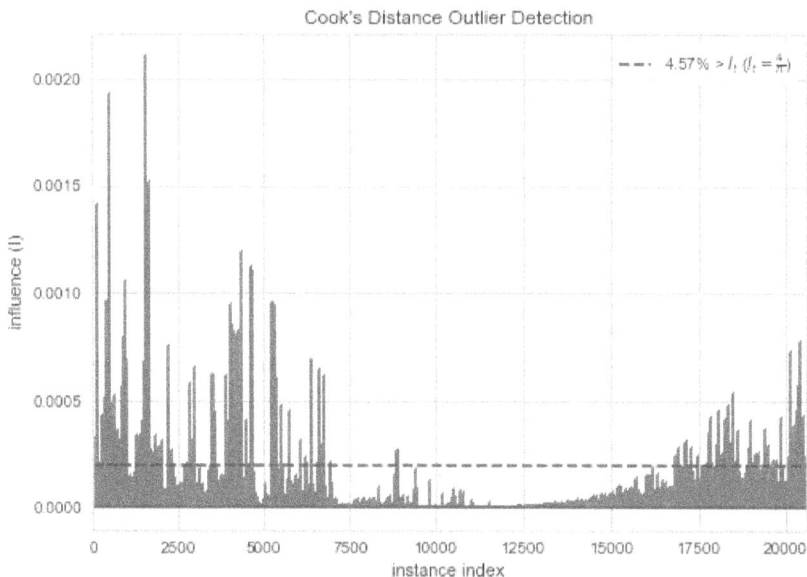

Figure 6.9 – The Cook's distance approach gives you specific places to look for anomalies

Cook's distance reduces the work you need to perform to determine where anomalies occur so that you aren't searching entire datasets for them. If this were a data stream, the `y` value could actually be time increments so that you'd know when the anomalies occur. By slicing and dicing your data in specific ways, you can quickly check for anomalies in a number of useful ways.

### Relying on the z-score

Obtaining the z-score for data points in a dataset can help you detect specific instances of outliers. There are a number of ways to do this, but the basic idea is to determine when a particular data point value falls outside of a specific range, normally beyond the third deviation. Fortunately, the math for performing this task is simplified by using NumPy (https://numpy.org/), rather than calculating it manually.

Using seaborn (https://seaborn.pydata.org/) greatly reduces the amount of work you need to do to visualize your data distribution. This example relies on seaborn version 0.11 and above. If you see an error message stating that module 'seaborn' has no attribute 'displot', it means that you have an older version installed. You can update your copy of seaborn using this command:

```
pip install --upgrade seaborn
```

The following example shows a technique for actually listing the records that fall outside of the specified range and understanding how the records in the dataset fall within a particular distribution:

1.  Import the required dependencies:

    ```
    import numpy as np
    import seaborn as sns
    ```

2.  Obtain an updated copy of the California housing data for median income:

    ```
    X = california.data["MedInc"]
    ```

3.  Determine the mean and standard deviation for the data:

    ```
    mean = np.mean(X)
    std = np.std(X)
    print('Mean of the dataset is: ', mean)
    print('Standard deviation is: ', std)
    ```

    You'll see an output like this when you run the code:

    ```
    Mean of the dataset is:  3.8706710029070246
    Standard deviation is:  1.899775694574878
    ```

4.  Create a list of precisely which records fall outside the third deviation:

    ```
    threshold = 3
    record = 1
    z_scores = []
    for i in X:
        z = (i - mean) / std
        z_scores.append(z)
        if z > threshold:
            print('Record: ', record, ' value: ', i)
        record = record + 1
    ```

*Figure 6.10* shows the results of performing this part of the task. Notice the method for calculating the z-score, z, is simplified by using NumPy. The resulting data tells you where each record falls in the expected range. The output is the records that fall outside the third standard deviation and may be an outlier:

```
Record:  132  value:  11.6017
Record:  410  value:  10.0825
Record:  511  value:  11.8603
Record:  512  value:  13.499
Record:  513  value:  12.2138
Record:  515  value:  12.3804
Record:  924  value:  9.7194
Record:  978  value:  10.9506
Record:  987  value:  10.3203
Record:  1542  value:  9.5862
Record:  1562  value:  9.7037
Record:  1564  value:  10.3345
Record:  1565  value:  12.5915
Record:  1567  value:  15.0001
Record:  1575  value:  9.8708
Record:  1583  value:  10.7372
Record:  1584  value:  13.4883
Record:  1587  value:  12.2478
Record:  1592  value:  10.4549
```

Figure 6.10 – You now have a precise list of the records that could contain outliers

5.  Plot the standard distribution of the z-scores:

```
axes = sns.displot(z_scores, kde=True)
axes.fig.suptitle('Z-Scores for MedInc')
axes.set(xlabel='Standard Deviation')
```

*Figure 6.11* shows you how this data fares. This graph shows that not many of the values are significantly out of range (the third deviation, either plus or minus), but that a few are and that some of them are quite a bit outside the range (up to six deviations):

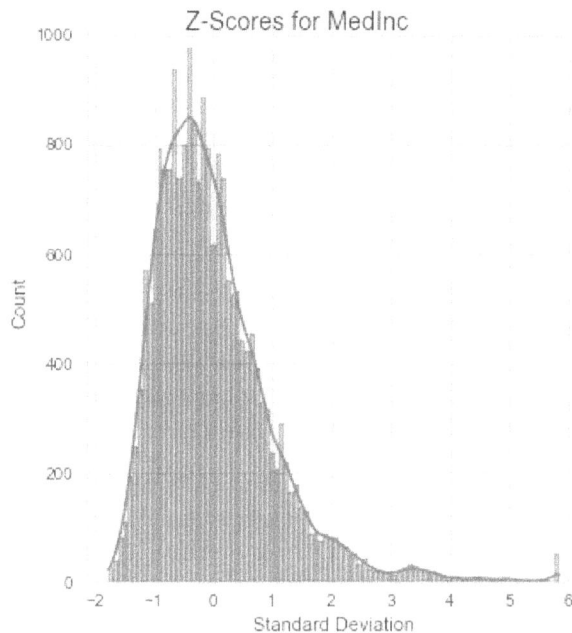

Figure 6.11 – Viewing the distribution plot can tell you just how
much the data is outside of the expected range

By now, you should be seeing a progression with the various examples presented in this and the three previous sections. A simple box plot can tell you whether your data is solid, PCA can tell you where the data is outside of the range, the Cook's distance can help you narrow down the record areas and show patterns of outliers, and the z-score can be very specific in telling you precisely which records are potential outliers. The danger in all of this analysis is that you are faced with too much information of any sort. You don't want to become mired in too much detail, so use the level of detail that satisfies the need at the time.

## Forecasting potential anomalies example

As mentioned earlier in the chapter, it's not really possible to predict anomalies. For example, you can't write an application that predicts that a certain number of users will enter incorrect data in a particular way today. You also can't create an application that will automatically determine what a hacker will attack today. However, you can predict, within a reasonable amount of error, what will likely happen in certain situations.

When you start a new project, you should be able to predict certain outcomes. If the prediction turns out to be completely false, then there might be something more than bad luck causing problems; you might be encountering outside influences, a lack of training, bad data, bad assumptions, or something else, but it all boils down to dealing with some sort of anomaly. One way to predict the future is to rely on a product such as BigML (`https://bigml.com/`) that provides an easy method of creating predictive models for all sorts of uses, such as those shown at `https://bigml.com/gallery/models`. Of course, you have to pay for the privilege of using such a product in many cases (there are also free models that may fit your needs).

Time series data provides a particularly interesting problem because the data is serialized. What occurs in the past affects the now and the future. However, because each data point in a time series is individual, it also violates many of the rules of statistics. From a security perspective, this individuality makes it hard to determine whether the increase in user input today, as contrasted to yesterday or last week on the same day, is a matter of an anomaly, a general trend, or simply random variance.

Fortunately, you can forecast many problems without using exotic solutions when you have historical data. The example in this section forecasts future passenger levels on an airline based on historical passenger levels. The `MLSec; 06; Forecast Passengers.ipynb` file contains source code and some detailed instructions for performing predictions on your data.

### Obtaining the data

If you use the downloadable source, the code automatically downloads the data for you (as shown here) and places it in the code folder for you. Otherwise, you can download it from `https://raw.githubusercontent.com/jbrownlee/Datasets/master/airline-passengers.csv`. The file is only 2.18 KB, so the download won't take long:

```
import urllib.request
import os.path


filename = "airline-passengers.csv"
if not os.path.exists(filename):
    url = "https://raw.githubusercontent.com/\
        jbrownlee/Datasets/master/airline-passengers.csv"
    urllib.request.urlretrieve(url, filename)
```

### Viewing the airline passengers data

One of the things that make the airline passengers dataset so useful for seeing how predictions can work is that it shows a definite cycle in values. The following steps show how to see this cycle for yourself:

1. Import the required dependencies. Note that you may see a warning message when working with this example and using `matplotlib` versions greater than 3.2.2. You can ignore these warnings as they don't affect the actual output:

```
import pandas as pd
import numpy as np
import matplotlib.pyplot as plt
import matplotlib.ticker as ticker

%matplotlib inline
```

2. Read the dataset into memory and obtain specific values from it:

```
apDataset = pd.read_csv('airline-passengers.csv')
passengers = apDataset['Passengers']
months = apDataset['Month']
```

What the example is most interested in is the number of passengers per month. To put this into security terms, you could use the same technique to see the number of suspect API calls per month or the number of invalid logins per month. It also doesn't have to be a monthly interval. You can use any equally spaced interval you want, perhaps hourly, perhaps by the minute.

3. Plot the data on screen:

```
fig = plt.figure()
ax = fig.add_axes([0.0, 0.0, 1.0, 1.0])
ax.plot(passengers)
ticks = months[0::12]
ticks = pd.concat([pd.Series([' ']), ticks])
ax.xaxis.set_major_locator(ticker.MultipleLocator(12))
ax.set_xticklabels(ticks, rotation='vertical')
start, end = ax.get_ylim()
ax.yaxis.set_ticks(np.arange(100, end, 50))
ax.grid()
plt.show()
```

The output is a simple line plot using the `ax.plot(passengers)` call. Here, `ticks` are simply the first month of each year, which are displayed on the *x* axis. The *y* axis starts at a value of 100 and ends at 650 using `ax.yaxis.set_ticks(np.arange(100, end, 50))`.

To make the values easier to read, the code calls `ax.grid()` to display grid lines. As you can see from *Figure 6.12*, the data does have a specific cycle to it as the number of passengers increases year by year.

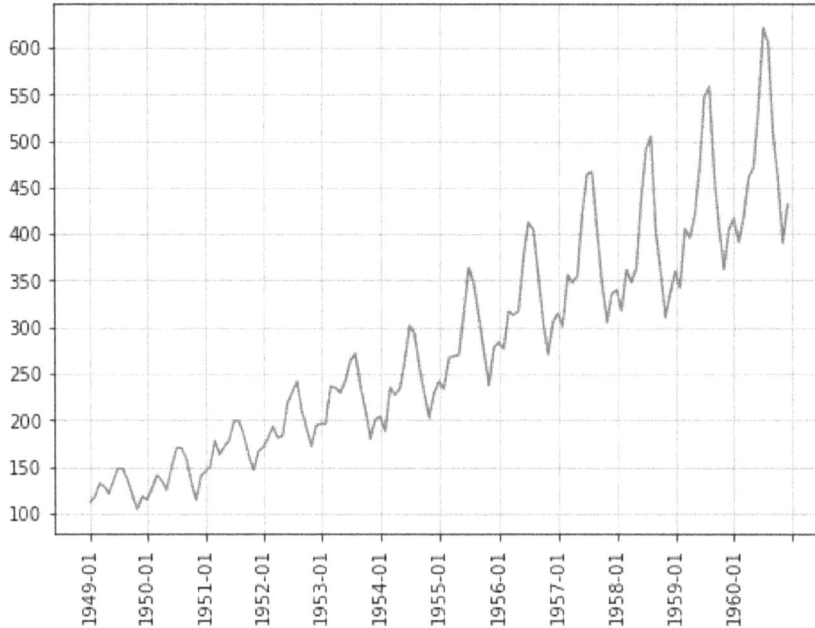

Figure 6.12 – A line chart of the airline passengers dataset shows a definite pattern

This is the sort of pattern you need to see in security data as well to make useful predictions. There has to be some sort of pattern, or it will be quite hard to create any sort of accurate prediction. The better the pattern, the better the prediction.

### Showing autocorrelation and partial autocorrelation

This part of the example comes in two parts: autocorrelation and partial autocorrelation. Both help you make predictions about the future state of data based on both historical and current information.

**Autocorrelation** is a statistical measure that shows the similarity between observations as a function of time lag. It looks at how an observation correlates with a time **lag** (previous) version of itself. Using autocorrelation helps you see patterns in the data so that you can understand the data better. This example reviews the autocorrelation for 3 years (36 months) of airline passenger data. The time interval is a month in this case, so each lag is 1 month, and there are 36 lags. Since all of the points in the graph are above 0, there is a positive correlation between all of the entries. The pattern shows that the data is seasonal. Here are the steps needed to perform an autocorrelation with the air passenger dataset:

1.  Import the required dependencies. Note that NumPy may complain about certain data types in the `statsmodels` package (`https://www.statsmodels.org/stable/index.html`). There is nothing you can do about this warning and can safely ignore it. The `statsmodels` package developer should fix the problem soon:

    ```
    from statsmodels.graphics import tsaplots
    ```

2.  Create the plot:

    ```
    fig = tsaplots.plot_acf(passengers, lags=36)
    fig.set_size_inches(9, 4, 96)
    plt.show()
    ```

    The code relies on a basic plot found in the `statsmodels` package, `plot_acf()`. All you supply is the data and the number of lags. *Figure 6.13* shows the autocorrelations and associated confidence cone:

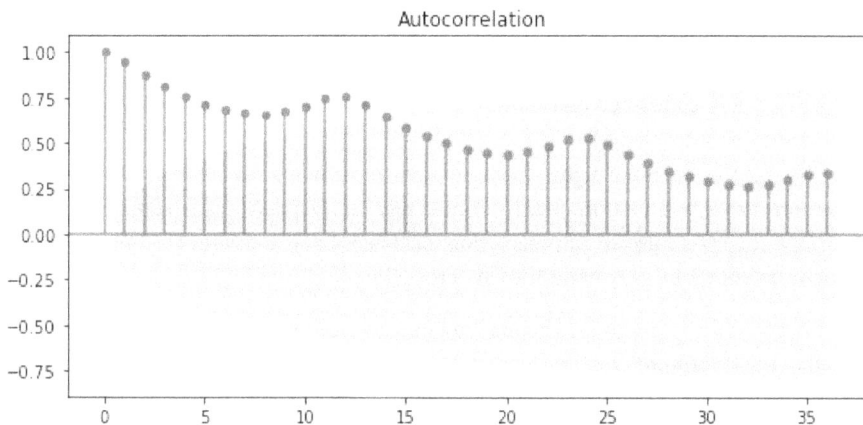

Figure 6.13 – An example of an autocorrelation plot with a confidence cone

The cone drawn as part of the plot is a **confidence cone** and shows the confidence in the correlation values, which is about 80 percent in this case near the end of the plot. More importantly, data with a high degree of autocorrelation isn't a good fit for certain kinds of regression analysis because it fools the developer into thinking there is a good model fit when there really isn't.

The **partial autocorrelation** measurement describes the relationship between an observation and its lag. An autocorrelation between an observation and a prior observation consists of both direct and indirect correlations. Partial autocorrelation tries to remove the indirect correlations. In other words, the plot shows the amount of each data point that isn't predicted by previous data points so that you can see whether there are obvious extremes (data points that don't quite fit properly). Ideally, none of the data points after the first two or three should exceed 0. You use the following code to perform the partial autocorrelation using another standard `statsmodel` package plot, `plot_pacf()`:

```
fig = tsaplots.plot_pacf(passengers, lags=36)
fig.set_size_inches(9, 4, 96)
plt.show()
```

When you run this code, you will see the output shown in *Figure 6.14*:

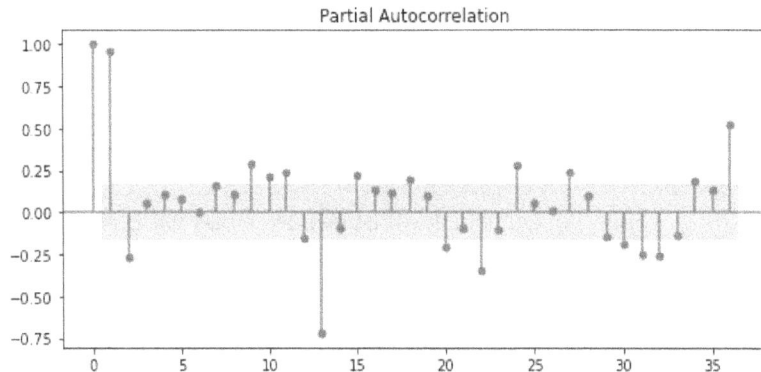

Figure 6.14 – A partial autocorrelation plot shows the relationships between an observation and its lag

The plot shows two lags that are quite high, which provides the term for any autoregression you perform of 2 (any predictions made by an autoregressive model rely on the previous two data points). If there were three high lags, then you'd use an autoregression term of 3. Look at month 13: it has a higher value due to some inconsistency. The gray band shown across the plot is the 95 percent confidence band.

By the time you've got to this point in the chapter, you may be wondering how you'll ever write enough code to cover all of the techniques you find here, much less study the output of all of the applications you seemingly need to write. The fact is that no one can write that much code or study that much output. The next section of the chapter is essential because it helps you put everything into perspective so that you can create an effective detection and mitigation strategy that you'll actually be able to follow.

## Making a prediction

Now that you've looked at the autocorrelation for the data, it's time to use the data to create a predictive model. The following steps show how to perform this process using the statsmodel package used earlier (there are obviously many other ways to perform the same task using other techniques, such as XGBoost):

1.  Import the required dependencies:

    ```
    from statsmodels.tsa.ar_model import AutoReg
    from sklearn.metrics import mean_squared_error
    from math import sqrt
    ```

2.  Import the data and split it into training and testing data:

```
series = pd.read_csv('airline-passengers.csv',
    header=0, index_col=0,
    parse_dates=True, squeeze=True)
X = series.values
train, test = X[1:len(X)-7], X[len(X)-7:]
print("Training Set: ", train)
print("Testing Set: ", test)
```

The data is read using a different technique this time because you need to format it in a different way for analysis. The resulting data appears as simply a list, as shown in *Figure 6.15*:

```
Training Set:  [118 132 129 121 135 148 148 136 119 104 118 115 126 141 135 125 149 170
 170 158 133 114 140 145 150 178 163 172 178 199 199 184 162 146 166 171
 180 193 181 183 218 230 242 209 191 172 194 196 196 236 235 229 243 264
 272 237 211 180 201 204 188 235 227 234 264 302 293 259 229 203 229 242
 233 267 269 270 315 364 347 312 274 237 278 284 277 317 313 318 374 413
 405 355 306 271 306 315 301 356 348 355 422 465 467 404 347 305 336 340
 318 362 348 363 435 491 505 404 359 310 337 360 342 406 396 420 472 548
 559 463 407 362 405 417 391 419 461 472]
Testing Set:  [535 622 606 508 461 390 432]
```

Figure 6.15 – The data is formatted using a different approach to meet the needs of the model software

3.  Create a model based on the training data:

```
model = AutoReg(train, lags=29)
model_fit = model.fit()
print('Coefficients: %s' % model_fit.params)
```

Remember that you've split the 36 lags defined in earlier sections into a training set of 29 lags and a testing set of 7 lags, so you only have 29 lags to train the model. The example prints the **coefficients** for the model, which indicate the direction of the relationship between a predictor variable and the response variable, as shown in *Figure 6.16*:

```
Coefficients: [ 4.40999915  0.54377634  0.31837725 -0.02344532 -0.09328385  0.23085169
 -0.15420365  0.09373805 -0.09291185  0.32015708 -0.46668114  0.05145669
  0.73075956 -0.25128548 -0.33381625  0.242229   -0.15796594 -0.09420813
 -0.08006763  0.05655165 -0.06648343 -0.14659542  0.29515085  0.24214347
  0.09150113 -0.02924769 -0.36689735 -0.01890586  0.12110732  0.0568705 ]
```

Figure 6.16 – The coefficients for the predictive model

4.  Make a prediction and compare that prediction to the test data:

```
predictions = model_fit.predict(start=len(train),
    end=len(train)+len(test)-1,
    dynamic=False)
for i in range(len(predictions)):
    print('predicted=%f, expected=%f'
        % (predictions[i], test[i]))
rmse = sqrt(mean_squared_error(test, predictions))
print('Test RMSE: %.3f' % rmse)
```

The example computes the **root-mean-square error** (**RMSE**) of the difference between the predicted value and the actual value and then outputs it, as shown in *Figure 6.17*. In this case, the RMSE is 18.165 percent, which is a little high but would be better with more data:

```
predicted=552.536526, expected=535.000000
predicted=632.633224, expected=622.000000
predicted=629.130658, expected=606.000000
predicted=527.457734, expected=508.000000
predicted=432.252914, expected=461.000000
predicted=399.961090, expected=390.000000
predicted=424.926456, expected=432.000000
Test RMSE: 18.165
```

Figure 6.17 – Show the predictions, the expected values, and the RSME

5.  Plot the prediction (in red) versus the actual data (in blue):

```
plt.plot(test)
plt.plot(predictions, color='red')
plt.show()
```

*Figure 6.18* shows that despite a somewhat high RSME, the prediction follows the test data pretty well:

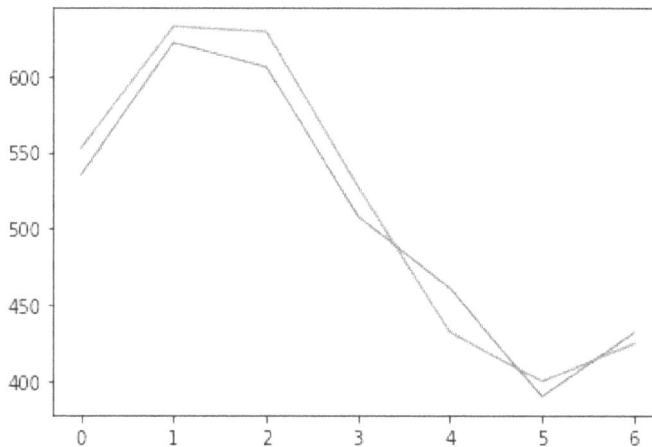

Figure 6.18 – The model's ability to predict the future closely matches the actual data

Now that you've created a model and tested it, you can use it to predict the future within a certain degree of accuracy, as defined by the RSME, and also take into account that the prediction will become less accurate the further you go into the future. The next section will take the next steps and discuss how you can use anomaly detection in an ML environment in an effective way. It's not really enough to simply detect what may appear as anomalies; they need to be identified as true anomalies instead of novelties or possibly noise.

# Using anomaly detection effectively in ML

Everyone has their own opinion of how to work effectively with ML; they can even back up their opinion with favorable statistics. Making things worse, you can find new techniques appearing on a daily basis, adding to the already burgeoning pile of strategies that will likely work within a certain range of probability. The one word that you need to keep in mind is *effective*. An anomaly detection strategy is only effective if you can use it regularly, and therein lies the problem for most overworked security professionals. So, here are some methods you can employ to make whatever ML strategy you use to detect anomalies effective:

- Ensure you actually use the strategy on a regular basis; daily is best
- Use the simplest approach that will work for your organization and you as an individual
- Look for anomalies that are actually likely to affect your organization
- Keep in mind that most anomalies will end up being novelties that you can hand off to someone else and keep your focus on hackers
- Rely on pre-built models only when the model is built specifically for your industry by professionals that understand it

Ultimately, this list is about people and not necessarily about software. The software will do what you tell it to do, computers are faithful in that manner. What you need to do is come up with the correct questions for the software to answer, which means knowing as much about potential anomalies as possible and understanding how these anomalies can lead you to hacker activities.

The next section of the chapter considers other mitigation techniques. A major problem with most approaches to security is that the people involved are looking for a simple fix and security is anything but simple. There is no amount of automation that will solve every issue, which is why you keep seeing reports of breaches in the news. These other techniques may seem to be outside the purview of the security professional and are definitely things you can't write an application to accomplish, but they're still necessary, and, unfortunately, they're neglected far too often.

## Considering other mitigation techniques

The *Using anomaly detection versus supervised learning* and *Using and combining anomaly detection and signature detection* sections of this chapter look at anomaly detection when combined with supervised learning techniques and signature detection. These two sections broach the topic of finding a way to create a defense in-depth strategy for your infrastructure. Developing multiple layers of detection is a strategy that most security experts see as crucial for stemming the tide of hacker attacks, at least to some extent. However, it's also important to understand that combining ML anomaly detection with other software strategies won't completely fix the problem because the issue is one of automation. In order to have the greatest chance of success, you need humans to help see the patterns in data creation, usage, and modification that are anomalous in nature. When considering anomaly detection, also include these human-based observations and mitigation strategies:

- Track behavioral changes. Security personnel should be involved with the people at your workplace enough that they can see patterns in how people act, and they should be authorized to question changes in established patterns.

- Review dashboard data frequently for changes in the way that users, customers, trusted third parties, and the like interact with your system.

- Ensure that security checks actually are run and that references actually are checked. If you don't make the required checks, then you have no idea who is accessing the network or what they're doing with it.

- Set aside time to actually look for trouble rather than relying on it hitting the people you trust in the face.

- Ask everyone to look for potential problems. Security personnel will look for a certain class of problems, but they won't see it all. Someone working in accounting may see an anomaly that the security people may not even understand.

In 2020, Gartner established a new category of cybersecurity tools called **network detection and response** (**NDR**) (`https://www.gartner.com/en/documents/3986225/market-guide-for-network-detection-and-response`). Unfortunately, the report comes at a cost, so you either need to pay up or find an alternative, such as IronNet (`https://www.ironnet.com/what-is-network-detection-and-response`), where the essence of the report is explained. Essentially, it comes down to a third party monitoring your network (in addition to the monitoring you provide for yourself) and then acting when the vendor detects a threat. Some of the focuses of these vendors cover areas where your organization may not have the required personnel, such as monitoring your IoT devices. These organizations also keep up with the latest threats—something that your own personnel may not have time to do well.

## Summary

The important takeaway of this chapter is about being observant but not being paranoid. An anomaly is always unexpected, but it's not always malicious or an indicator of impending doom. Some anomalies are actually welcome because they're novelties that signify a trend toward something positive. The techniques that this chapter contains help you to differentiate between novelties and hacker attacks so that you don't waste time chasing data that doesn't matter in security matters.

A large part of this chapter focused on showing various techniques for discovering anomalies so that you can mitigate them. Even though the univariate approach may seem weak, it also has the benefit of being both fast and simple. You should first try the univariate approach before moving on to the more complex techniques used for multivariate analysis. When it comes to security, speed and simplicity do matter, and some advice you might find in data science texts for fully discovering your data may not apply as much when you need an answer now rather than allow that hacker time with your data, applications, and network.

Predicting anomalies isn't feasible. However, predicting when the conditions are ripe for an anomaly to occur can be done with some amount of accuracy. Just how accurate such a prediction will be depends on your reading of your security environment. However, it's essential to remember that just because conditions indicate that an anomaly *could* happen, doesn't mean that the event *will* occur. What it really means is that there is a need for additional vigilance on your part.

The next chapter moves on to malware, which is code that is designed to do something harmful to your data, applications, network, or personnel. Malware may not always make itself known. In fact, from the hacker's perspective, the longer the malware remains hidden, the better. Of course, there are exceptions, such as ransomware, when the hacker most definitely wants you to know something terrible has happened, but there is a fix for a price. *Chapter 7* will take more of a global view of malware rather than focus on one particular aspect, as is done in some books. Yes, monitoring individual machines for malware is important (and the chapter will provide you with resources to do so), but *Chapter 7* has a strong emphasis on web-based applications and users relying on multiple machines to perform their work today; a global approach is both valuable and necessary.

# Further reading

The following list will provide you with some additional reading that you may find useful in further understanding the materials in this chapter.

- A subset of the ImageNet dataset that is easier to use for experimentation: *Tiny ImageNet:* `https://paperswithcode.com/dataset/tiny-imagenet`

- This paper provides some interesting ideas on how to create an anomaly detection setup based on supervised methods: *Toward Supervised Anomaly Detection:* `https://arxiv.org/ftp/arxiv/papers/1401/1401.6424.pdf`

- This article provides additional information on the differences between MAD and STDEV: *Relationship Between MAD and Standard Deviation for a Normally Distributed Random Variable:* `https://blog.arkieva.com/relationship-between-mad-standard-deviation/`

- Understand the math behind PCA a little better: *A Step-by-Step Explanation of Principal Component Analysis (PCA):* `https://builtin.com/data-science/step-step-explanation-principal-component-analysis`

- Learn more about autocorrelation and time series: *Autocorrelation in Time Series Data:* `https://dzone.com/articles/autocorrelation-in-time-series-data`

- Understand the difference between autocorrelation and partial autocorrelation better: *What's The Difference Between Autocorrelation & Partial Autocorrelation For Time Series Analysis?:* `https://besmarv.medium.com/interpreting-autocorrelation-partial-autocorrelation-plots-for-time-series-analysis-23f87b102c64`

# 7
# Dealing with Malware

**Malware** encompasses a vast array of applications that are designed to disrupt, damage, gain illegal access to, spy on, and do all sorts of other unwanted things to networks, applications, data, and users. Trying to cover every potential kind of malware in all of its various forms in a single chapter, or even a single book, is impossible. Even limiting the topic to just the detection and mitigation of malware using ML techniques is impossible. So, this chapter is more of an overview of malware with some specific examples and references you can use to find additional details. No, you won't learn how to build your very own piece of malware for experimentation and the chapter will try to limit the potential damage to your system from any example code. A focus of this chapter is the use of safe techniques for learning the skills you need to tackle malware. With this in mind, the actual sample executable is benign, but the techniques shown are effective with any executable.

This chapter is all about helping you understand that malware is a serious threat, but that threat has morphed since systems have taken to the cloud, organizations have started to rely heavily on web applications, and users have chosen to use multiple systems to perform work. So, the approach used in this chapter will also be more modern than that found in some perfectly usable detailed articles and tomes written by others. You will gain an appreciation for just how large the problem of malware is today and an understanding of what you can do about it for your organization.

A defining characteristic of malware that this chapter studies in some depth is that it's always an application, which differentiates it from other kinds of attacks that involve doing things like making API calls. The malware may be compiled or interpreted, appear as part of a web application or on the local machine, or attack the network, local system, data store, or other locations, but it's *always* an application. This means that malware has specific features that you can analyze using ML techniques, and that detection is possible in an automated sort of way (versus trying to figure out how an anomaly is related to humans or nature, as was the case in the previous chapter). With these issues in mind, this chapter discusses these topics:

- Defining malware
- Generating malware detection features
- Classifying malware

# Technical requirements

You won't work with actual malware in this chapter because doing so requires a special virtual machine set up to quarantine the host system from any other connection using a sandbox setup. However, you will see some examples that use code that could interact with malware. This chapter requires that you have access to either Google Colab or Jupyter Notebook to work with the example code. The *Requirements to use this book* section of *Chapter 1, Defining Machine Learning Security*, provides additional details on how to set up and configure your programming environment. When testing the code, use a test site, test data, and test APIs to avoid damaging production setups and to improve the reliability of the testing process. Testing over a non-production network is highly recommended, and pretty much essential for this chapter because you don't want to let any of the malware you experiment with on your own get out. In fact, you may actually want to use a system that isn't connected to anything else. Using the downloadable source is always highly recommended. You can find the downloadable source on the Packt GitHub site at `https://github.com/PacktPublishing/Machine-Learning-Security-Principles` or on my website at `http://www.johnmuellerbooks.com/source-code/`.

# Defining malware

Besides the requirement that it be an application of some sort, which means compiled or interpreted executable code, malware takes on a lot of different forms that are consistent with the goals of the attacker. For example, **ransomware** (software that encrypts your data and then asks you to pay for a key to decrypt it) is quite vocal about its presence, while **keyloggers** (software that records your keystrokes in an attempt to gain access to sensitive data such as passwords) are quite stealthy. The goal of the following sections is to help you understand the various kinds of malware from an overview perspective so that it's possible later to understand how such software would have characteristics that you can turn into features for ML analysis.

> **Applications that aren't malware, but also behave badly**
>
> This chapter doesn't include discussions about applications that behave badly, but aren't necessarily dangerous, just annoying. This includes extremely common software categories such as **adware** (an application that is determined to sell you something you don't want), **riskware** (applications that may do something interesting, but also open your system up to security threats), and **pornware** (I'll leave this one to your imagination). Obviously, you want to keep these other categories of software off your business systems as well, but they're generally obvious and sometimes even come with their own uninstall programs, so they're not really the same thing as malware.

## Specifying the malware types

Some parts of the book already include a little information about malware. The *Describing the most common attack techniques* section of *Chapter 3, Mitigating Inference Risk by Avoiding Adversarial Machine Learning Attacks*, discusses the ML application attack element of these kinds of attack. In addition, malware types including spyware are discussed in the *Determining which features to track* section of *Chapter 5, Keeping Your Network Clean*. However, the following list is a more comprehensive view of the categories of malware than found anywhere else in the book because it treats the topic in a more general sense:

- **Worm**: Some people consider worms as a subset of viruses, but they act in a different manner. Worms always replicate themselves, but the goal is to install themselves on other machines that the host machine has contact with. This leaves just one instance of the worm on the host system. Specialists often classify worms by where they spread, which includes **Instant Messaging** (**IM**), **Peer-to-Peer** (**P2P**), and **Internet Relay Chat** (**IRC**).

- **Virus**: For many, the term virus encapsulates all malware of any kind, but that's not really what the term means. Instead, it means a kind of application that replicates and optionally morphs its signature on a single machine, so that removing one copy doesn't get rid of the virus software. In many cases, the term is augmented with a description of the virus target, such as a *macro virus* that attacks macros in products such as Microsoft Word.

- **Trojan**: This term comes from the Trojan horse built at the end of the Trojan War to allow the Greek army to gain entrance to Troy in Greek mythology (`https://www.greekmythology. com/Myths/The_Myths/The_Trojan_Horse/the_trojan_horse.html`). It's a perfect metaphor for this kind of malware because it speaks of deception and something appearing to be one thing, when it's really quite another. There are so many different kinds of trojans that each kind has its own special term as described in the list that follows. The main thing that defines all trojans is that they don't replicate like viruses and worms do. They're usually quite stealthy until ready to attack as well.

- **Backdoor**: A special type of trojan that allows the operator access to the user's machine, so this type is especially stealthy. The level of target system access varies, but the hacker generally looks for administrator-level access. Once the hacker has access, it's possible to do anything that an administrator can do: perform **Create, Read, Update, and Delete** (**CRUD**) operations on files, execute applications, change permissions, and so on.

- **Downloader**: A kind of application with trojan-like behavior that downloads content from a remote source, sometimes without user permission. A downloader can have positive uses, such as performing local system updates. However, hackers often use them to download malicious software onto the user's machine.

- **Dropper**: Similar to a downloader, except that the payload is included as part of the trojan software's payload. The dropper is normally benign to avoid detection by anti-virus software. It also usually entices the user to download it by offering something of value, at which point it delivers its payload to the user's machine. Droppers often contain multiple pieces of malicious software as the payload and are never used for benign purposes.

- **Password Stealing Ware** (**PSW**): The goal of this kind of malware is to steal user account information, including passwords when possible. However, the trojan does more than just record keystrokes or perform other types of background monitoring. It also looks for potential sources of account information, such as the system registry, configuration files, and other places where account data might appear in plain form. One method to detect this sort of malware is that it sends the data back to the hacker using email, an FTP site, or other means that is permanently recorded on the remote system, so firewall monitoring is helpful in this case.

- **Spyware**: A kind of trojan that remains as quiet as possible and installs itself in a location where antivirus applications are less likely to find it. Adware and other user-oriented applications, especially software that is obtained illegally, contain spyware. The purpose of spyware is to send data like passwords to the hacker. Some spyware will also track user behaviors to allow the hacker to better mimic the user during an attack.

- **Distributed Denial of Service** (**DDoS**): In this scenario, a trojan drops a payload that infects a group of machines, each of which performs a **Denial of Service** (**DoS**) attack on a particular target. One of the compromised systems normally coordinates the attack (often the first one compromised), so that the hacker's system remains invisible even if the coordinating system is discovered. Consequently, this attack differs from a DoS attack because multiple machines are involved and there is a coordinated effort.

- **Ransomware**: The payload for this trojan performs some task on the target system designed to allow the hacker to extort money or other goods from the victim. In general, ransomware encrypts most or all of the victim's hard drive. The software then displays a message to the victim essentially saying that the hacker wants money to decrypt the drive (although some hackers play the "cute" card by saying that the system is running illegal or dated software and that the victim needs to buy a new license). Although many virus detection applications can detect ransomware, the drive is usually compromised by the time they do so. The only way to protect your data is to back it up regularly, maintain several backups (in case one is compromised), and keep the backups in multiple locations in the cloud or off site.

- **GameThief**: As the name indicates, this kind of trojan was originally designed to steal the account information for online games (and it still does most of the time). The GameThief trojan targets mobile devices in addition to desktop systems. According to the article *Gaming, Banking Trojans Dominate Mobile Malware Scene* (`https://threatpost.com/gaming-banking-trojans-mobile-malware/178571/`) the number of mobile exploits is down, but they're becoming more sophisticated. The reason to be concerned about this particular trojan when thinking about your ML application is that most users today want to use their mobile device in addition to their desktop device. They may also have a laptop and a tablet they want to use (yes, some users rely on four different devices to access your application whether the devices are approved for use or not). Any attack vector that affects the user's games can also affect your ML application, so it's essential not to discount this kind of attack simply because it mostly targets games and mostly targets mobile devices.

- **Instant Messaging (IM)**: A type of trojan that steals IM credentials for sites such as Facebook Messenger, Skype, and Telegram. The hacker then poses as the user to perform social engineering attacks, which is the real focus of this particular trojan. This form of attack varies from an **IM-Flooder**, which is designed to clog IM channels with garbage messages. An IM-Flooder is more along the lines of an exploit (as described later in the list).

- **Banker**: A type of trojan that steals credentials for online banking systems, e-payment systems, and most kinds of credit/debit cards. The result is that the hacker makes a lot of expensive purchases in a hurry once the banking source is confirmed with a small purchase. When combined with ransomware that makes the account inaccessible to the user, the hacker can keep spending the user's money and the user can't directly do anything about it except hope that the financial institution is successful in stopping the attack. In addition, this kind of attack often includes the ability to thwart **Multi-Factor Authentication** (**MFA**) so that the hacker can pose as the user without any problem on multiple systems that the hacker has infected.

- **Short Message Service** (**SMS**): In this case, the trojan sends messages from the device to premium rate numbers, so the user gets charged for the calls without actually making them. The hacker likely gets a kickback from the effort. However, because the message appears to come from the user, a hacker could use this approach to send any SMS text anywhere for any reason. Because there is no voice communication involved, it's quite hard to prove that the user didn't make the call. Also, consider the effect on your ML model if it relies on text messages as part of its data input. You may suddenly find the results of most analysis skewed by a data stream that's virtually impossible to track down because it comes from a legitimate (recognized) source.

- **Clicker**: This trojan works by connecting to online resources as if it's the user by sending commands to the browser. It can also replace system files that specify standard web addresses so that user ends up going to an infected site, rather than the intended site. The thing that makes this particular exploit so dangerous for ML application is that it also works for automation used for screen scraping in search of data. The trojan would make it possible to obtain tainted data from a hacker site, rather than the trusted site that the automation intended to use. One way to keep this sort of exploit under control is to verify automation site lists regularly and to use the hashing techniques found in the *Detecting dataset modification* section of *Chapter 2, Mitigating Risk at Training by Validating and Maintaining Datasets*.

- **Proxy**: The trojan in this case acts as a proxy server that gives the attacker access to internet resources on the user's machine. This means that the attacker could have direct access to your ML application through the user's machine if it has a web interface, making illegal access tough to track down. The biggest issue in this scenario is one of authentication as described in the *Developing a simple authentication example* section of *Chapter 4, Considering the Threat Environment*. The trojan won't know the user's password for accessing your application unless it also includes a key logger (remember that the idea is to place many hurdles in the hacker's way so that it becomes a nuisance to attack your application). Authenticating the user before every new session will help mitigate the problem. Changing passwords regularly is also helpful.

- **Notifier:** One technique for keeping hackers at bay is to ensure users shut down their system when not in use. This way, there is a chance that the user will see any suspicious activity that would normally occur when the user isn't using the system. In addition, because the system isn't always on, the hacker will have to spend extra time figuring out when the system is available. A notifier trojan automatically notifies the hacker when the system becomes available through an email, special website access, or IM. Because this approach lacks much of the automation of other attack vectors, it is normally used in a multi-component scenario to let the attackers know that the other components of an attack are installed.

- **ArcBomb:** This is a particularly dangerous trojan for ML applications because it affects data archives and does things such as fill the hard drive with useless data so that a server freezes or its performance slows to a crawl. There are a number of useless data types employed for this exploit: malformed archive headers, repeating data, and multiple copies of identical files. Because of the techniques used, a 5 GB data load can appear in an archive as small as 200 KB. The way to avoid this exploit is to only download and use files from sources you know and ensure you scan the file for potential problems before attempting to use it.

- **Rootkit:** A kind of application that is used to hide something like an application, an object, data, or hacker activity from the user. Theoretically, a rootkit is normally harmless by itself, except for being incredibly hard to remove. It's the items that the rootkit is designed to hide that are the problem. Removing a rootkit requires special tools and it's a painstaking process at the best of times. Hackers hide code and other resources in all sorts of places including the following:

  - Alternate data streams

  - Drivers hooking the **System Service Descriptor Table (SSDT)**

  - Drivers hooking the **Interrupt Descriptor Table (IDT)**

  - Files

  - Inline hooks

  - **Master Boot Record (MBR)**

  - Modules

  - Processes

  - Registry keys

  - Services

  - Threads

- **Exploit**: An exploit is a special piece of code or carefully crafted data that takes advantage of a bug, error, or behavior (intended or not) of an application, operating system, or environment. Of the places where exploits are used, cloud-based exploits have the greatest potential to affect your ML application because they can affect every device that the user relies upon to access your application (as described in *7 Cloud Computing Security Vulnerabilities and What to Do About Them* at `https://towardsdatascience.com/7-cloud-computing-security-vulnerabilities-and-what-to-do-about-them-e061bbe0faee`). Here are some exploit categories to consider when securing your ML application:

  - **Constructor**: An application designed to create new viruses, trojans, and worms so that it's possible to morph an attack on the fly and take advantage of system vulnerabilities when located on the host system. Most constructors currently reside on Windows or macOS systems. Hackers also use constructors to create new classes of malware based on current research about system vulnerabilities.

  - **Denial of Service** (**DoS**): Used to hinder the normal operation of a website, server, desktop system, other devices, or any other resource. The most common way to carry out this attack is to overload the target in some manner so that it can't process incoming data. The best way to overcome this exploit is to look for significant increases in traffic of any sort or the appearance of invalid data.

  - **Spoofer**: The attacker replaces a real address with some other address in an effort to remain hidden. This exploit sees common use in user-oriented interactions, such as email, but it could also appear in message traffic to an application. If the application includes a whitelist of acceptable addresses, the hacker can spoof one of these addresses to obtain illegal access to a resource and perform tasks such as sending fake data. There are several methods to help you overcome this exploit that include using MFA and challenges to ensure the user and not some outsider is sending the data.

  - **Flooder**: A kind of DoS that directly affects network channels used for IM, email, SMS, and other communication. This kind of attack could feed false information to your ML application using spoofed addresses to bypass any filtering you have in place. The best way to overcome this particular kind of exploit is to look for unusual patterns in the message traffic.

  - **Hoax**: A hoax can take multiple forms, but it always contains some sort of fake information, usually in the form of a warning. For example, the user receives an email stating that system software has detected a virus on their system and that they should click a link to get rid of the problem. Of course, they click the link and now have the virus. The best way to avoid hoaxes is through vigorous, mandatory, user training. Unfortunately, this is a kind of social engineering type attack the users find very difficult to resist.

  - **VirTool**: This is a hacker management aid that helps direct, modify, and otherwise interact with any sort of malware that a hacker has placed on your system. The goal is to keep the malware hidden from antivirus software and the user so that it can keep working in the background to corrupt the system, steal data, and perform other tasks the hacker may have in mind.

- **HackTool**: Provides the means to perform clean-up behind any malware on a system so that it's harder to detect and clean the malware up. A HackTool commonly performs tasks such as adding new users to a system so that the hacker can have a dedicated account, clean system logs, and analyze network activity.

As you can see, the list is rather long and the definitions seem to cover everything except possibly what happens when you sneeze or yawn. Making things more difficult is the fact that hackers often combine vectors when making an attack to increase the likelihood of a successful attack. The idea is that you might be looking for one sort of malware but not another on your network, local systems, servers, and hosted cloud applications and servers. In addition, the hacker is hoping that you haven't been comprehensive or consistent in your coverage. Any chink in your security armor is enough to give the hacker an advantage of some sort, even if that advantage only leads to another attack.

> **Watch out for the online model!**
>
> You may think that hacks only happen to games or other consumer applications, or that only users encounter issues with their online viewing habits. However, developers can encounter these problems as well. For example, the **pickle** library used to serialize objects in Python (`https://docs.python.org/3/library/pickle.html`) can cause you serious woe. Watch the YouTube video *The hidden dangers of loading open-source AI models* at `https://www.youtube.com/watch?v=2ethDz9KnLk` for a detailed account of how developers can be taken in just as easily as any other person (I apologize in advance for the commercial you'll have to sit through, but the video really is worth watching). All of the attack vectors listed in this section do apply directly to developers, especially those who are new to working with models and are experimenting heavily to discover what does and doesn't work. *Never* experiment on a production system, and use a virtual platform so that you can just destroy the setup without personal loss should the configuration become contaminated.

The classification keywords used in the list are important because you often see them used when describing a particular kind of malware. For example, `VirTool:Win32/Oitorn.A` attacks certain Windows systems (`https://www.microsoft.com/en-us/wdsi/threats/malware-encyclopedia-description?Name=VirTool:Win32/Oitorn.A`). The term VirTool identifies the category of the attack vector. You can use this information to better prepare yourself for pertinent malware, while ignoring other possibilities that have a lower chance of success with your particular setup.

## Understanding the subtleties of malware

Like any software developer, hackers constantly seek to improve the effectiveness of the malware available. It's not just a matter of extorting money, data, or resources from a target. For the hacker, creating a particularly difficult attack is a matter of pride in their work *and* of staying out of jail. Much of the malware discussed in this chapter so far is incredibly subtle. More often than not, the hacker wants to remain unobserved until the attack is under way. In fact, many attacks require that the hacker remain permanently unknown and that the attack itself remains under the radar. Unlike the application software designer, a hacker has a considerable number of good reasons to design applications that are pretty much invisible or can be made invisible through technologies such as rootkits. Oddly enough, application developers could learn a few things from hackers about being subtle and making software perform a task without being seen.

Locating malware of all types involves understanding the subtleties of malware design. Sometimes it's a matter of hiding in plain sight, other times it's a matter of being discrete or simply clever. Consider the malware that hides bugs in source code used to create applications. This malware lurks on developer systems and creates application holes that have nothing to do with the coder's abilities. You can read about this particular exploit at *'Trojan Source' Hides Invisible Bugs in Source Code* (`https://threatpost.com/trojan-source-invisible-bugs-source-code/175891/`). It's in the hacker's best interest to keep this particular kind of malware completely hidden from view so that it can continue to do its job of corrupting applications in a way that the hacker can then exploit once the application moves into production. Of course, this article points out the need to constantly scan developer systems and possibly keep their machines free from connectivity outside the organization.

### Understanding the command and control server

One long-established technique for locating malware is to look for applications with a large amount of compressed data. A special piece of code in the malware application unpacks this data into the payload that the malware carries. By installing the malware in a special environment and watching how it performs this unpacking process, it's possible to determine malware behaviors that help identify the malware category and point to methods of mitigation. Now, however, many pieces of malware rely on a remote **Command and Control** (**C2**) server to perform the unpacking. This means that the malware does nothing in the lab when someone tries to study it. You only get to see it in action as it infects the target device. The C2 strategy is often compounded by a waterfall effect in the malware where one module unpacks and then provides the address needed to unpack the next module. Consequently, unpacking one module doesn't tell security professionals about the malware as a whole. This is also the reason that older techniques for detecting and mitigating malware may not work when performing your own research.

An essential element in working through the subtleties of malware is not only deciding precisely why a hacker is making the attack, but also determining a creative way to perform the task. The rather long list in the *Specifying the malware types* section doesn't begin to cover all of the approaches that hackers use. You also have to consider that hackers usually combine attacks to improve their chances of success so that each attack begins to take on a unique appearance. This is the reason why you want to locate software that will help you in your detection task using unbiased statistical sources whenever possible. Avoiding sites that are also trying to sell you the software is usually a good idea because they're hardly unbiased. Look for articles such as *The Best Antivirus of 2022: A data-driven comparison* (`https://www.comparitech.com/antivirus/`) that provide a numeric basis after running tests to defend their decisions. Note that this site provides you with its testing criteria so you know how the comparison is made after testing is performed (it's not simply an opinion). When looking for anti-malware tools, these features will help keep malware at bay:

- **Boot activation**: The software starts before the operating system does so that it's possible to monitor the environment for applications such as rootkits.

- **Behavior analysis**: Looking for certain application behaviors can clue you in to a potential piece of malware, even when other monitoring aids fall short.

- **Signature analysis**: Using ML techniques allows for learning the malware's signature, even when the developer attempts to hide it from view. This approach also helps locate malware that doesn't use traditional techniques such as particular file extensions.

- **Content filtering**: Anything that looks like it could be a problem, such as those ads that make it sound like you can get rich quick, are also likely sources of malware, so getting rid of them helps everyone.

- **Suspicious link tagging**: If a user is in a hurry, a link with a small misspelling error is likely to go unnoticed, so it's important to locate any link that doesn't quite match expectations.

- **Offsite backups**: Making continuous offsite backups can be helpful, but only if the backups are kept separate so that the backup doesn't become contaminated with the very malware it's supposed to avoid. Having malware software checking the backup stream as you make the backup adds additional insurance against infection.

- **Sandboxing**: Using sandboxing techniques for all efforts with any new code helps keep malware at bay. This includes the use of new libraries or other developer tools.

- **Stronger firewalls**: The operating systems used by various user devices likely come with some level of firewall protection, which may not be enough. In days gone by, castles didn't rely on just one wall. They had walls within walls so that an enemy would have to breach each wall in turn to gain entrance. Likewise, malware detection and prevention tools you use should provide multiple layers of protection in addition to the protection offered by the operating system.

These features are more or less mandatory even if you create your own custom solution for locating and destroying malware. Using tools created by experts to assist in locating malware saves you time and ultimately money. The tools you build should focus on the issues that malware software doesn't cover; issues that are unique to your business.

> **Feature comparison versus tested results**
>
> Articles or other resources that provide feature comparisons are doing just that, comparing features, which is helpful if done in an unbiased manner because it's all too easy for a site to cherry-pick features that give one piece of software an advantage over another. One such feature site (even though you need to consider whether the resource is properly vetted) is on Wikipedia at `https://en.wikipedia.org/wiki/Comparison_of_antivirus_software`. It's important to realize that you are getting a list of features on Wikipedia, but that no one has tested those features to see how well they work or whether they work at all. A feature site normally lists the features of many kinds of software, so it makes a good starting point for your selection process. However, once you have a list of candidates, then you need to find a testing site to tell you how well those features work to achieve your particular goals.

## Determining malware goals

This chapter demonstrates that malware doesn't always have the goal of damaging a system or extorting money from a user. Malware sometimes hides from view and records keystrokes or uses the system as a zombie when it detects the user isn't there to stop it. Malware need not affect the current user at all. If you're a developer, the malware may hide in the background and add known bugs to the software you're developing so that the hacker knows how to attack your application once it's in production. Consequently, part of the thought process of tackling malware is to determine what a hacker might want from you or your organization and it always benefits you to think outside the box.

Of course, you also don't want to waste lots of time figuring out goals that have no chance of success or little interest to the hacker. With this in mind, think about these factors as part of the determination of malware goals:

- Organizational assets
- Devices in use
- User habits
- Resources of all types (physical, digital, data, and so on)
- Interactions with others
- Sources of potential grudges or other reasons for revenge

If you can think of a particular goal that a hacker might have in mind for attacking your organization as a whole or individuals in particular, it's possible start narrowing your search for malware and also narrow the types of malware that a hacker is most likely to use. Most people associate malware with executable files, but malware comes as part of scripts and finds its way in all sorts of other places. Malware in images presents a particular problem because most people don't see images as harmful, yet images are often used to install a dropper on the user's system while the user is looking at the image. More importantly, images appear on websites that are accessible to any device that users in your organization might use.

As with any ML application, you need to determine which features to use to train a model that you can use to detect malware. Security professionals have a lot of different views on the matter. They ask a lot of questions that involve the kinds of malware, where the malware will operate, and the environment in which the malware will attack. The next section of the chapter delves into selecting malware features for analysis, which can be very helpful in choosing various malware detection solutions and building models.

## Generating malware detection features

In ML, features are the data that you use to create a model. You analyze features to look for patterns of various sorts. The *Checking data validity* section of *Chapter 6*, *Detecting and Analyzing Anomalies*, shows you one kind of analysis. However, in the case of the *Chapter 6* example and all of the other examples in the book so far, you were viewing data that humans can easily understand. This section talks about a new kind of data hidden in the confines of malware. Consequently, you're moving from the realm of human-recognizable data to that of machine-recognizable data. The interesting thing is that your ML model won't care about what kind of data you use to build a model, the only need is for enough data of the right kind to build a statistically sound model to use to locate malware.

> **Working with a first step example**
>
> To actually work with malware, you need a system that has appropriate safety measures in place, such as a **virtual machine** (`https://www.howtogeek.com/196060/beginner-geek-how-to-create-and-use-virtual-machines/`) and a **sandbox** configuration (`https://blog.hubspot.com/website/sandbox-environment`). In fact, it's just a generally good idea to separate the test machine from any other machine, including the Internet. Many discussions of malware throw you into the deep end of the pool without helping you build any skills. This means that it's likely that you'll destroy your machine and every machine around you. The approach taken in this book is to provide you with that first step so that you can move on to books that do demonstrate examples using real malware, such as *Malware Data Science*, by Joshua Saxe with Hillary Sanders, No Starch Press (the next step up that I would personally recommend).

### Getting the required disassembler

The act of dissecting an executable file of any kind is called **disassembly**. You turn the machine code or byte code contained in the executable file back into something a human can at least interpret. Disassembly is never completely accurate in returning code to the same form as the developer of the executable created. What you get instead is something that's close enough that you can determine how the executable works, but nothing else.

The goal of disassembly isn't to completely replicate the original source code anyway. You disassemble an executable to find data you can use to create an ML model. Executable files contain all sorts of statistical information, such as the number of compressed or encrypted data sections. You can find strings of all sorts in any executable that can provide you with clues as to how the executable works. Sometimes the executable contains images that you can pull out and view to see what sort of presentation the executable will provide. In short, the executable contains a lot of hidden information that you must then pull together into a set of features that help you recognize malware.

The example in this section uses the **Portable Executable File** (**PEFile**) disassembler (`https://pypi.org/project/pefile/` and `https://github.com/erocarrera/pefile`) to break down the Windows PEFile format. The following steps show how to obtain this package if necessary and install it on your system. You can also find this code in the `MLSec; 07; View Portable Executable File.ipynb` file of the downloadable source. Let's begin:

1.  Set a search variable, `found`, to `False`:

    ```
    found = False
    ```

2.  Obtain a list of installed packages on the system and search it for the required PEFile package:

    ```
    packages = !pip list
    for package in packages:
        if "pefile" in package:
            found = True
            break
    ```

3.  If the package is missing, then install it. Otherwise, print a message saying that the package was found:

    ```
    if not found:
        print("Package is missing, installing...")
        !pip install pefile
    else:
        print("PEFile package found.")
    ```

When you run this code, you will either see a message stating `PEFile package found` or the code will install the PEFile package for you. When the installation process finishes, you see output similar to that shown in *Figure 7.1*.

```
Package is missing, installing...
Collecting pefile
  Downloading pefile-2022.5.30.tar.gz (72 kB)
    -------------------------------------- 72.9/72.9 kB 3.9 MB/s eta 0:00:00
  Preparing metadata (setup.py): started
  Preparing metadata (setup.py): finished with status 'done'
Requirement already satisfied: future in c:\users\john\anaconda3\lib\site-packa
ges (from pefile) (0.18.2)
Building wheels for collected packages: pefile
  Building wheel for pefile (setup.py): started
  Building wheel for pefile (setup.py): finished with status 'done'
  Created wheel for pefile: filename=pefile-2022.5.30-py3-none-any.whl size=693
62 sha256=a930d47d59917a48cad471ae8249e0a7fe875e10f6907004a2ac19614941dc71
  Stored in directory: c:\users\john\appdata\local\pip\cache\wheels\c7\ca\2c\b2
bc3360e75954899f344606ea58a307d8f3a7060899b7b7fd
Successfully built pefile
Installing collected packages: pefile
Successfully installed pefile-2022.5.30
```

Figure 7.1 – The installation process shows which version of PEFile you have installed

Now that you have a disassembler to use, you can actually begin working with an executable file. The executable file used for this example isn't harmful in any way, so you don't have to worry about it.

## Collecting data about any application

The example in this section shows you how to dissect a Windows PEfile. You can use this technique on any PE file, including malware, but working with files that you know are benign is a good starting point if you don't want to mistakenly infect your system and all of the systems around you. The code for this example appears in the `MLSec; 07; View Portable Executable File.ipynb` file of the downloadable source. Use these steps to dissect your first file. The example uses `Notepad. exe` because you can find it on every Windows system.

### Checking for the PE file

Before you can disassemble a PE file, you need to know that it actually exists. The following steps show how to check for a file on a system even if you don't precisely know where the file is:

1.  Import the required methods:

    ```
    from os.path import exists, expandvars
    ```

2.  Create an expanded path variable and then display the path on screen so you know where to look later:

```
path = "$windir/notepad.exe"
exppath = expandvars(path)
print(exppath)
```

3.  Ensure that the file does actually appear on the system:

```
if exists(exppath):
    print("Notepad.exe is available for use.")
else:
    print("Use another Windows PE file.")
```

This simple check works on any system. Note the use of the `windir` Windows environment variable that specifies the location of the Windows directory on a host system (it's created by default during installation). If you wanted to expand this variable at the command prompt, you'd use `%windir%`, but in Python you use `$windir`. Note that Windows comes with a host of environment variables that you can display at the command prompt by typing `set` and pressing *Enter*.

### Loading the Windows PE file

At this point, you have a special package installed on your system that will disassemble Windows PE files and you know whether you can use `Notepad.exe` as a target for your disassembly. It's time to load the Windows PE file for examination using the following steps:

1.  Load the required functionality:

```
import pefile
```

2.  Load the PE file:

```
exe_file = pefile.PE(exppath)
```

Now you have `Notepad.exe` loaded into memory where you can now examine it in some detail. The `exe_file` variable won't tell you much if you just print it out. You need to become a detective and look for specific features.

### Looking at the executable file sections

Despite the fact that executable files are really just long lists of numbers that only your computer processor can really understand, they're actually highly organized (or else the processor would run amok and no one wants that). One of the ways in which to organize executable files is in sections. There are specialized sections in an executable file for every need. The following steps detail how you can look at the sections in an executable file:

1. Display a list of all of the section information:

```
for section in exe_file.sections:
    print(section)
```

2. Choose specific section information to focus on during your detective work:

```
for section in exe_file.sections:
    print(section.Name)
```

The first step is to look at what the executable has to offer in the way of information. When you perform this step, you see something like the output shown in *Figure 7.2* for each section within the file. That's a lot of really hard-to-understand information, but that's how your executable is organized.

```
[IMAGE_SECTION_HEADER]
0x208      0x0    Name:                         .text
0x210      0x8    Misc:                         0x247FF
0x210      0x8    Misc_PhysicalAddress:         0x247FF
0x210      0x8    Misc_VirtualSize:             0x247FF
0x214      0xC    VirtualAddress:               0x1000
0x218      0x10   SizeOfRawData:                0x24800
0x21C      0x14   PointerToRawData:             0x400
0x220      0x18   PointerToRelocations:         0x0
0x224      0x1C   PointerToLinenumbers:         0x0
0x228      0x20   NumberOfRelocations:          0x0
0x22A      0x22   NumberOfLinenumbers:          0x0
0x22C      0x24   Characteristics:              0x60000020
```

Figure 7.2 – Each section output contains a wealth of information that is useful for statistical analysis

The second step refines the output by just looking at the Name value of each section. Note that the entries are case sensitive, so name (it does exist) is not the same thing as Name. In this case, you see the output shown in *Figure 7.3*.

```
b'.text\x00\x00\x00'
b'.rdata\x00\x00'
b'.data\x00\x00\x00'
b'.pdata\x00\x00'
b'.didat\x00\x00'
b'.rsrc\x00\x00\x00'
b'.reloc\x00\x00'
```

Figure 7.3 – The section names help you see the organization of the file

Well, those names are singularly uninformative. What precisely is a .text section? The *Special Sections* section of the *PE Format* documentation at https://docs.microsoft.com/en-us/windows/win32/debug/pe-format tells you what all of these names mean. However, here is the meaning for all of the sections found in Notepad.exe:

- .text: Executable code

- .rdata: Read-only initialized data

- .data: Read/write initialized data

- .pdata: Exception information

- .didat: Delayed import table (not found in the table but does appear in the *PE Format* documentation)

- .rsrc: Resource directory

- .reloc: Image relocations

In most cases, you won't use all of the sections unless you're involved in detailed research that is well beyond the scope of this book. For example, you really don't care where the image gets relocated at this point, so the .reloc section isn't very helpful. On the other hand, looking at the .rdata and .data sections can be quite illuminating.

### Examining the imported libraries

Executable files contain directories of items needed to execute. These directories list specific **Dynamic Link Libraries** (**DLLs**) that the executable accesses and the specific methods within those DLLs that it calls. When you review enough executable files, you begin to develop a feel for which calls are common and which aren't. You also start to be able to tell when a particular DLL could be downright dangerous to access. For example, you might ask yourself whether the executable really needs to fiddle around in the registry. If not, importing registry methods may be a pointer to malware. The PEFile package makes a number of directories accessible and the following code shows you how to identify them and then access the list of imported libraries:

1. Display the list of accessible directories. Note that you remove the IMAGE_ part of the entry to use it in code:

   ```
   pefile.directory_entry_types
   ```

2. Obtain a list of imported DLLs to use for continued analysis:

   ```
   entries = []
   for entry in exe_file.DIRECTORY_ENTRY_IMPORT:
       print (entry.dll)
       entries.append(entry)
   ```

3.  Examine the calls within a target DLL that are used by the executable application:

```
print(entries[0].dll)
for function in entries[0].imports:
    print(f"\t{function.name}")
```

The list of accessible directory entries for PEFile appear in *Figure 7.4*. As you can see, the list is rather extensive, but as someone who is just looking for potential malware markers, some of the entries stand out, such as IMAGE_DIRECTORY_ENTRY_IMPORT:

```
[('IMAGE_DIRECTORY_ENTRY_EXPORT', 0),
 ('IMAGE_DIRECTORY_ENTRY_IMPORT', 1),
 ('IMAGE_DIRECTORY_ENTRY_RESOURCE', 2),
 ('IMAGE_DIRECTORY_ENTRY_EXCEPTION', 3),
 ('IMAGE_DIRECTORY_ENTRY_SECURITY', 4),
 ('IMAGE_DIRECTORY_ENTRY_BASERELOC', 5),
 ('IMAGE_DIRECTORY_ENTRY_DEBUG', 6),
 ('IMAGE_DIRECTORY_ENTRY_COPYRIGHT', 7),
 ('IMAGE_DIRECTORY_ENTRY_GLOBALPTR', 8),
 ('IMAGE_DIRECTORY_ENTRY_TLS', 9),
 ('IMAGE_DIRECTORY_ENTRY_LOAD_CONFIG', 10),
 ('IMAGE_DIRECTORY_ENTRY_BOUND_IMPORT', 11),
 ('IMAGE_DIRECTORY_ENTRY_IAT', 12),
 ('IMAGE_DIRECTORY_ENTRY_DELAY_IMPORT', 13),
 ('IMAGE_DIRECTORY_ENTRY_COM_DESCRIPTOR', 14),
 ('IMAGE_DIRECTORY_ENTRY_RESERVED', 15)]
```

Figure 7.4 – PEFile provides access to a number of PE file directories

In fact, the IMAGE_DIRECTORY_ENTRY_IMPORT entry is the focus of the next step, which is to list the DLLs that are used by Notepad.exe as shown in *Figure 7.5*. Some of the DLL names are pretty interesting. For example, you might wonder what api-ms-win-shcore-obsolete-11-1-0.dll is used for (see the details at https://www.exefiles.com/en/dll/api-ms-win-shcore-obsolete-11-1-0-dll/). Oddly enough, you may find yourself needing to fix this file, so it's helpful to know it's in use.

```
b'KERNEL32.dll'
b'GDI32.dll'
b'USER32.dll'
b'api-ms-win-crt-string-l1-1-0.dll'
b'api-ms-win-crt-runtime-l1-1-0.dll'
b'api-ms-win-crt-private-l1-1-0.dll'
b'api-ms-win-core-com-l1-1-0.dll'
b'api-ms-win-core-shlwapi-legacy-l1-1-0.dll'
b'api-ms-win-shcore-obsolete-l1-1-0.dll'
b'api-ms-win-shcore-path-l1-1-0.dll'
b'api-ms-win-shcore-scaling-l1-1-1.dll'
b'api-ms-win-core-rtlsupport-l1-1-0.dll'
b'api-ms-win-core-errorhandling-l1-1-0.dll'
b'api-ms-win-core-processthreads-l1-1-0.dll'
b'api-ms-win-core-processthreads-l1-1-1.dll'
b'api-ms-win-core-profile-l1-1-0.dll'
b'api-ms-win-core-sysinfo-l1-1-0.dll'
b'api-ms-win-core-interlocked-l1-1-0.dll'
b'api-ms-win-core-libraryloader-l1-2-0.dll'
b'api-ms-win-core-winrt-string-l1-1-0.dll'
b'api-ms-win-core-synch-l1-1-0.dll'
b'api-ms-win-core-winrt-error-l1-1-0.dll'
b'api-ms-win-core-string-l1-1-0.dll'
b'api-ms-win-core-winrt-l1-1-0.dll'
b'api-ms-win-core-winrt-error-l1-1-1.dll'
b'api-ms-win-eventing-provider-l1-1-0.dll'
b'api-ms-win-core-synch-l1-2-0.dll'
b'COMCTL32.dll'
```

Figure 7.5 – The list of imported DLLs can prove interesting even for benign executables

The example examines a far more common DLL however, KERNEL32.dll, in the third step. The output shown in *Figure 7.6* tells you that this DLL sees a lot of use in Notepad.exe (the list shown in *Figure 7.6* isn't complete, it's been made shorter to fit in the book).

```
b'KERNEL32.dll'
        b'GetProcAddress'
        b'CreateMutexExW'
        b'AcquireSRWLockShared'
        b'DeleteCriticalSection'
        b'GetCurrentProcessId'
        b'GetProcessHeap'
        b'GetModuleHandleW'
        b'DebugBreak'
        b'IsDebuggerPresent'
        b'GlobalFree'
        b'GetLocaleInfoW'
        b'CreateFileW'
        b'ReadFile'
        b'MulDiv'
        b'GetCurrentProcess'
        b'GetCommandLineW'
        b'HeapSetInformation'
        b'FreeLibrary'
        b'LocalFree'
        b'LocalAlloc'
        b'FindFirstFileW'
        b'FindClose'
```

Figure 7.6 – Looking at specific method calls can tell you want the executable is doing

The list of calls from KERNEL32.dll is extensive and well documented (https://www.geoffchappell.com/studies/windows/win32/kernel32/api/index.htm). You can drill down as needed to determine what each function call does. By breaking the various imports down, you start to understand what the executable is doing. You've discovered all of the various sections contained within the executable and what it imports from other sources. Just these two bits of information are enough to start getting a feel for how the executable works and what it's doing on your system.

## Extracting strings from an executable

It's only possible to garner so much information by looking at what an executable is doing. You can also look at the strings contained in an executable. In this case, you need to use something like the **Strings utility** for Windows (https://docs.microsoft.com/en-us/sysinternals/downloads/strings) or the same command on Linux (https://www.howtogeek.com/427805/how-to-use-the-strings-command-on-linux/). The Windows version requires that you download and install the product. Oddly enough, both versions use the same command line switch, -n, which is needed for the example in this section. To extract the strings, you need to add a command to your code, something like !strings -n 10 %windir%\Notepad.exe. This is the Windows version of the command; the Linux command would be similar. *Figure 7.7* shows the output for Notepad.exe if you limit the string size to ten characters or more.

```
UATAUAVAWH
ADVAPI32.dll
Unknown exception
bad allocation
bad array new length
COMDLG32.dll
PROPSYS.dll
SHELL32.dll
WINSPOOL.DRV
urlmon.dll
%hs(%u)\%hs!%p:
(caller: %p)
%hs(%d) tid(%x) %08X %ws
Msg:[%ws]
CallContext:[%hs]
[%hs(%hs)]
kernelbase.dll
RaiseFailFastException
onecore\internal\sdk\inc\wil\opensource\wil\resource.h
```

Figure 7.7 – Reviewing the strings in an executable can reveal some useful facts

This output is typical of any executable, so if you were looking for specific information, then you'd need to start manipulating the list to locate it using Python features. As you can see, there are error messages, references to DLLs, format strings, and all sorts of other useful information. However, a specific example is helpful. Perhaps you're interested in the format strings contained in a file, so you might use code like this:

```
exe_strings = !strings -n 10 %windir%\Notepad.exe
for exe_string in exe_strings:
    if "%" in exe_string:
        print(exe_string)
```

The output in *Figure 7.8* shows that you still don't get only format strings, but it's a lot shorter and you have a better chance of finding what you need.

```
%hs(%u)\%hs!%p:
(caller: %p)
%hs(%d) tid(%x) %08X %ws
Msg:[%ws]
CallContext:[%hs]
[%hs(%hs)]
Local\SM0:%d:%d:%hs
%s%c*.txt%c%s%c*.*%c
%s\%s.autosave
%081X-%04X-%04x-%02X%02X-%02X%02X%02X%02X%02X%02X
```

Figure 7.8 – Using Python features it's possible to make the list of string candidates shorter

In this case, you can see that some format strings are standalone, while others are part of messages. The point is that you now have a basis for performing additional work with the executable, without actually loading it. Obviously, loading malware to see it run is something best done in a lab.

## Extracting images from an executable

How you look for images in an executable depends on the image type and the platform that you're using. For example, two of the most popular tools for performing this task in Linux are `wrestool` (`https://linux.die.net/man/1/wrestool`) and `icotool` (`https://linux.die.net/man/1/icotool`). Windows users have an entirely different list of favorites (`https://www.alphr.com/extract-save-icon-exe-file/`), some of which come with the **Windows Software Development Kit** (**SDK**). Some tools are quite specialized and only look for a particular image type. To give you some idea of what is involved, the following steps locate icons used with `Notebook.exe`:

1.  Download a copy of `IconsExtract` from `https://www.nirsoft.net/utils/iconsext.html`. The product doesn't require any installation; all you need to do is unzip the file and double-click the executable to start it.

2.  Open `IconsExtract` and you see a **Search For Icons** dialog box like the one shown in *Figure 7.9*. This is where you enter the name of the file you want to check. However, you'll quickly find that `Notepad.exe` doesn't contain any icons. Instead, you need to look through the list of DLLs that `Notebook.exe` loads to find the icons.

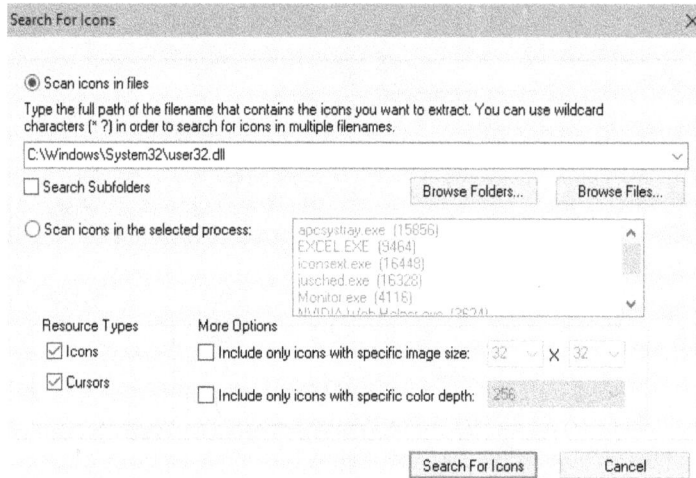

Figure 7.9 – Specify where to look for the images you want to see

3.  Locate the `user32.dll` file on your system. This is one of the files you see listed in *Figure 7.5*.

4.  Click **Search For Icons**. You see the output shown in *Figure 7.10*.

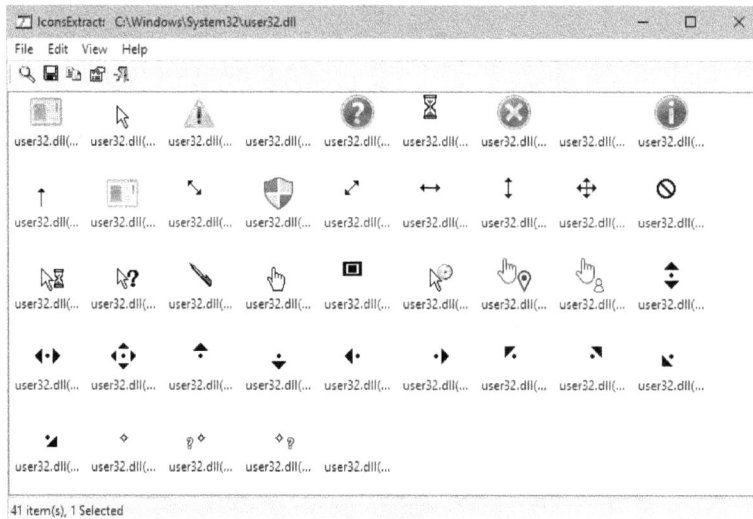

Figure 7.10 – The display shows icons that Notepad.exe may import from user32.dll

When working with malware, you generally want to find images that indicate some type of fake presentation. Perhaps you see a ransom message or other indicator that the executable is malware and not benign. The point is that this is just one of many ways to make the required determination. You shouldn't rely on graphics alone as the only method to perform your search.

## Generating a list of application features

In order to create a model to detect malware that may be trying to sneak onto your system or may already appear on your system, you need to define which features to look for. A problem with many models is that the developers haven't really thought through which features are important. For example, looking for applications that use the `ReadFile()` function of `Kernel32.dll` really won't do anything for your detection possibilities. In fact, it will muddy the waters. What you need to do is figure out which features are likely to distinguish the malware target and then build a model around those features. The examples in this chapter should bring up some useful ideas:

- Making unusual or less used method calls, such as interacting with the registry or writing to configuration files

- Executables that contain a great deal of compressed or encrypted data

- Executables or scripts that call on libraries, packages, DLLs, or other external code sources that you don't recognize and can't find documented somewhere online

- Any file, including things such as sound files and images, that contain strings or other unexpected data

- Strings that contain spelling errors

- Any suspected use of **steganography** (the hiding of data or code in a container, such as an image, that looks normal otherwise) in any file, especially images

- Anything out of the ordinary for your particular organization (such as finding patient information out in the open for a hospital)

The list of application features that you choose to use has to reflect the particulars of your organization, rather than the generalized list that you might find on a security website because the hacker is most likely attacking you or your organization as opposed to a generalized attack. When you do want to use the list of generalized features, then creating custom software is likely not the best option – you should go with off-the-shelf software designed by a reputable security company that has already taken these generalized features into account.

## Selecting the most important features

There are a considerable number of ways to determine which features are most important in any dataset. The method you use depends as much on personal preference as the dataset content. It's possible to categorize feature selection techniques in four essential ways:

- **Filter methods**: This approach uses univariate statistics to filter the feature set to locate the most important features based on some criterion. The advantage of this approach is that it's less computationally intensive, which is important when working with a system that may not include a high-end **Graphics Processing Unit (GPU)** to speed the computations. These methods also work well with data that has high dimensionality, which is what security data often has because you need to monitor so many different inputs to find a hacker. Because these approaches are so well suited to security needs, they're used for the examples in this section.

- **Wrapper methods**: This approach performs a search of the entire feature set looking for high-quality subsets. This is an ML approach that relies on a classifier to make feature selection determinations. The method relies on greedy algorithms to accomplish the task of determining which set of features best matches the evaluation criterion. The advantage of this approach is that the output provides better predictive accuracy than using filter methods. However, it's also a time-consuming approach that requires a lot of resources and a system that has a GPU if you want to get the results in a timely manner. This is probably the worst approach to use for any sort of security situation that requires real-time analysis. You would get fewer false positives, but not in a timeframe that meets security needs.

- **Embedded methods**: This is a combination of the filter and wrapper methods. It is less computationally intensive, but the iterative approach can make using it for most security needs less than helpful. This is the method you might use to analyze security logs after the fact, not in real time, to ascertain how a hacker gained entrance to your system with a higher degree of precision than would be allowed by filter methods. The algorithms that appear to work best for security needs are **LASSO Regularization (L1)** and **Random Forest Importance**.

- **Hybrid methods**: This is an **approach** that uses the result of multiple algorithms to mine data. Generally, it isn't used for security needs because the landscape changes too fast to make it effective. This is the sort of approach that a medical facility might use to mine a dataset for new knowledge needed to treat a disease. These methods rely heavily on instance learning and have the goal of providing consistency in feature selection. However, it could be useful if applied to historical security data in looking for particular trends.

Feature selection is an essential part of working with data of any sort. Otherwise, the problem of creating a model would become resource expensive, time consuming, and not necessarily accurate. When thinking about feature selection, consider these goals:

- Making the dataset small enough to interact with in a reasonable timeframe

- Reducing the computational resources needed to create and use a model

- Improving the understandability of the underlying data

The examples in this section are both filtering methods because you use filtering methods most often for security data. To make the data more understandable, the examples rely on the same California Housing dataset used for the examples in *Chapter 6*. The important thing to remember with this example is that you're looking for features to use and the example shows you how to do this in an understandable manner. Feature selection is critical whether you're building a malware detection model or looking for data in the California Housing dataset, using the California Housing dataset is simply easier to understand.Of course, you use the dataset in a different way in this chapter. You can the example code in the MLSec; 07; Feature Selection.ipynb file of the downloadable source.

## Obtaining the required data

Both feature selection examples rely on the same dataset, so you only have to obtain and manipulate the data once. The following steps show how to perform the required setup:

1. Import the required packages and methods. The imports include the California Housing dataset, one of the analysis methods, plotting packages, and pandas (https://pandas.pydata. org/) for use in massaging the data:

```
from sklearn.datasets import fetch_california_housing
from sklearn.feature_selection import mutual_info_classif
import matplotlib as plt
import seaborn as sns
import pandas as pd

%matplotlib inline
```

2. Fetch the data and configure it for use:

```
california = fetch_california_housing(as_frame = True)
X = california.data
X = pd.DataFrame(X, columns=california.feature_names)
print(X)
```

The result of fetching the data and massaging it is a DataFrame that will act as input for both of the filtering methods. *Figure 7.11* shows the DataFrame used in this case.

```
        MedInc  HouseAge  AveRooms  AveBedrms  Population  AveOccup  Latitude  \
0       8.3252      41.0  6.984127   1.023810       322.0  2.555556     37.88
1       8.3014      21.0  6.238137   0.971880      2401.0  2.109842     37.86
2       7.2574      52.0  8.288136   1.073446       496.0  2.802260     37.85
3       5.6431      52.0  5.817352   1.073059       558.0  2.547945     37.85
4       3.8462      52.0  6.281853   1.081081       565.0  2.181467     37.85
...        ...       ...       ...        ...         ...       ...       ...
20635   1.5603      25.0  5.045455   1.133333       845.0  2.560606     39.48
20636   2.5568      18.0  6.114035   1.315789       356.0  3.122807     39.49
20637   1.7000      17.0  5.205543   1.120092      1007.0  2.325635     39.43
20638   1.8672      18.0  5.329513   1.171920       741.0  2.123209     39.43
20639   2.3886      16.0  5.254717   1.162264      1387.0  2.616981     39.37

       Longitude
0        -122.23
1        -122.22
2        -122.24
3        -122.25
4        -122.25
...          ...
20635    -121.09
20636    -121.21
20637    -121.22
20638    -121.32
20639    -121.24

[20640 rows x 8 columns]
```

Figure 7.11 – It's necessary to massage the data before filtering the features

Now that you've obtained the required data and massaged it for use in feature selection, it's time to look at two approaches for filtering the data. The first is the **Information Gain** technique, which relies on the information gain of each variable in the context of a target variable. The second is the **Correlation Coefficient** method, which is the measurement of the relationship between two variables (as one changes, so does the other).

### Using the Information Gain technique

As mentioned earlier, the Information Gain technique uses a target variable as a source of evaluation for each of the variables in a dataset. You chose a feature set based on the target you want to use for analysis. In a security setting, you might choose your target based on known labels, such as whether the input is malicious or benign. Another approach might be to view API calls as common or rare. The target variable must be categorical in nature, however, or you will receive an error message saying that the data is continuous (which really doesn't tell you what you need to know to fix it).

The California Housing dataset doesn't actually include a categorical feature, which makes it a good choice for this example because security data often lacks a categorical feature as well. The following steps show how to create the required categorical feature using the MedInc column. This is a good example to play with by changing the category conditions or even selecting a different column, such as AveRooms or HouseAge:

1.  Create a categorical variable to use for the analysis:

    ```
    y = [1 if entry > 7 else 0 for entry in X['MedInc']]
    print(y)
    ```

2. Perform the required analysis and display the result on screen:

```
importances = mutual_info_classif(X, y)
imp_features = pd.Series(importances,
    california.feature_names)
imp_features.plot(kind='barh')
```

The example uses a list comprehension approach to creating the categorical variable, which is very efficient and easy to understand. The result creating the categorical variable is a y variable containing a list of 0s and 1s as shown in *Figure 7.12*. A value of 1 indicates that MedInc is above $70,000 and a value of 0 indicates that the MedInc value is equal to or less than $70,000. The point is to categorize the data according to some criterion.

```
0, 0, 0, 0, 0, 0, 0, 0, 0, 0, 0, 0, 0, 0, 0, 0, 0, 0, 0, 0, 0, 0, 0, 0, 0, 0,
0, 0, 0, 0, 0, 0, 0, 0, 0, 0, 0, 0, 0, 0, 0, 0, 0, 0, 0, 0, 0, 0, 0, 0, 0, 0,
0, 0, 0, 0, 0, 0, 0, 0, 0, 0, 0, 0, 0, 0, 0, 0, 0, 0, 0, 0, 0, 0, 0, 0, 0, 0,
0, 0, 0, 0, 0, 0, 0, 0, 0, 0, 0, 0, 0, 0, 0, 0, 0, 0, 0, 0, 0, 0, 0, 0, 0, 0,
0, 0, 0, 0, 0, 0, 0, 0, 0, 0, 0, 0, 0, 0, 0, 0, 0, 0, 0, 0, 0, 0, 0, 0, 0, 0,
0, 0, 0, 0, 0, 0, 0, 0, 0, 0, 0, 0, 0, 0, 0, 0, 0, 0, 0, 0, 0, 0, 0, 0, 0, 0,
0, 0, 0, 0, 0, 0, 0, 0, 0, 0, 0, 0, 0, 0, 0, 0, 0, 0, 0, 0, 0, 0, 0, 0, 0, 0,
0, 0, 0, 0, 0, 0, 0, 0, 0, 0, 0, 0, 0, 0, 0, 0, 0, 0, 0, 0, 0, 0, 0, 0, 0, 0,
0, 0, 0, 0, 0, 0, 0, 0, 0, 0, 0, 0, 0, 0, 0, 0, 0, 0, 0, 0, 0, 0, 0, 0, 0, 0,
0, 0, 0, 0, 0, 0, 0, 0, 0, 0, 0, 1, 0, 0, 0, 0, 0, 0, 0, 0, 0, 0, 0, 0, 0, 0,
0, 0, 0, 0, 0, 0, 0, 0, 0, 0, 0, 0, 0, 0, 0, 1, 0, 0, 0, 0, 0, 0, 0, 0, 0, 0,
0, 0, 1, 0, 1, 0, 0, 0, 0, 0, 0, 0, 0, 0, 0, 1, 0, 0, 0, 0, 0, 0, 0, 0, 0, 0,
0, 0, 1, 0, 0, 1, 0, 0, 1, 0, 0, 0, 0, 0, 0, 0, 0, 0, 0, 0, 1, 1, 0, 1, 1, 1,
0, 0, 1, 0, 1, 1, 0, 1, 0, 1, 1, 1, 0, 0, 0, 1, 0, 0, 1, 1, 1, 1, 1, 1, 1, 1,
1, 1, 1, 1, 1, 1, 1, 0, 1, 1, 0, 0, 0, 0, 1, 0, 0, 1, 0, 0, 0, 0, 0, 0, 0, 0,
0, 0, 1, 1, 0, 1, 1, 1, 0, 1, 1, 0, 1, 0, 1, 1, 0, 1, 1, 1, 1, 0, 0, 1, 1, 1,
1, 1, 1, 0, 0, 0, 0, 1, 1, 1, 1, 0, 0, 0, 0, 0, 1, 1, 1, 1, 0, 1, 1, 0, 1, 0,
0, 0, 0, 0, 0, 0, 0, 0, 0, 0, 0, 0, 0, 0, 0, 0, 0, 0, 0, 0, 0, 0, 0, 0, 0, 0,
0, 0, 0, 0, 0, 0, 0, 0, 0, 0, 0, 0, 0, 0, 0, 0, 0, 0, 0, 0, 0, 0, 0, 0, 0, 0,
```

Figure 7.12 – Create a categorical variable to use as the target

Once you have the data in the required form, you can perform the required analysis, which produces the horizontal bar chart shown in *Figure 7.13*. Obviously, there is going to be a high degree of correlation between the target variable and MedInc, so you can safely ignore that bar.

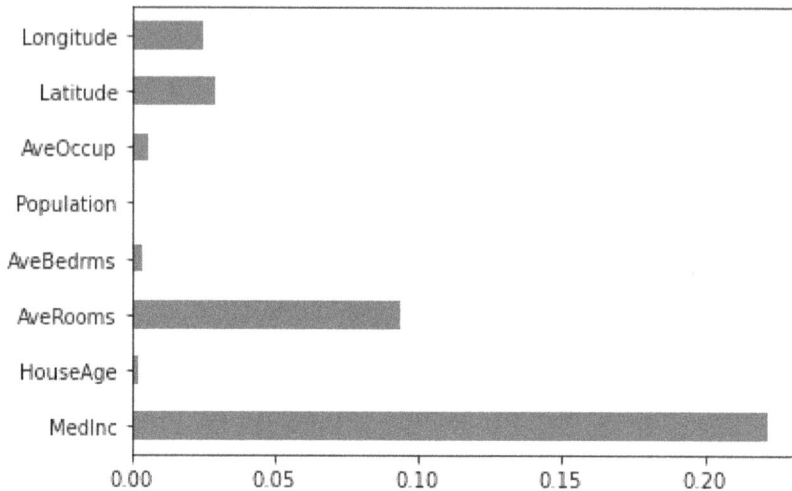

Figure 7.13 – The results are interesting because they show a useful correlation

What is interesting in the result is the high correlation between MedInc and AveRooms. Far less important is HouseAge – the results show you could likely eliminate the Population feature and not even notice its absence. However, you have to remember that these results are in the context of the target variable selection. If you change the target variable, the results will also change, so you can't simply assume that deleting Population from the original dataset is a good idea. What you need to do is create a new dataset that lacks the Population feature for this particular analysis.

### Using the Correlation Coefficient technique

Of the filtering techniques, the Correlation Coefficient technique is probably the least code-intensive and doesn't actually require any variable preparation. This approach simply compares the correlation between two (and sometimes more) variables. The example uses the default settings for the corr() function, but you can try other approaches as documented at https://pandas.pydata.org/docs/reference/api/pandas.DataFrame.corr.html. The following code shows the Correlation Coefficient technique as provided by pandas:

```
cor = X.corr()
plt.figure
sns.heatmap(cor, annot=True)
```

Note that the actual analysis only requires one step. The second and third steps are used to plot the results shown in *Figure 7.14*.

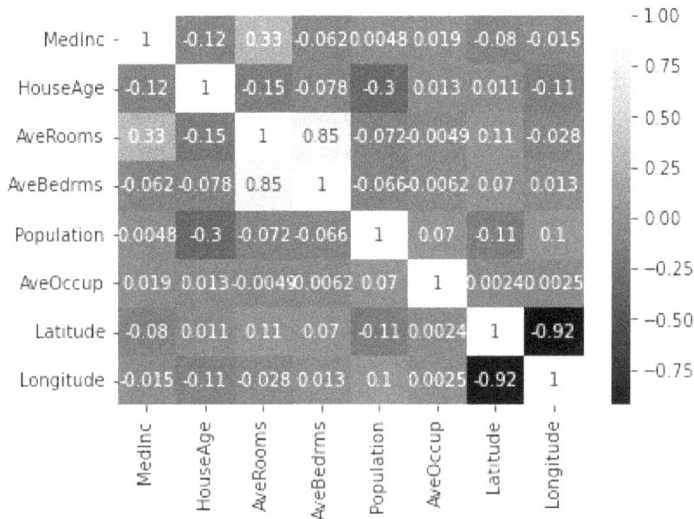

Figure 7.14 – The Correlation Coefficient technique shows some
interesting relationships between variables

This is a **heatmap plot**, which is a Cartesian plot with data shown as colored rectangular tiles where the color designates a level of correlation in this case. You see the correlation levels on the right side of the plot as a bar where lighter colors represent higher levels of correlation and darker colors indicate lower levels of correlation. Since MedInc (horizontally) has a 100-percent-degree of correlation with MedInc (vertically), this square receives the lightest color and a value of 1. Not shown in the bar on the right is that black indicates no correlation at all. So, for example, there is no correlation between Latitude and Longitude in the California Housing dataset.

There are some interesting correlations in this case. Notice that AveRooms has a high degree of correlation with AveBedrms. This plot also corroborates the result in *Figure 7.13* in that there is a moderate level of correlation between MedInc and AveRooms.

Not corroborated in this case is the correlation between MedInc, Latitude, and Longitude shown in *Figure 7.13*. This is due the different method used to filter the feature selection. The output in *Figure 7.14* isn't considering the amount of MedInc as a factor in feature selection, so now you understand that the Information Gain approach helps you target a specific criterion for filtering, while the Correlation Coefficient method is more generalized. These approaches are both helpful in filtering security features because sometimes you don't actually know what to target and the Correlation Coefficient approach will present you with some ideas.

## Considering speed of detection

When performing security analysis using ML techniques, and with malware in particular, detection is extremely time critical. This is the reason that you need to choose datasets used to create models with care, target the kind of threats you want to address with your business in mind, and layer your defenses to reduce the load on any one defense so it doesn't slow down. This chapter has discussed a number of malware threat types, feature selection, and detection techniques that will help you create a useful security model for your organization. However, here are some things to consider when creating your model and tuning it for the performance needed for usable security:

- Err in favor of false positives, rather than false negatives, because you can always take benign data out of the virus vault after verification, but letting malware through will always cause problems.

- Reduce your feature set to ensure that you're not creating a model that will get mired in useless detail. Smaller feature sets mean faster and more targeted training.

- Use faster algorithms that provide good enough analysis because you generally can't afford the time required by models that provide highly precise output.

- Decide at the outset that humans will be involved in the mitigation process because automation will always produce false positives that someone needs to verify. It also isn't necessarily possible to use lessons learned to create a more precise model because the new model will run slower.

- Ensure that any virus vault you create to hold suspected malware is actually secure and that only authorized personnel (your security professionals and no one else) can access it. It's usually better to lose some small amount of data (having it sent again from the source) than to allow any malware to enter your system.

The essential factor in all of these bullets is speed. Getting precise details about a threat after the threat has already passed is useless. Security issues won't wait on your model to make a decision. This need for speed doesn't mean creating a sloppy model. Rather it means creating a model that will always err on the side of being too safe and allowing a human to make the ultimate decision about the malicious or benign nature of the malware later.

## Building a malware detection toolbox

The example in the *Collecting data about any application* section tells you how to dissect just one kind of executable file, a Windows PE file. If all you ever work on is Windows systems that are possibly connected to each other, but nowhere else, and none of your employees use their smartphones and other devices to perform their work, then you may have everything you need to start studying malware. However, this is unlikely to be the case. To really get involved in malware detection, you need to know how to disassemble and review every type of executable for every device that interacts with your system in any way. It's really quite a job, which is why this chapter has focused so hard on using off-the-self solutions when possible, rather than trying to build everything from scratch on your own.

A malware detection toolbox needs to consist of the assortment of items needed to analyze the malware you want to target. This includes hardware that will keep any malware contained, which usually means using sandboxing techniques (`https://www.barracuda.com/glossary/sandboxing`) on virtual machines (`https://www.vmware.com/topics/glossary/content/virtualized-security.html`). Note that these hardware techniques don't work well in real time. For example, sandboxing is time intensive, so you couldn't attach a sandbox to your network and expect that the network will continue to provide good throughput. In addition, some types of malware evade sandbox setups by remaining dormant (appearing benign) until they leave the sandbox and enter the network.

Once you have decided on which malware features to detect, you need to work with a data source to understand how to classify the malware. The classification process leads to detection, avoidance, and mitigation. Knowing how your adversary works is essential to eventual victory over them. The next section discusses the malware classification process and provides advice on how to use this classification to perform detection and possible avoidance.

# Classifying malware

Even though this chapter prepares you to disassemble and analyze malware, nothing replaces actual experience. The best option to start with is to disassemble and analyze benign software of the kind you eventually want to work with before you attempt to work with any actual malware. Otherwise, you may find yourself the target of whatever malware you're studying at the time. The following sections provide you with some additional insights into classifying malware that may target your particular setup.

## Obtaining malware samples and labels

There are a lot of malware sites online where you can download live malware. The problem with live malware is that it can suddenly turn on you if you're not prepared. A good alternative is to download and study disabled malware first, which is what you find at `https://github.com/sophos/SOREL-20M`. This site also provides detailed instructions for working with the dataset with as much safety as working with malware can allow. The *Frequently Asked Questions* section of the site tells you how the malware is disarmed.

Once you have gotten past the early learning stages with malware, you may want to start looking at samples from other sites, such as VirusTotal (`https://www.virustotal.com/gui/intelligence-overview`). The datasets from these organizations usually come with a mix of malware and benign samples. There is no guarantee that the malware is disarmed, so you need to proceed with caution. In addition, you may have to pay a membership fee on some sites or jump through other hurdles to obtain samples. Obviously, these sites don't want to make live malware available to just anyone.

## Development of a simple malware detection scenario

It's helpful to have existing source code to try when starting your malware detection efforts. Trying to come up with code of your own would definitely be a bad idea. Even though the site contains older versions of malware, the *Microsoft Malware Classification Challenge (BIG 2015)* at `https://www.kaggle.com/c/malware-classification` does contain some useful techniques and freely downloadable data. The nice thing about this site is that you can find a number of solutions to the malware problem presented by viewing the submitted code (`https://www.kaggle.com/competitions/malware-classification/code`).

If you're looking for malware samples and sometimes code specifically for research, you can find a list at *Free Malware Sample Sources for Researchers* (`https://zeltser.com/malware-sample-sources/`). Another good place to look is *Contagio Malware Dump* (`http://contagiodump.blogspot.com/2010/11/links-and-resources-for-malware-samples.html`). This second site is older, but still contains a lot of useful links and resources for you to use (some of which do contain newer exploits).

## Summary

This chapter focuses on malware, but not malware in a single location. Most hackers realize that users don't rely on a single machine anymore and many users employ four or even more systems to interact with your ML application. Consequently, securing just one system is sort of like locking the barn with a truly impressive lock, but then forgetting to close the shackle. It's essential to consider the bigger picture and ensure that you have looked into securing all of the systems that a user may own, including personal systems.

The most important takeaway from this chapter is that classifying malware is a difficult process best left to security professionals. The job of the administrator, DBA, manager, data scientist, or other ML expert is to convey the potential risks to a security professional and come up with a good solution to secure the application as a whole from whatever location the user might access it.

Classifying malware means ensuring that you understand how current threats will combine attacks in order to overcome any prevention measures you have in place. A trojan will likely contain multiple payloads designed to work together to achieve specific hacker goals. These payloads may not all appear at once and may actually remain encrypted and compressed until a C2 server provides the correct instructions to start their attack.

Many of the pieces of malware explored in this chapter are designed to steal credentials so that the hacker can do something in the user's stead. Often, this amounts to some sort of fraud as described in the next chapter, *Locating Potential Fraud*. Committing fraud in the user's name lets the hacker off the hook until the authorities become convinced that the user really isn't to blame, at which point it's usually too late to do anything. *Chapter 8* provides you with ideas on how to overcome potential fraud that could create chaos for your ML application.

# Further reading

The following bullets provide you with some additional reading that you may find useful to advance your understanding of the materials in this chapter:

- See the method used to classify various kinds of malware: *Rules for classifying* (`https://encyclopedia.kaspersky.com/knowledge/rules-for-classifying/`)

- Locate potential malware dataset sources: *Top 7 malware sample databases and datasets for research and training* (`https://resources.infosecinstitute.com/topic/top-7-malware-sample-databases-and-datasets-for-research-and-training/`)

- Discover techniques for discovering behavior tracking spyware: *Behavior-based Spyware Detection* (`https://sites.cs.ucsb.edu/~chris/research/doc/usenix06_spyware.pdf`)

- Learn how combined attacks are becoming significantly more common and some of them also thwart MFA: *This Android banking malware now also infects your smartphone with ransomware* (`https://www.zdnet.com/article/this-android-banking-trojan-malware-can-now-also-infect-your-smartphone-with-ransomware/`)

- See a list of tools that can remove rootkits from the MBR of a hard drive, among other places: *Best Rootkit Scanners for 2022* (`https://www.esecurityplanet.com/networks/rootkit-scanners/`)

- Get an overview of the Windows **NT File System** (**NTFS**) version of alternate data streams: *NTFS File Streams – What Are They?* (`https://stealthbits.com/blog/ntfs-file-streams/`)

- Discover how images can install dropper trojans on user systems: *Malware in Images: When You Can't See "the Whole Picture"* (`https://blog.reversinglabs.com/blog/malware-in-images`)

- Understand how techniques like steganography pose a real risk to your organization: *What is steganography?* (`https://www.techtarget.com/searchsecurity/definition/steganography`)

- Create any custom malware detection aids with less effort: *11 Best Malware Analysis Tools and Their Features* (`https://www.varonis.com/blog/malware-analysis-tools`)

- Obtain a graphical presentation of the PE file format: *Portable Executable Format Layout* (`https://drive.google.com/file/d/0B3_wGJkuWLytbnIxY1J5WUs4MEk/view?resourcekey=0-n5zZ2UW39xVTH8ZSu6C2aQ`)

- See more feature selection techniques: Feature Selection Techniques in Machine Learning (`https://www.analyticsvidhya.com/blog/2020/10/feature-selection-techniques-in-machine-learning/`)

- Gain a better understanding of hybrid feature selection methods: *Hybrid Methods for Feature Selection* (`https://digitalcommons.wku.edu/cgi/viewcontent.cgi?article=2247&context=theses`)

- Learn more about virtualized machines: *What is a virtual machine (VM)?* (`https://azure.microsoft.com/en-us/resources/cloud-computing-dictionary/what-is-a-virtual-machine/`)

# 8
# Locating Potential Fraud

**Fraud** is deception that is perpetrated with either financial or personal gain in mind. Many people think about identity theft or credit card theft when thinking about fraud, but that's only a very small part of the picture. A person claiming or being credited with another's accomplishments or qualities is another form of fraud. The key word when thinking about fraud is *deception*, which takes many forms – for example, disinformation and hypocrisy. Consequently, it's important to come up with a solid definition of what fraud is when working with ML applications and data, which is the goal of the first part of this chapter.

Determining fraud sources is essential with ML because fraud sources generate data – deceptive data. Yes, the data looks just fine, but when you study it in depth, it contains one or more of the five mistruths of data described in the *Defining the human element* section of *Chapter 1*, *Defining Machine Learning Security*. Using this data will cause hard-to-locate issues within your ML models and make it difficult to ensure a secure environment.

In the ML realm, fraud either occurs in the background or in real time. **Background fraud** is the type that a hacker or another individual hopes will go unnoticed while the perpetrator makes small changes a little at a time. This is the kind of fraud that eventually improves sales for one company, while starving another, because the data used for comparisons is just a little off or resources are being siphoned off. **Real-time fraud** (the kind that most people think about when the term fraud is mentioned) occurs quickly with the intent of obtaining a fast gain, such as identity theft. With these issues in mind, this chapter discusses these topics:

- Understanding the types of fraud
- Defining fraud sources
- Considering fraud that occurs in the background
- Considering fraud that occurs in real time
- Building a fraud detection example

## Technical requirements

This chapter requires that you have access to either Google Colab or Jupyter Notebook to work with the example code. The *Requirements to use this book* section of *Chapter 1, Defining Machine Learning Security*, provides additional details on how to set up and configure your programming environment. When testing the code, use a test site, test data, and test APIs to avoid damaging production setups and to improve the reliability of the testing process. Testing over a non-production network is highly recommended, but not necessary. Using the downloadable source is always highly recommended. You can find the downloadable source on the Packt GitHub site at `https://github.com/PacktPublishing/Machine-Learning-Security-Principles` or my website at `http://www.johnmuellerbooks.com/source-code/`.

## Understanding the types of fraud

Fraud entails deception. The kind of deception depends on the fraud being perpetrated. For example, a compelling product message could result in identity theft, stolen credentials, or other resource gains for the perpetrator. Creating a condition in which one entity receives the blame for another entity's actions is another form of fraud. When considering fraud, it pays to have a Machiavellian mindset because the deception can become quite complex. However, the majority of fraud is quite simple: someone is deceived into giving someone else a resource the other wants for no apparent return. It amounts to a kind of theft.

There are many types of fraud. Some are watched by professionals, while others aren't, possibly because no one thinks to monitor them. With this in mind, here are a few common types of fraud that professionals do monitor:

- **Bank account takeover:** The common way to perpetrate this fraud when using ML techniques is to install a key logger on a system that someone uses to interact with their bank account over a computer. However, it's also possible to gain access to the information through the mail system, using a social engineering attack, or by hacking an unsecured Wi-Fi connection. The goal is to drain the person's bank account or to perpetrate other forms of fraud using the person's credentials. Constant monitoring can help with this kind of fraud and, interestingly enough, you could build an application to provide automated alerts.

- **Charity and disaster:** This particular kind of fraud often affects businesses with a benevolent urge directly. The disaster or charity is real, but the organization that is collecting funds or goods does the least that it can, so that little (if anything) of what is donated ends up going to the charity or disaster. An application can help in locating information about the organization doing the collecting. However, this is one case where human intervention is needed to keep the fraudsters at bay.

- **Debit and credit card:** There are many ways to perpetrate this fraud, including the use of card skimmers and persons of dubious character at restaurants and other locations making copies of your card. A man-in-the-middle attack works well for computer-based fraud. ML techniques can also be used to break into business databases to steal the information there. Having a chipped card does reduce the potential for fraud, but can't eliminate it. Some sites recommend using a third-party purchasing agent such as PayPal, which does provide a measure of protection, but only if PayPal isn't broken into by a hacker.

- **Driver's license:** A driver's license provides proof of identity for many situations, such as voting or boarding a plane. Often, you must provide a driver's license when checking into a hotel or engaging in an activity. Recent advances in driver's license construction have made copying a driver's license harder, but not impossible. The Real ID (`https://www.dhs.gov/real-id`) initiative also helps. However, the technology doesn't protect the actual number when used for online transactions, so there is still significant risk.

- **Elderly people:** Some fraud specifically targets the elderly because they're considered easy prey, especially when it comes to navigating computer systems. Because elderly people leave breadcrumbs that ML algorithms can use to create a pattern for fraud, perpetrators have more than the usual amount of information available for a social engineering attack. One way around this problem would be to use ML methodologies to give everyone a greater level of protection from prying eyes.

- **Healthcare:** Someone can use someone else's medical identification to obtain their care. ML could help thwart this type of attack by looking for inconsistent usage patterns. Two (or possibly more) people using the same ID will leave different usage trails, something that a human auditor might miss, but an ML application could be trained to see.

- **Internet:** Most of this book has explored the many ways hackers gain some advantage illegally through the internet. Many forms of fraud involve stealing user credentials and then using those credentials to make purchases. The products are then sold for cash. Sometimes, a hacker doesn't even take that many steps – the stolen credentials are used to empty bank accounts or cause other problems for the user.

- **Mail:** It's easy to use ML techniques to create a profile about people using data sources such as Facebook. The profile is then used to create a targeted letter that is sent through the mail system to fraudulently obtain money or other goods from the recipient. Because people don't quite get the connection between paper mail and the internet, this particular kind of fraud can be quite successful. Another form of this kind of fraud is to create a profile of somebody who receives money in the mail and then steal the check. However, this approach is becoming far less common with the use of techniques such as a direct deposit.

- **Phishing:** The internet provides many methods for a smart fraudster to gain access to an entity's personal or financial information through the use of social engineering attacks. It may be easy to think that this is an individual's problem until it affects an organization in a big way when the individual loses security clearance, is sent to jail, blamed for another's action in a way that reflects badly on the business, and so on. Unfortunately, phishing also affects organizations as a whole, so this isn't just something that individuals need to worry about. Training does help with this sort of fraud, but keeping the sources of phishing out of the organization in the first place is better. The ML techniques mentioned throughout this book can help you look for patterns that indicate someone is employing phishing that you can avoid.

- **Insurance:** Insurance fraud is one in which someone makes a false claim. Damage may have occurred, but the claim is exaggerated or perhaps an accident is staged and hasn't happened for real. In some cases, a third party will create a situation where some damage occurs, but the damage is self-inflicted, such as someone jumping in front of a car to make an insurance claim.

---

**Email is fraud's friend**

This chapter could become a book if it were to cover all of the creative ways in which fraud occurs through email. One of the most famous forms of fraud is Nigerian Fraud. There are whole websites devoted to just this one form of fraud (see `https://consumer.georgia.gov/consumer-topics/nigerian-fraud-scams` as an example). There are also debt elimination, advanced fee, cashier check, high-yield investment, and a great many other email-based fraud attacks. They all have one thing in common: they use email to perform their tasks to take money or other resources from people. As such, monitoring the email stream to your organization using a combination of off-the-shelf and customized ML applications can reduce the problem by reducing these emails. Of course, there still has to be a mechanism in place for dealing with false positives, but targeting your defenses is a good way to reduce false positives as well. Oddly enough, one of the suggested methods for dealing with this problem that seems effective is to mark all emails that come from outside the organization with the word [EXTERNAL] in the subject line.

---

- **Stolen tax refund:** This is one of those weird, never would have thought of it, types of fraud. Someone steals credentials from someone else, fills out their taxes as soon as possible, and then claims the refund for themselves. When the real filer sends in their tax information, the IRS rejects it because the person has already filed. Automation can help prevent this problem by allowing a person to file their taxes early (cutting off any attempt at a fraudulent return).

- **Voters:** This is a kind of fraud that many people don't fully understand. It's more than simply getting the wrong person into office or spending money on bonds that no one wants. Voter fraud can and does shape all sorts of issues that affect everyone, such as new taxes or drawing new voting district lines. ML techniques could at least help in locating votes by people who have passed away or are currently too young to vote.

This list doesn't include some types of common fraud that are unlikely to affect a business or its applications directly, such as adoption fraud, holiday scams, or money mules. However, because these kinds of fraud do affect your employees and make them less efficient because they're thinking about something other than work, you also need to be aware of them and provide the required support. There are many other kinds of fraud that can take place and it's important to remember that the people who perpetrate fraud can be quite creative. The simple rule to follow is that if something seems too simple, too good to be true, or simply too convenient, then it's likely a fraud of some sort. A fraudster uses human weaknesses to their advantage.

Now that you have a better understanding of fraud types, it's time to look at the source of fraud. Often, the source of fraud is the entity that is most trusted by the target. It's easier to perpetrate fraud when there is a trust relationship. However, it's important to understand that organizations have trust relationships with other organizations, organizations trust a particular application or service, and people trust other people. To understand the scope of fraud sources, it's important to stretch the concept of trust to fit all sorts of scenarios.

## Defining fraud sources

A **fraud source** is an entity that is generating deception. People are most often considered fraud sources, but organizations and even **non-entities** (things that someone would normally not consider an **entity**, such as a smart device) can become fraud sources as well. An application that is fed misinformation can become a fraud source by outputting the incorrect analysis of data input. A user is deceived into believing one thing is true when another thing is actually true. The deception isn't intentional, but fraud occurs just the same. However, if the application contains a bug, then it's broken and needs to be fixed; it isn't a fraud source in this case. Organizational fraud usually involves a group of people working together to develop this deception. For example, an organization could try to attract investors for a product that doesn't exist yet and will never exist (**vaporware** is specific to the computer industry because it involves non-existent hardware or software).

---

### Distinguishing between types of fraud and fraud sources

Most articles online on this topic cover the different kinds of fraud. For example, identity theft is a **type of fraud** because it focuses on what an entity is doing. Unfortunately, without some knowledge of the potential sources for identity theft, the task of locating the culprit and ending the activity is made much harder. A fraudster is the primary perpetrator of identity theft, followed by hackers. You already have detailed knowledge of other organizations, company insiders, and customers (or fraudsters posing as customers; you don't know until you ferret out their intentions), so the potential for identity theft from these sources is much less (or possibly non-existent) when compared to other sources. So, it's important to know the type of fraud, but also important to know the potential fraud source so that you don't spend a lot of time looking for a particular type of fraud from a source that's unlikely to commit it.

## Considering fraudsters

A **fraudster** is someone who perpetrates fraud solely for some type of gain, usually immediate gain. Fraud is just business to them. They aren't out for revenge, to gain access to your organization, or any of the other motivations that can and do affect others who commit fraud. A fraudster uses some technique to gain a person's or organization's trust and then robs them of everything possible. However, some fraudsters take years to set up their fraud or they engage in it continually. There is a form of fraud called long-term fraud where a fraudster places small orders with a business and promptly pays for the goods to win the business's trust. The fraudster then places one or more large orders and absconds with the goods without paying. ML techniques can look for patterns in the fraudster's buying habits. For example, would any one organization need the eclectic assortment of goods that the fraudster is buying? Anomalies (as discussed in *Chapter 6, Detecting and Analyzing Anomalies*) can help detect these sorts of frauds because the fraudster isn't buying for a business, but rather to create a reputation.

Of the kinds of input that these schemes produce, the Ponzi scheme (`https://www.investor.gov/protect-your-investments/fraud/types-fraud/ponzi-scheme`) is the most common for businesses because it affects people managing the businesses' assets directly. The use of virtual currencies has made such Ponzi schemes incredibly easy to perpetrate and it may take a long time for anyone to see that there is any sort of problem. ML can help avoid this particular type of fraud by automating a paperwork review. A Ponzi scheme often provides payments that are too regular and hides details in copious paperwork that might be hard to analyze without a computer. Given that a human would take too long to spot the necessary patterns, ML is the only solution to augment human capability when dealing with lots of high-speed transactions using virtual currency.

## Considering hackers

Hackers have a different set of goals from fraudsters when it comes to fraud. As discussed previously in this book, a hacker is usually trying to gain entry to your organization to perform some task, such as installing ransomware or stealing credentials. The hacker's motivations can include personal gain or it may simply be business on the part of someone else who wants to access your business. Unlike fraudsters, you won't have any actual contact with a hacker in most cases. The organization may receive an email, but that email won't be personalized for the most part unless the hacker is perpetrating some type of personal attack. The goal is to blanket an organization in the hopes that someone will take the bait and leave your organization wide open. While a fraudster usually makes targeted attacks based on social engineering, a hacker can use any of the techniques described in *Chapter 5, Keeping Your Network Clean,* through *Chapter 7, Dealing with Malware,* to commit fraud based on automation, hitting as many targets as possible in the shortest time possible. These differences between fraudsters are clues that you can use when trying to detect a fraud source, whether you're dealing with a fraudster or a hacker.

## Considering other organizations

When dealing with fraud from other organizations, the prime motivators are espionage, sabotage, or stealing trade secrets (sometimes, there is a bit of revenge involved as well). The organization is using fraud to obtain something that your organization has that gives you some sort of advantage or that can be sold to someone else for a profit. Fraud between organizations tends to be more personalized than that provided by fraudsters because the members of each organization know each other. With this fact in mind, the kind of fraud is usually very personal based on what one party knows about the other.

Considering only electronic interactions (and setting aside personal interactions), it's possible to use ML to look for patterns in data exchanges with the other organization. In this case, the main means of detection is fake data. The other organization will have enough information to avoid potential issues with anomalies or making randomized purchases. However, it's still possible to use ML to detect fake data that the other organization generates to skew the target organization's perception of it. This is one of the situations where using unsupervised learning techniques may be necessary, as described in *Unsupervised Deep Learning for Fake Content Detection in Social Media* at `https://scholarspace.manoa.hawaii.edu/items/6d7560aa-2aff-4439-a884-35994e242c06`. The same techniques work when reviewing other sorts of textual data exchanged between organizations.

## Considering company insiders

Oddly enough, company insiders are often the most eager perpetrators of fraud. A company insider generally has good access to company assets and is familiar enough with company policies and procedures to avoid detection. However, the kinds of fraud that the company insiders perform are different because the focus of the fraud is different. For example, a company insider is unlikely to try to use someone else's credit card to make a purchase and may not make any purchases at all (depending on what the company sells). Consequently, consider these kinds of fraud when looking at company insiders as a potential source:

- **Asset misappropriation:** Any time that someone steals something from the organization, it amounts to asset misappropriation. The most commonly stolen asset is money. For example, someone who is allowed to create and sign checks may create and sign a check for themselves. Likewise, an employee on a trip could misrepresent the costs for items purchased at the company's expense while on the trip. ML can significantly automate any checks made to verify the actual amounts that employees are paid or spend, reducing the workload on auditors and ensuring accuracy. Difficulties can arise when third parties become involved in asset misappropriation because the third party (such as another organization) is better able to hide details that an ML application would need to detect. However, it's still possible to locate potential sources of asset misappropriation through behavior monitoring and reviewing the change in individual or organizational patterns.

- **Corruption schemes:** This kind of fraud normally involves bribery, conflicts of interest, and kickbacks that somehow involve your organization, especially your organization's reputation or monetary state. The problem with this sort of fraud is that it relies on personal relationships that don't leave a paper trail. So, even behavior analysis may not reveal the information you need (at least, not completely). In this particular case, ML may provide some facts, but you need a human investigator to complete the task.

- **Financial or other statement issues:** This particular kind of fraud is a deliberate misrepresentation of an organization's financial strength or abilities. While the previous kinds of fraud are commonly performed by employees, this kind of fraud is common to managers or owners. The methodology for the financial form of this kind of fraud includes manipulating revenue figures, delaying the recognition of expenses, or overstating inventory (among a great many other ways). Other types of statement fraud can include stating that an organization has a certain number of employees with a particular certification or has access to specific kinds of equipment. In general, someone outside the organization will expose this type of fraud. ML can help research and correlate facts, but this form of fraud requires a human investigator. One approach to ferreting out the required information is to perform comparisons between the target organization and its competitors to look for odd differences in performance or ability.

As you can see, these are still kinds of fraud, but they're different because they would be harder for someone outside the organization to perpetrate due to a lack of knowledge and/or access. Make sure that you also consider the role third parties can play in company insider fraud. In addition, employees can be blackmailed into committing fraud by a third party, which complicates the situation even further. Given modern ML methods, it's entirely possible that the target of a fraud investigation isn't even involved, but is being blamed for the fraud activity by someone else (so, look for things such as misuse of credentials).

## Considering customers (or fraudsters posing as customers)

Customers are the lifeblood of any business, so learning how to avoid fraud when dealing with customers is essential. A customer that wants to perpetrate fraud against your business will usually do so in the form of a scam. The idea is to get whatever is desired as quickly as possible and then disappear without a trace. With this in mind, here are typical scammer tactics you need to be aware of:

---

**Customer or fraudster?**

Some fraudsters start as customers and only perpetrate fraud when they see weakness on the part of the business, so there is a great amount of ambiguity here: you don't always know when someone goes from being a customer to being a fraudster; a well-prepared business will avoid certain kinds of fraud through a show of strength. The addition of a sign saying *"You're on candid camera!"* immediately below a surveillance camera has been shown to reduce theft. That camera could be attached to an ML application with facial recognition capabilities to make it easier for you to detect repeat offenders. With this ambiguity in mind, the following section uses the term *customer* because what you see is a customer until you find out otherwise.

- **Trust environment:** The customer will try to create some level of trust as quickly as possible so that you lower your defenses.

- **Urgency:** The customer will often create a situation in which you need to make a decision quickly. ML applications can help with this issue by reducing the time to perform in-depth research about the customer from multiple sources and also to provide some type of reading on the probability of a particular situation occurring.

- **Intimidation and fear:** The customer will make it seem as if something bad will happen if you don't take advantage of an offer. For example, the customer may intimate that they'll provide whatever it is to your competition instead. An expert system can provide a sanity check by providing a worst-case scenario thought process that will help reduce the potential for fear taking the place of reason.

- **Untraceable payment methods:** Once a deal is struck, the customer may want to use alternative payment methods such as wire transfers, reloadable cards, or gift cards to make payments. The right application can help you trace the payment method and help you determine whether it's legitimate. Otherwise, this is probably one situation where you can just let a potential opportunity pass you by.

There are potential ways to use ML to help reduce the potential for customer fraud in your organization. One of the most important methods is to create and maintain a database of customer contacts with detailed information. A customer who wants to perpetrate fraud will often try multiple employees, looking for one that is more open to creating a trust environment. You can also use this database to search for odd customer contact patterns that don't match other customers that you deal with. ML applications can also validate customers in several ways:

- Ensuring that the person on the phone is actually who they say they are because their caller ID can easily be faked.

- Verifying that email addresses and websites are legitimate. Typos in addresses are a common method of misdirecting searches and checks. Avoid opening such sites because they often download malware to the caller's system.

- Securing every aspect of your network, APIs, and other computer resources using ML techniques found in previous chapters and then monitoring these resources constantly using other ML applications.

- Researching the customer by checking online reviews, rating systems, and so on. Calling colleagues who may have dealt with the customer is always a good addition to automated checks.

- Checking whether the service, goods, or other items of interest are free. It's the old idea of someone trying to sell someone else the Brooklyn Bridge.

- Performing **onboarding**, which relies on a series of both human and automated controls to verify and validate customers.

The actual techniques that customers use to defraud your business vary, but they tend to fall into one of several areas. The following list provides ideas of what you should look for:

- **Fake invoices:** The customer provides an invoice for a service or product that your organization uses, except that you didn't receive the invoiced product or service. A good database design wouldn't allow payment for products or services without a corresponding request. However, you can also rely on ML to analyze the database logs to look for scam patterns where a customer employs more than one person to commit the scam.

- **Things you didn't order:** The customer calls and asks whether you want a free item such as a catalog. You accept the offer and receive it, along with the merchandise you didn't order, in the mail. The customer then tries to high-pressure you into paying for the goods. A good database application will reveal that you didn't create a purchase order for the merchandise. However, given the numerous methods that businesses now employ to make purchases, you may need an ML application to search all of the potential places in which an order could have been made.

- **Directory listings and advertising:** The customer tries to interest you in a directory listing or advert that does not actually exist. Banner ads are especially prone to this scam because it's hard to verify that they're being displayed. Screen scraping techniques and an ML application designed to analyze the related advertising can ensure that you are getting the amount of advertising or the directory listings that you're paying for.

- **A utility company, government agent, or other professional contact imposters:** Customers will often pose as someone that they're not to create a trust relationship. Any method that you can use to verify the person's identity will help reduce the potential for fraud.

- **Tech support:** There are several ways to perpetrate this particular fraud, but it comes down to convincing someone in your organization that they need some type of technical support to fix a problem that doesn't exist. The scam itself may consist of a one-time payment to fix this problem, or possibly a maintenance plan to keep it from happening again. Often, the goal is not the payment or the maintenance plan, but rather to gain access to sensitive information on your network, which is worth a lot more.

- **Coaching and training scams:** The customer promises to provide you with coaching or training that will greatly improve your organization's income. They use false testimonials and other methods to convince you that the techniques work, except that they don't. This is an especially prevalent scam for new businesses. Finding a service, usually ML-based, to help you separate the wheat from the chaff is the only way to get past this particular scam if you do feel that coaching or training will be helpful.

- **Payment processing or acceptance scams:** This category includes things such as credit card processing and equipment scams and the use of fake checks. Most of these scams focus on getting customers to pay what they owe or obtaining payment for lower processing costs. Possibly the best advice here is the idea that **There Is No Such Thing As A Free Lunch** (**TINSTAAFL**) (see https://www.investopedia.com/terms/t/tanstaafl.asp for details).

Whether it's creating a model for your ML application, developing a new product, selling to customers, or getting paid, there is a cost involved and no one gets by for free.

As you can see, customers use several highly successful approaches to defrauding businesses. This list doesn't even include the social engineering, phishing, and ransomware attacks discussed in other chapters. Many of these scams can be tracked down with the assistance of an ML application. In addition, using expert systems to create smart scripts so that employees know what process to follow when working with customers can make a huge difference in ensuring that things go smoothly and don't necessarily require a lot of management time to solve.

## Obtaining fraud datasets

To create effective ML models for your organization, you need to train the model using real data or something that approximates real data. The best choice for a dataset is one that your organization gathers, but this is a time-consuming process and you may not generate enough data to produce a good dataset. With this in mind, you can use any of several fraud datasets available online. One of the best places to get this sort of dataset if you need several different types is the *FDB: Fraud Dataset Benchmark* at `https://github.com/amazon-research/fraud-dataset-benchmark`. This one download provides you with access to nine different fraud datasets as listed on the website. You can read the goals of creating this dataset at `https://www.linkedin.com/posts/groverpr_fdb-fraud-dataset-benchmark-activity-6970921322067427328-fNCo`.

It's tough, perhaps impossible, to create a single grouping of datasets that answers all fraud detection needs. The *Building a fraud detection example* section of this chapter uses an entirely different dataset to demonstrate actual credit card fraud based on sanitized data from a real-world dataset. Any public dataset you get will likely be sanitized already (and you need to exercise care in downloading or using any dataset that isn't already sanitized). Here are some other dataset sites you should visit:

- **Amazon Fraud Detector (fake data that looks real):** `https://docs.aws.amazon.com/frauddetector/latest/ug/step-1-get-s3-data.html`

- **data.world:** `https://data.world/datasets/fraud`

- **Machine Learning Repository (Australian Credit Approval):** `https://archive.ics.uci.edu/ml/datasets/Statlog+%28Australian+Credit+Approval%29`

- **Machine Learning Repository (German Credit Data):** `https://archive.ics.uci.edu/ml/datasets/statlog+(german+credit+data)`

- **OpenML (CreditCardFraudDetection):** `https://www.openml.org/search?type=data&sort=runs&id=42175&status=active`

- **Papers with Code (Amazon-Fraud):** `https://paperswithcode.com/dataset/amazon-fraud`

- **Papers with Code (Yelp-Fraud):** `https://paperswithcode.com/dataset/yelpchi`

- **Synthesized (Fraud Detection Project):** `https://www.synthesized.io/data-template-pages/fraud-detection`

- **US Treasury Financial Crimes Enforcement Unit:** `https://www.fincen.gov/fincen-mortgage-fraud-sar-datasets`

Note that older datasets are often removed without much comment from the provider. Two examples that are often cited by researchers and data scientists are the Kaggle dataset at `https://www.kaggle.com/dalpozz/creditcardfraud` and the one at `http://weka.8497.n7.nabble.com/file/n23121/credit_fruad.arff`. Both of these datasets are gone and it isn't worth pursuing them because there are others. The reason this is important is that the previous list may become outdated at some point, but because fraud is such a common and consistent problem, there will be others to take their place.

Now that you've considered all of the kinds of fraud sources and perhaps reviewed a few fraud datasets for use in your research, it's time to look at the slow type of fraud that occurs in the background. A background fraud scenario is the most dangerous kind of fraud because it happens slowly over a certain period. In some cases, this sort of fraud is never found out except accidentally as part of an audit or other research that has nothing to do with the fraud in question. Of course, this kind of fraud also requires patience on the part of the deceiver, so it can also be the hardest kind of fraud to perpetrate successfully.

## Considering fraud that occurs in the background

Niccolo Machiavelli, the person from whose name the term *Machiavellian* is derived, is one of those individuals who observed human nature with great patience and in great depth in many cases. Unlike the perception that many have of him being a scoundrel of the worst sort, he was a philosopher who saw human nature as it was at the time when it came to politics. Background fraud is often Machiavellian in nature. It has an "ends justifies the means" view of the world and can be utterly immoral in its approach to obtaining some goal through the use of deception. In other words, yelling that something isn't fair is unlikely to garner any sort of response in this situation. The following sections define the kinds of background fraud and how to detect it.

### Detecting fraud that occurs when you're not looking

There are many different definitions for the difference between background, or long-term, fraud and real-time, or short-term, fraud. When viewing long-term fraud strictly from a financial perspective, many experts view it as a scenario where a fraudster makes lots of small purchases and pays for them to build trust and then makes one or two large purchases, but doesn't pay for them. By the time anyone thinks to look for the fraudster, the products have been sold for cash, and the fraudster is gone.

Because the term long-term fraud has a specific meaning in many financial circles, this book uses the term **background fraud** to indicate fraud that happens over weeks or months in a manner that the fraudster hopes no one will notice until it's too late to do anything about it. The key difference between any sort of fraud and an attack is that fraud relies on deception – it appears to be one thing when it's another. For example, the *Using supervised learning example* section of *Chapter 5, Keeping Your Network Clean*, represents a kind of attack because no deception is employed during the probing process. To change that example into a kind of fraud, the hacker would need to make completely legitimate calls exclusively during the learning process from what appears to be a legitimate caller, and then commit an act that would derive some sort of tangible benefit once the learning process is done by using API calls to manipulate the database in some manner (as an example). Perhaps the hacker would obtain company secrets from the database to sell to a competitor. Consequently, the technique shown in *Chapter 5, Keeping Your Network Clean*, would have a slim chance of working because the model wouldn't necessarily have time to recognize a change in pattern.

Detecting background fraud often involves more detailed detective work that relies on some sort of targeted knowledge. Looking again at the example from *Chapter 5, Keeping Your Network Clean*, a form of fraud might be detected by reviewing the IP addresses making calls to the API and comparing them to a list of IP addresses that could make legitimate requests. As an alternative, it might be possible to trace the IP addresses to a specific caller. Reviewing the history and background of the caller might reveal some anomalies that would point to fraud. Yes, automation plays a very big role in helping a human track the caller down and check into the caller's information, but eventually, a human has to decide on the viability of the caller because that's something automation can't do.

## Building a background fraud detection application

There are some types of fraud that can be tracked using pattern recognition models such as the one described in *Chapter 5, Keeping Your Network Clean*. For example, if a fraudster wanted to slowly manipulate the price of a product to obtain financial gain, the act of manipulating the price would create a pattern of some sort that an ML application could detect, given a model that is trained using enough examples (there are a lot of caveats here, so make sure you pursue this approach with care).

It's also important to think small in some cases when it comes to fraud. For example, loyalty rewards may seem like a very small sort of fraud to commit, but according to Kount (`https://kount.com/blog/how-to-prevent-loyalty-fraud/`), the value of just the unredeemed loyalty rewards in a given year may amount to $160 billion. Fraudsters perform an **Account Takeover** (**ATO**) attack to gain access to royalty rewards using stolen credentials in many cases. The reason that this is such perfect fraud is that many people never check their royalty rewards information; they just assume that the business will automatically apply a loyalty reward at the cash register when one is available. Consequently, the fraudster has a significantly reduced chance of being caught when perpetrating this form of fraud. An ML application could help in this case by flagging customer accounts that usually have no access, but are currently experiencing a surge in access. In addition, the ML application could monitor the customer's loyalty rewards for large withdrawals of benefits in a short time.

Now that you have a better idea of how background fraud works, it's time to look at something with a faster turnaround: real-time fraud. Real-time fraud is marked by impatience; the quick use of deception to make a fast gain, even if that gain isn't substantial. The idea is that even a small gain is worthwhile if there are enough small gains.

# Considering fraud that occurs in real time

**Real-time fraud** is marked by a certain level of impatience and often requires quick thinking to pull off. Real-time fraud usually infers a kind of interaction that is performed by an entity that can perpetrate the fraud and then become inaccessible (usually by changing venue). In addition, real-time fraud normally relies on social engineering, a lack of knowledge, or some type of artifice. The following sections provide insights into real-time fraud and its detection.

## Considering the types of real-time fraud

Real-time fraud, a term this book uses to indicate a kind of fraud that occurs within hours or possibly days (or sometimes even seconds), targets quick gains with little effort on the part of the fraudster. Here are a few real-time fraud types to consider:

- **Email phishing:** The fraudster commonly sends what appears to be a legitimate email to a user with some method of invoking fraud that antivirus or antispam software can't detect. One such approach is the use of a malformed URL to take the user to the fraudster's site where the user enters confidential information, such as credentials. The best way to avoid this kind of fraud is to ignore the message, but users will often click on the link in an email without thinking about it first. Consequently, the best way to avoid this fraud is to keep the user from seeing the email in the first place, which normally involves two levels of protection:

- **Authentication checks:** Where the identity of the sender is verified. The *Developing a simple authentication example* section of *Chapter 4, Considering the Threat Environment*, provides some ideas on how to implement this solution.

- **Network-level checks:** The use of filters to keep the message out of the site. The most common approaches are the use of whitelists, blacklists, and pattern matching. The example in the *Developing a simple spam filter example* section of *Chapter 4, Considering the Threat Environment*, provides you with some ideas on how to implement this approach.

- **Payment fraud:** The fraudster somehow obtains a payment card or its information. In many cases, this type of fraud requires the use of stolen or counterfeit cards. The use of chipped cards has helped reduce this form of fraud, but no method of protection is absolute or foolproof. Pattern matching and logical checks can also detect this sort of fraud. For example, it's physically impossible for someone to use a card in one location, and then use it in another location 500 miles away just a few minutes later. Humans monitoring the system can't pick up clues such as this; you need an ML application and a database with specific triggers set on it to look for illogical entries and then look for a pattern that indicates the fraud source.

- **ID document forgery:** It's hard to detect fraud when the ID used to verify a person's identity appears legitimate because a fraudster has expertly copied the original. There are techniques in place to detect this kind of fraud based on the ID itself, but fraudsters are continually improving their skill at creating fake IDs. One possible way around this kind of fraud would be to have cameras at the site where the ID is used to capture the person's image and then compare it to the legitimate image on file. Of course, fraudsters are already looking at ways around facial recognition, as we will see in *Chapter 10, Considering the Ramifications of Deepfakes*. It's also possible to use pattern recognition to look for behavioral anomalies. The problem is that trying to get some of these techniques to work in real time is going to be quite hard. Customers balk over the smallest delays in any transaction, so making the check promptly is important. What you need to consider is your business model and the kinds of checks that it can support. For example, a bank could more easily accommodate a facial recognition check than a grocery store, which would likely rely on some type of pattern matching and possibly historical data.

- **Identity theft:** Identity theft differs from ID document forgery in that the fraudster focuses on the user's accounts in some way. Yes, credentials are stolen, but the goal is to gain access to some second-tier level of access, such as a bank account. There are three common types of identity theft: real name theft (where the fraudster acts in the victim's stead), ATO (where the fraudster actually takes over the user's account and usually seals the user out), and synthetic (where the fraudster combines real information with fake information to create a new identity).

There are many other kinds of real-time fraud. Fraudsters are extremely skilled at coming up with the next **confidence trick** (**con**) to play on unwitting victims. So, the people that are protecting others from the fraudster have to be equally skilled and quite fast. The next section provides some ideas on the tools that someone can use to aid in detection.

## Detecting real-time fraud

The act of detecting real-time fraud as it occurs is really hard because everything happens so quickly. Humans have developed intuitive approaches to knowing when fraud is occurring based on clues that ML has yet to pick up on (and may never pick up on). For example, there is some evidence to suggest that odors can indicate that a person is lying. According to the article, *Artificial networks learn to smell like the brain* (`https://news.mit.edu/2021/artificial-networks-learn-smell-like-the-brain-1018`), this approach is still a work in progress. Gamblers often rely on a person's tell (behavior change) to know what appears in the person's hand. What this amounts to is a kind of facial recognition over time. *Chapter 10* will tell you that ML is getting closer to good facial recognition, but this too is a work in progress. So, what can ML do for you today to help mitigate real-time fraud?

- **Physical inspection:** Humans can look at an ID and not see any problem with it. A scanned image provided to an ML application can locate small (possibly imperceptible) problems and flag the ID as fake. The same technique also holds for any other kind of document, including money.

- **Smart scripting based on the situation:** Rule-based systems are inflexible because they require hardcoded rules to determine a course of action. However, they can provide a basis for smart scripting where a rattled employee can remember the recommended process for dealing with fraud based on the current situation.

- **Constantly upgraded models using unsupervised learning:** Normally, you want to use labeled data when building a model because of the chances for an increase in accurate analysis. However, unsupervised models are better with constantly changing situations because they learn without using labels (which means the learning is more automatic). The downside is that an unsupervised model will present more false positives.

The set of features commonly used for real-time fraud detection is reduced compared to those used for background fraud detection, partly because less data is available and partly because time is of the essence. Here are the four features most commonly used to detect real-time fraud:

- **Identity:** The main method for avoiding real-time fraud is to ensure that the entity is who they say they are. Most of the time, you look for an ID, but online transactions require confirmation of the credentials provided. Something as small as a purchase from a new device or the use of a new address can provide a clue that the credentials need further investigation, which will be provided by your ML application.

---

**The effect of using multiple devices**

The use of different devices by a single individual today commonly triggers an email to the authorized person, an indicator that tracking the device does work.

---

- **Amount:** Purchases generally follow a routine of sorts. Trips to the grocery store usually follow within a few dollars of each other unless there is some sort of special event (and even then, the total won't be outrageously different). Time at the coffee shop or a night at a restaurant also tends to follow patterns. It's when the amount of a purchase is much higher or much lower than normal that you need to check further. An amount can also mean the number of items or the type of items purchased. Consider the amount as being a quantifier, rather than a specific value.

- **Location:** Even online purchases are normally performed from the same location. Any time that the buyer starts changing locations radically, it could be an indicator of fraud (or that they simply moved to a new house). Some credit card companies will contact a purchaser about a change in location to verify that the purchase was made by them from a different location. Of course, since someone can't be in more than one location at a time, purchases from more than one location in a given timeframe are almost certainly indicative of fraud.

- **Frequency:** Habits are often considered bad, but they can also protect a potential victim. A fraudster can grab credentials and try to use them successfully, but a change in credential usage frequency can often signal potential fraud.

As you can see from this list, it's not a perfect setup for detecting absolutely every kind of fraud because there is a potential for fraudsters to slip through the cracks. A highly motivated fraudster could verify that the identity used is perfect, make purchases only within the range that the target would make, spoof the location of the purchase, and ensure that they followed the person's habits within a reasonable range. Of course, that's a lot of ifs to consider, but it could happen. The point is that observing these four characteristics as part of an ML application will greatly reduce the potential for fraud.

Now that you have some basis for understanding the nature of fraud detection, the next section looks at a specific fraud detection example. In this case, you will see how to detect fraud in credit card purchases.

# Building a fraud detection example

This section will show you how to build a simple fraud detection example using real sanitized credit card data available on Kaggle. The transactions occurred in September 2013 and there are 492 frauds out of 284,807 transactions, which is unbalanced because the number of frauds is a little low for training a model. The data has been transformed by **Principal Component Analysis (PCA)** using the techniques demonstrated in the *Relying on Principle Component Analysis* section of *Chapter 6, Detecting and Analyzing Anomalies*. Only the Amount column has the original value in it. The Class column has been added to label the data. You can also find the source code for this example in the MLSec; 08; Perform Fraud Detection.ipynb file of the downloadable source.

## Getting the data

The dataset used in this example appears at https://www.kaggle.com/datasets/mlg-ulb/creditcardfraud?resource=download. The data is in a 69 MB .zip file. Download the file manually and unzip it into the source code directory. Note that you must obtain a Kaggle subscription if you don't already have one to download this dataset. Getting a subscription is easy and free; check out https://www.kaggle.com/subscribe.

## Setting the example up

This example begins with importing the data, which requires a little work in this case. The following steps show you how to do so:

1.  Import the required packages:

    ```
    import pandas as pd
    from sklearn.preprocessing import StandardScaler
    ```

2.  Import the creditcard.csv data file. Note that this process may require a minute or two, depending on the speed of your system. This is because the unzipped file is 150.83 MB:

    ```
    cardData=pd.read_csv("creditcard.csv")
    ```

3.  Check the dataset's statistics:

```
total_transactions = len(cardData)
normal = len(cardData[cardData.Class == 0])
fraudulent = len(cardData[cardData.Class == 1])
fraud_percentage = fraudulent/normal
print(f'Total Number Transactions:
    {total_transactions}')
print(f'Normal Transactions: {normal}')
print(f'Fraudulent Transactions: {fraudulent}')
print(f'Fraudulent Transactions Percent: ' \
        f'{fraud_percentage:.2%}')
```

The number of actual fraudulent transactions is smaller, accounting for only 0.17% of the transactions, as shown in *Figure 8.1*:

```
Total Number Transactions: 275663
Normal Transactions: 275190
Fraudulent Transactions: 473
Fraudulent Transactions Percent: 0.17%
```

Figure 8.1 – Output showing the transaction statistics

4.  Look for any potential null values. Having null values in the data will skew the model and cause other problems:

```
cardData.info()
```

*Figure 8.2* shows that there are no null values in the dataset. If there had been null values, then you would need to clean the data by replacing the null values with a specific value, such as the mean of the other entries in the column. The same thing applies to **missingness**, which is missing data in a dataset and sometimes indicates fraud. You need to replace the missing value with some useful alternative:

```
<class 'pandas.core.frame.DataFrame'>
Int64Index: 275663 entries, 0 to 284806
Data columns (total 30 columns):
 #   Column  Non-Null Count   Dtype
---  ------  --------------   -----
 0   V1      275663 non-null  float64
 1   V2      275663 non-null  float64
 2   V3      275663 non-null  float64
 3   V4      275663 non-null  float64
 4   V5      275663 non-null  float64
 5   V6      275663 non-null  float64
 6   V7      275663 non-null  float64
 7   V8      275663 non-null  float64
 8   V9      275663 non-null  float64
 9   V10     275663 non-null  float64
 10  V11     275663 non-null  float64
 11  V12     275663 non-null  float64
 12  V13     275663 non-null  float64
 13  V14     275663 non-null  float64
 14  V15     275663 non-null  float64
 15  V16     275663 non-null  float64
 16  V17     275663 non-null  float64
 17  V18     275663 non-null  float64
 18  V19     275663 non-null  float64
 19  V20     275663 non-null  float64
 20  V21     275663 non-null  float64
 21  V22     275663 non-null  float64
 22  V23     275663 non-null  float64
 23  V24     275663 non-null  float64
 24  V25     275663 non-null  float64
 25  V26     275663 non-null  float64
 26  V27     275663 non-null  float64
 27  V28     275663 non-null  float64
 28  Amount  275663 non-null  float64
 29  Class   275663 non-null  int64
dtypes: float64(29), int64(1)
memory usage: 65.2 MB
```

Figure 8.2 – Dataset output showing a lack of missing or null values

5.  Determine the range of the Amount column. The Amount column is the only unchanged column from the original dataset:

```
print(f'Minimum Value: {min(cardData.Amount)}')
print(f'Mean Value: ' \
        f'{sum(cardData.Amount)/total_transactions}')
print(f'Maximum Value: {max(cardData.Amount)}')
print(cardData['Amount'])
```

If the `Amount` column has too great a range, as shown in *Figure 8.3*, then the model will become skewed. In this case, the standard practice is to scale the data to obtain better results:

```
Minimum Value: 0.0
Mean Value: 88.34961925087359
Maximum Value: 25691.16
0           149.62
1             2.69
2           378.66
3           123.50
4            69.99
           . . .
284802       0.77
284803      24.79
284804      67.88
284805      10.00
284806     217.00
Name: Amount, Length: 284807, dtype: float64
```

Figure 8.3 – Output showing variance in the Amount column

6.  Remove the `Time` column. For this analysis, time isn't a factor because the model isn't concerned with when credit card fraud occurs; the concern is detecting it when it does occur. The output of this code, `(284807, 30)`, shows that the `Time` column is gone:

```
cardData.drop(['Time'], axis=1, inplace=True)
print(cardData.shape)
```

7.  Scale the dataset to avoid negative values in any of the columns. Some classifiers work well with negative values, while others don't. For example, if you want to use a `DecisionTreeClassifier`, then negative values aren't a problem. On the other hand, if you want to use a `MultinomialNB`, as was done in the *Developing a simple spam filter example* section of *Chapter 4, Considering the Threat Environment*, then you need positive values. Selecting a scaler is also important. This example uses a `MinMaxScaler` to allow the minimum and maximum range of values to be set:

```
scaler = MinMaxScaler(feature_range=(0,1))
col = cardData.columns
cardData = pd.DataFrame(
    scaler.fit_transform(cardData),
    columns = col)
print(cardData)
```

*Figure 8.4* shows the partial results of the scaling operation:

```
              V1        V2        V3        V4        V5        V6        V7  \
0       0.935192  0.766490  0.881365  0.313023  0.763439  0.267669  0.266815
1       0.978542  0.770067  0.840298  0.271796  0.766120  0.262192  0.264875
2       0.935217  0.753118  0.868141  0.268766  0.762329  0.281122  0.270177
3       0.941878  0.765304  0.868484  0.213661  0.765647  0.275559  0.266803
4       0.938617  0.776520  0.864251  0.269796  0.762975  0.263984  0.268968
...          ...       ...       ...       ...       ...       ...       ...
284802  0.756448  0.873531  0.666991  0.160317  0.729603  0.236810  0.235393
284803  0.945845  0.766677  0.872678  0.219189  0.771561  0.273661  0.265504
284804  0.990905  0.764080  0.781102  0.227202  0.783425  0.293496  0.263547
284805  0.954209  0.772856  0.849587  0.282508  0.763172  0.269291  0.261175
284806  0.949232  0.765256  0.849601  0.229488  0.765632  0.256488  0.274963
```

Figure 8.4 – Presentation of data values after scaling

8.  Remove any duplicates. Removing the duplicates ensures that any entries in the dataset that replicate another entry don't remain in place and cause the model to overfit and it won't generalize well. This means that it won't catch as many fraudulent transactions from data it hasn't seen. The output of (275663, 30) shows that there were 9,144 duplicated records:

```
cardData.drop_duplicates(inplace=True)
print(cardData.shape)
```

9.  Determine the final Amount column's characteristics:

```
print(f'Minimum Value: {min(cardData.Amount)}')
print(f'Mean Value: ' \
        f'{sum(cardData.Amount)/total_transactions}')
print(f'Maximum Value: {max(cardData.Amount)}')
print(cardData['Amount'])
```

Now that the dataset has been massaged, it's time to see the result, as shown in *Figure 8.5*:

```
Minimum Value: 0.0
Mean Value: 0.003341246851336739
Maximum Value: 1.0
0          0.005824
1          0.000105
2          0.014739
3          0.004807
4          0.002724
             ...
284802     0.000030
284803     0.000965
284804     0.002642
284805     0.000389
284806     0.008446
Name: Amount, Length: 275663, dtype: float64
```

Figure 8.5 – Output of the Amount column after massaging

A lot of ML comes down to ensuring that the data you use is prepared properly to create a good model. Now that the data has been prepared, you can split it into training and testing sets. This process ensures that you have enough data to train the model, and then test it using data the model hasn't seen before so that you can ascertain the goodness of the model (something you will see later in this process).

## Splitting the data into train and test sets

Splitting the data into training and testing sets makes it possible to train the model on one set of data and test it using another set of data that the model hasn't seen. This approach ensures that you can validate the model concerning its goodness in locating credit card fraud (or anything else for that matter). The following steps show how to split the data in this case:

1.  Import the required packages:

    ```
    from sklearn.model_selection \
        import train_test_split
    ```

2.  Split the dataset into data (X, which is a matrix, so it appears in uppercase) and labels (y, which is a vector, so it appears in lowercase):

    ```
    X = cardData.drop('Class', axis=1).values
    y = cardData['Class'].values
    print(X)
    print(y)
    ```

    When you print the result, you will see that X is indeed a matrix and y is indeed a vector, as shown in *Figure 8.6*:

    ```
    [[9.35192337e-01 7.66490419e-01 8.81364903e-01 ... 4.18976135e-01
      3.12696634e-01 5.82379309e-03]
     [9.78541955e-01 7.70066651e-01 8.40298490e-01 ... 4.16345145e-01
      3.13422663e-01 1.04705276e-04]
     [9.35217023e-01 7.53117667e-01 8.68140819e-01 ... 4.15489266e-01
      3.11911316e-01 1.47389219e-02]
     ...
     [9.90904812e-01 7.64079694e-01 7.81101998e-01 ... 4.16593177e-01
      3.12584864e-01 2.64215395e-03]
     [9.54208999e-01 7.72855742e-01 8.49587129e-01 ... 4.18519535e-01
      3.15245157e-01 3.89238944e-04]
     [9.49231759e-01 7.65256401e-01 8.49601462e-01 ... 4.16466371e-01
      3.13400843e-01 8.44648509e-03]]
    [0. 0. 0. ... 0. 0. 0.]
    ```

Figure 8.6 – Contents of the training and testing datasets

3.  Divide X and y into training and testing sets. There are many schools of thought as to what makes for a good split. What it comes down to is creating a split that provides the least amount of variance during training, but then tests the model fully. In this case, there are plenty of samples, so using the 80:20 split that many people use will work fine:

```
X_train, X_test, y_train, y_test = \
    train_test_split(X, y, test_size=0.2, random_state=1)
print(f"X training data size: {X_train.shape}")
print(f"X testing data size: {X_test.shape}")
print(f"y training data size: {y_train.shape}")
print(f"y testing data size: {y_test.shape}")
```

*Figure 8.7* shows that the data is split according to the 80:20 ratio and that the training and testing variables have the same number of entries:

```
X training data size: (220530, 29)
X testing data size: (55133, 29)
y training data size: (220530,)
y testing data size: (55133,)
```

Figure 8.7 – The output shows that the data is split according to an 80:20 ratio

Now that the data is in the correct form, you can finally build a model. Of course, that means selecting a model and configuring it. For this example, you will use `DecisionTreeClassifier`. However, there are a wealth of other models that could give you an edge when working with various kinds of data.

> **Considering the importance of testing model goodness**
>
> The whole issue of data splitting, selecting the correct model, configuring the model correctly, and so on, comes down to getting a model with very high accuracy. This is especially important when doing things such as looking for malware or detecting fraud. Unfortunately, no method or rule of thumb provides a high level of accuracy in every case. The only real way to tweak your model is to change one item at a time, rebuild the model, and then test it.

## Building the model

As you've seen in other chapters, building the model involves fitting it to the data. The following steps show how to build the model using a minimal number of configuration changes:

1.  Import the required packages:

```
from sklearn.tree import DecisionTreeClassifier
```

2.  Fit the model to the data. Notice that the one configuration change here is to limit the max_depth setting to 5. Otherwise, the classifier will keep churning away until all of the leaves are pure or they meet the min_samples_split configuration setting requirements:

```
dtc = DecisionTreeClassifier(max_depth = 5)
dtc.fit(X_train, y_train)
```

You'll know that the process is complete when you see the output shown in *Figure 8.8*:

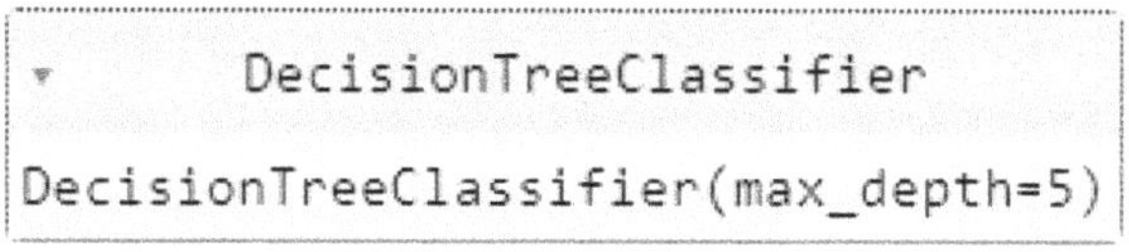

```
▼              DecisionTreeClassifier

DecisionTreeClassifier(max_depth=5)
```

Figure 8.8 – Description of the DecisionTreeClassifier model

You now have a model to use to detect credit card fraud. Of course, you have no idea of how good that model is at its job. Perhaps it's not very generalized and overfitted to the data. Then again, it might be underfitted. The next section shows how to verify the goodness of the model in this case.

## Performing the analysis

Having a model to use means that you can start detecting fraud. Of course, you don't know how well you can detect fraud until you test it using the following steps:

1.  Import the required packages:

```
from sklearn.metrics import accuracy_score
from sklearn.metrics import \
    precision_recall_fscore_support
from sklearn.metrics import confusion_matrix
from sklearn.metrics import plot_confusion_matrix
import matplotlib.pyplot as plt
```

2.  Perform a prediction using the X_test data that wasn't used for training purposes. Use this prediction to verify the accuracy by comparing the actual labels in y_test (showing whether a record is fraudulent or not) to the predictions found in yHat ($\hat{y}$):

```
dtc_yHat = dtc.predict(X_test)
print(f"Accuracy score: ' \
        f'{accuracy_score(y_test, dtc_yHat)}")
```

*Figure 8.9* shows that the accuracy is very high:

```
Accuracy score: 0.9993651714943863
```

Figure 8.9 – Output of the DecisionTreeClassifier model accuracy

3.  Determine the **precision** (ratio of correctly predicted positive observations to the total of the predictive positive observations), **recall** (ratio of correctly predicted positive observations to all observations in the class), **F-beta** (the weighted harmonic mean of precision and recall), and **support** scores (the number of occurrences of each label in y_test). These metrics are important in fraud work because they tell you more about the model given that the dataset has a few positives and a lot of negatives, making the dataset unbalanced. Consequently, accuracy is important, but these measures are also quite helpful for fraud work:

```
precision, recall, fbeta_score, support = \
    precision_recall_fscore_support(y_test, dtc_yHat)
print(f"Precision: {precision}")
print(f"Recall: {recall}")
print(f"F-beta score: {fbeta_score}")
print(f"Support: {support}")
```

4.  *Figure 8.10* shows the output from this step. Ideally, you want a precision as close as possible to 1.0. The output shows that the precision is 0.9994551 for predicting when a transaction isn't fraudulent and a value of 0.93506494 when a transaction is fraudulent. For recall, there are two classes: not fraud and fraud. So, the ratio of correctly predicting fraud against all of the actual fraud cases is only 0.70588235. The F-beta score is about weighting. The example uses a value of 1.0, which means that recall and precision have equal weight in determining the goodness of a model. If you want to give more weight to precision, then you use a number less than 1.0, such as 0.5. Likewise, if you want to give more weight to recall, then you use a number above 1.0, such as 2.0. Finally, the support output simply tells you how many non-fraud and fraud entries there are in the dataset:

```
Precision: [0.9994551  0.93506494]
Recall: [0.99990914 0.70588235]
F-beta score: [0.99968207 0.80446927]
Support: [55031    102]
```

Figure 8.10 – Output of the precision, recall, F-beta, and support statistics

5.  Print a confusion matrix to show how the prediction translates into the number of entries in the dataset:

```
print(confusion_matrix(y_test,
        dtc_yHat,
        labels=[0, 1]))
```

*Figure 8.11* shows the output. The number of true positives appears in the upper-left corner, which has a value of 55025. Only 29 of the records generated a false positive. There were 73 true negatives and six false negatives in the dataset. So, the chances of finding credit card fraud are excellent, but not perfect:

$$[[55025 \qquad 6]$$
$$[ \quad 29 \qquad 73]]$$

Figure 8.11 – The confusion matrix output for the DecisionTreeClassifier model prediction

6.  Create a graphic version of the confusion matrix to make the output easier to understand. Using graphic output does make a difference, especially when discussing the model with someone who isn't a data scientist:

```
matrix = plot_confusion_matrix(dtc,
        X=X_test,
        y_true=y_test,
        cmap=plt.cm.Blues)
plt.title('Confusion Matrix for Fraud Detection')
plt.show(matrix)
plt.show()
```

*Figure 8.12* shows the output in this case:

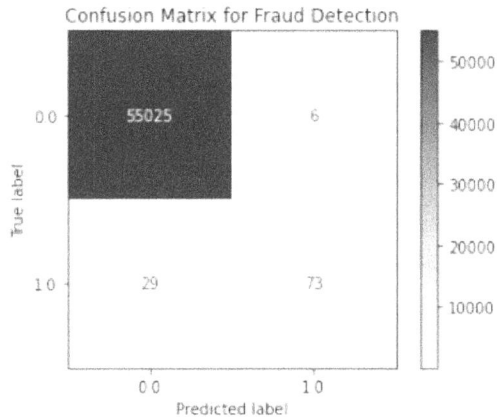

Figure 8.12 – A view of the graphic version of the confusion matrix

This section has shown you one complete model building and testing cycle. However, you don't know that this is the best model to use. Testing other models is important, as described in the next section.

## Checking another model

The decision tree classifier does an adequate job of separating fraudulent credit purchases from those that aren't, but it could do better. A random forest classifier is a group of decision tree classifiers. In other words, you put multiple algorithms to work on the same problem. When the classification process is complete, the trees vote and the classification with the most votes wins.

In the previous example, you used the max_depth argument to determine how far the tree should go to reach a classification. Now that you have a whole forest, rather than an individual tree, at your disposal, you also need to define the n_estimators argument to define how many trees to use. There are a lot of other arguments that you can use to tune your model in this case, as described in the documentation at https://scikit-learn.org/stable/modules/generated/sklearn.ensemble.RandomForestClassifier.html. The following steps help you create a random forest classifier model so that you can compare it to the decision tree classifier used earlier:

1.  Import the required packages:

```
from sklearn.ensemble \
    import RandomForestClassifier
```

2. Fit the model to the data and display its accuracy. Note that this step will take a while to complete because you're using a whole bunch of trees to complete it. If your system is taking too long, you can always reduce either or both of the max_depth and n_estimators arguments. Changing either of them will affect the accuracy of your model. Of the two, max_depth is the most important for this particular example, but in the real world, you'd need to experiment to find the right combination. The purpose of the n_jobs=-1 argument is to reduce the amount of time to build the model by using all of the processors on the system:

```
rfc = RandomForestClassifier(max_depth=9,
    n_estimators=100,
    n_jobs=-1)
rfc.fit(X_train, y_train)
rfc_yHat = rfc.predict(X_test)
print(f"Accuracy score: " \
        f"{accuracy_score(y_test, rfc_yHat)}")
```

*Figure 8.13* shows that even though it took longer to build this model (using all of the processors no less), it performs only slightly better than the decision tree classifier. In this case, the contributing factors are the small dataset and the fact that the number of fraud cases is small. However, even a small difference is better than no difference at all when it comes to fraud and you need to consider that a real-world scenario will be dealing with far more entries in the dataset:

```
Accuracy score: 0.9994014474089928
```

Figure 8.13 – The accuracy of the random forest classifier

3. Calculate the precision, recall, fbeta_score, and support values for the model:

```
precision, recall, fbeta_score, support = \
    precision_recall_fscore_support(y_test, rfc_yHat)
print(f"Precision: {precision}")
print(f"Recall: {recall}")
print(f"F-beta score: {fbeta_score}")
print(f"Support: {support}")
```

As with accuracy, the differences (as shown in *Figure 8.14*) between the two models are very small, but important when dealing with fraud. After all, you don't want to claim a customer has committed fraud unless it's true:

```
Precision: [0.99949141 0.93670886]
Recall: [0.99990914 0.7254902 ]
F-beta score: [0.99970023 0.81767956]
Support: [55031    102]
```

Figure 8.14 – The precision, recall, F-beta, and support scores for the random forest classifier

4.  Plot the confusion matrix for this model:

```
matrix = plot_confusion_matrix(rfc,
    X=X_test,
    y_true=y_test,
    cmap=plt.cm.Blues)
plt.title('Confusion Matrix for RFC Fraud Detection')
plt.show(matrix)
plt.show()
```

*Figure 8.15* shows what it all comes down to in the end. The random forest classifier has the same prediction rate for transactions that aren't fraudulent as the decision tree classifier in this case. However, it also finds one more case of fraud, which is important:

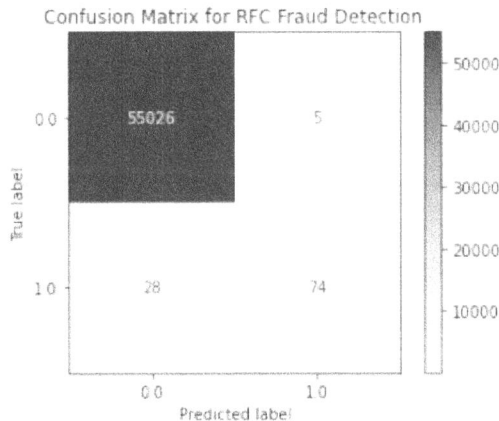

Figure 8.15 – The confusion matrix for the random forest classifier

This example has looked at credit card fraud, but the same techniques work on other sorts of fraud as well. The main things to consider when looking for fraud are to obtain a large enough amount of data, train the model using your best assumptions about the data, and then test the model for accuracy. Tweak the model as needed to obtain the required level of accuracy. Of course, no model is going to be completely accurate. While ML will greatly reduce the burden on human detectives looking for fraud, it can't eliminate the need for a human to look at the data entirely.

## Creating a ROC curve and calculating AUC

A **Receiver Operating Characteristic (ROC)** curve and **Area Under the Curve (AUC)** calculation help you determine where to set thresholds in your ML model. These are methods of looking at the performance of a model at all classification thresholds. The $X$-axis shows the false positive rate, while the $Y$-axis shows the true positive rate. As the true positive rate increases, so does the false positive rate. The goal is to determine where to place the threshold for determining whether a particular sample is fraudulent or not based on its score during analysis. A score indicates the model's confidence as to whether a particular sample is fraud or not, but the model doesn't determine where to place the line between fraud and legitimate; that line is the **threshold**. Therefore, a ROC curve helps a human user of a model determine where to set the threshold, and where to say that it's best to detect fraud.

The **True Positive Rate (TPR)** defines the ratio between true positives (TP) (as shown in *Figure 8.11*, *Figure 8.12*, and *Figure 8.15*) and false negatives (FN): TPR = TP / TP + FN. The **False Positive Rate (FPR)** defines the ratio between the false positives (FP) and the true negatives (TN): FPR = FP / FP + TN. Essentially, what you're trying to determine is how many true positives are acceptable for a given number of false positives when working with a model.

As part of plotting a ROC curve, you also calculate the AUC, which is essentially another good measure for the model. It's a measure of overall performance against all classification thresholds; the higher the number, the better the model. The AUC is a probability measure that tells you how likely it is that the model will rank a random positive example higher than a random negative example. Consequently, the MultinomialNB classifier used in an earlier example would have an AUC of 0, which means it never produces a correct detection. With all of these things in mind, use the following steps to create two ROC curves comparing DecisionTreeClassifier to the RandomForestClassifier classifier used earlier:

1.  Import the required packages:

    ```
    from sklearn.metrics import roc_curve
    from sklearn.metrics import auc
    from numpy import argmax
    from numpy import sqrt
    ```

2.  Perform the required DecisionTreeClassifier analysis. Notice that this is a three-step process of calculating the probabilities for each value in X_test, determining the FPR and TPR values, and then estimating the AUC value. dtc_thresholds will be used in a later location to determine the best-calculated place to put the threshold for this model:

    ```
    dtc_y_scores = dtc.predict_proba(X_test)
    dtc_fpr, dtc_tpr, dtc_thresholds = \
        roc_curve(y_test, dtc_y_scores[:, 1])
    dtc_roc_auc = auc(dtc_fpr, dtc_tpr)
    ```

3.  Perform the required `RandomForestClassifier` analysis. Notice that this is the same process we followed previously:

```
rfc_y_scores = rfc.predict_proba(X_test)
rfc_fpr, rfc_tpr, rfc_thresholds = \
    roc_curve(y_test, rfc_y_scores[:, 1])
rfc_roc_auc = auc(rfc_fpr, rfc_tpr)
```

4.  Calculate the best threshold for each of the models and display the **generic mean** (**G-mean**) for each one. The best-calculated threshold represents a balance between sensitivity, which is the TPR, and specificity, which is 1 - FPR. To calculate this value, the code must perform the analysis for each threshold; then, a call to `argmax()` will report the best option based on the calculation:

```
dtc_gmeans = sqrt(dtc_tpr * (1-dtc_fpr))
dtc_ix = argmax(dtc_gmeans)
print('Best DTC Threshold=%f, G-Mean=%.3f'
        % (dtc_thresholds[dtc_ix],
            dtc_gmeans[dtc_ix]))

rfc_gmeans = sqrt(rfc_tpr * (1-rfc_fpr))
rfc_ix = argmax(rfc_gmeans)
print('Best RFC Threshold=%f, G-Mean=%.3f'
        % (rfc_thresholds[rfc_ix],
            rfc_gmeans[rfc_ix]))
```

*Figure 8.16* shows the result of the calculation for each model:

```
Best DTC Threshold=0.062500, G-Mean=0.885
Best RFC Threshold=0.010000, G-Mean=0.939
```

Figure 8.16 – Calculated best threshold and G-mean values for each model

5.  Create the ROC curve plot and display the AUC values:

```
plt.title('Receiver Operating Characteristic')
plt.plot(dtc_fpr, dtc_tpr, 'g',
        label = 'DTC AUC = %0.2f' % dtc_roc_auc)
plt.plot(rfc_fpr, rfc_tpr, 'b',
        label = 'RFC AUC = %0.2f' % rfc_roc_auc)
plt.plot([0, 1], [0, 1],'r--',
```

```
            label = 'No Skill')
    plt.scatter(dtc_fpr[dtc_ix], dtc_tpr[dtc_ix],
            marker='o', color='g',
            label='DTC Best')
    plt.scatter(rfc_fpr[rfc_ix], rfc_tpr[rfc_ix],
            marker='o', color='b',
            label='RFC Best')
    plt.legend(loc = 'lower right')
    plt.xlim([0, 1])
    plt.ylim([0, 1])
    plt.ylabel('True Positive Rate')
    plt.xlabel('False Positive Rate')
    plt.title('ROC Curve Comparison DTC to RFC')
    plt.show()
```

The output of this example appears in *Figure 8.17*. Notice that the RFC model outperforms the DTC in this particular case by a small margin. In addition, the plot shows where you'd place the threshold for each model. Given the data and other characteristics of this example, once the model has achieved a maximum value, there is little advantage in increasing the threshold further:

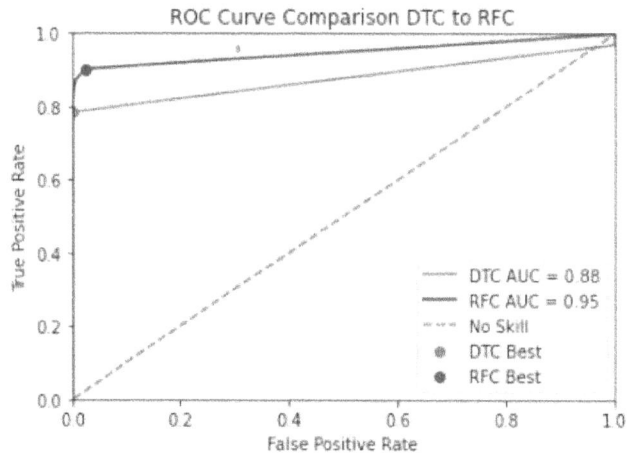

Figure 8.17 – The ROC curve and AUC calculation for each model

Of course, you won't likely always see this result. The main takeaway from this example is that you need to compare models and settings to determine how best to configure your ML application to detect as much fraud as possible without creating an overabundance of false positives.

# Summary

This chapter introduced you to the topic of fraud as it applies to ML. The key takeaway from this chapter is that fraud involves deception for some type of gain. Often, this deception is completely hidden and subtle; sometimes, the gain is even hard to decipher unless you know how the gain is used. Fraud affects ML security by introducing flawed data into the dataset, which produces unreliable or unpredictable results that are skewed to the perpetrator's goals. In addition, because the data is unreliable, it also presents a security risk.

When reviewing the security needs of an organization, it's important to consider both background and real-time fraud. Depending on your organization, one form of fraud or the other may take precedence. For example, a marketing company with no direct consumer interaction would need to consider background fraud more strongly. Likewise, an online seller would need to consider real-time fraud more strongly. Tailoring the type of fraud detection used is incredibly important to make detection both precise and efficient.

*Chapter 9*, *Defending Against Hackers*, moves on to defending against direct hacker attacks. Previous chapters have considered what might be termed hacker agents, such as network interference, the introduction of anomalies, the reliance on malware, and now the use of fraud in this chapter. In *Chapter 9*, the hacker will become actively engaged in what could be termed either sabotage or espionage. It's the sort of attack seen in movies, but not in the way that movies depict them. You may be amazed at just how subtle a hacker's machinations against your organization can become.

# Further reading

The following bullets provide you with some additional reading that you may find useful in understanding the materials in this chapter:

- Discover how deep learning is already being used to extract information from ID cards: *ID Card Digitization and Information Extraction using Deep Learning – A Review*: `https://nanonets.com/blog/id-card-digitization-deep-learning/`

- Understand how organizations suffer phishing attacks: *How to Prevent Phishing on an Organizational Level*: `https://www.tsp.me/blog/cyber-security/how-to-prevent-phishing-on-an-organizational-level/`

- Uncover Ponzi schemes using ML techniques: *Data-driven Smart Ponzi Scheme Detection* at `https://arxiv.org/pdf/2108.09305.pdf` and *Evaluating Machine-Learning Techniques for Detecting Smart Ponzi Schemes* at `https://ieeexplore.ieee.org/document/9474794`

- Read about the man who did sell the Brooklyn Bridge and other monuments: *Meet the Conman Who Sold the Brooklyn Bridge — Many Times Over*: `https://history.howstuffworks.com/historical-figures/conman-sold-brooklyn-bridge.htm`

- Learn about the no free lunch theorem as it applies to ML: *No Free Lunch Theorem for Machine Learning*: `https://machinelearningmastery.com/no-free-lunch-theorem-for-machine-learning/`

- Learn more about the historical basis of Niccolo Machiavelli: *Italian philosopher and writer Niccolo Machiavelli born* at `https://www.history.com/this-day-in-history/niccolo-machiavelli-born` and *Machiavelli* at `https://www.history.com/topics/renaissance/machiavelli`

- Gain an understanding of how AI is used to detect government fraud: *Using AI and machine learning to reduce government fraud*: `https://www.brookings.edu/research/using-ai-and-machine-learning-to-reduce-government-fraud/`

- See how accuracy, precision, and recall relate: *Accuracy, Precision, Recall & F1 Score: Interpretation of Performance Measures*: `https://blog.exsilio.com/all/accuracy-precision-recall-f1-score-interpretation-of-performance-measures/`

- Gain a better understanding of the F-beta measure: *A Gentle Introduction to the Fbeta-Measure for Machine Learning*: `https://machinelearningmastery.com/fbeta-measure-for-machine-learning/`

- Consider how long-term fraud detection and protection improve business productivity and profit: *How long-term digital fraud protection can increase profitability*: `https://kount.com/blog/long-term-fraud-protection-benefits/`

- Pick up techniques to help determine when someone is lying: *Become a Human Lie Detector: How to Sniff Out a Liar*: `https://www.artofmanliness.com/character/behavior/become-a-human-lie-detector-how-to-sniff-out-a-liar/`

- Discover credit card statistics that demonstrate why fraud detection is so important: *42 credit card fraud statistics in 2022 + steps for reporting fraud*: `https://www.creditrepair.com/blog/security/credit-card-fraud-statistics/`

# 9

# Defending against Hackers

Previous chapters have addressed a wide variety of threats and an even wider variety of threat sources. Most people would be having full-on bouts of paranoia about now! The fact is that these threats and threat sources are real, but most of them *become* a threat source as a secondary matter. For example, an employee doesn't normally join your organization with the thought of stealing as much as possible from you and then running away. The very few employees that actually do steal something and then run away do it later after they have been in your business for a while. Hackers, on the other hand, start out with the idea of stealing, damaging, or monitoring something in your business. They've never had any other idea in mind. The orientation and priority of the attack are why defending against hackers is different from defending against other threats and why the separate treatment in this chapter is important.

Of course, hackers are not superhuman, nor are they necessarily evil all of the time. Despite what you may read in the media, they are simply humans with the same flaws and desires as everyone else. You can thwart hacker attacks using a variety of methods, including behavioral approaches that **Machine Learning** (**ML**) methods can help you create. Once you become aware that an attack is about to happen or is happening right now, simply trying to shut the hacker down may not prove very effective. Some hackers will take your attempts to shut them down as a challenge, and they'll come up with ever better methods of overcoming your defenses. So, part of this chapter deals with strategies that will thwart the hacker attack through a variety of means, not all of them necessarily meant to build your defensive walls higher. With these issues in mind, this chapter discusses the following topics:

- Considering hacker targets
- Defining hacker goals
- Monitoring and alerting
- Improving security and reliability

# Technical requirements

This chapter requires that you have access to either Google Colab or Jupyter Notebook to work with the example code. The *Requirements to use this book* section of *Chapter 1, Defining Machine Learning Security*, provides additional details on how to set up and configure your programming environment. When testing the code, use a test site, test data, and test APIs to avoid damaging production setups and to improve the reliability of the testing process. Testing on a non-production network is highly recommended but not absolutely necessary. Using the downloadable source is always highly recommended. You can find the downloadable source on the Packt GitHub site at `https://github.com/PacktPublishing/Machine-Learning-Security-Principles` or my website at `http://www.johnmuellerbooks.com/source-code/`.

# Considering hacker targets

One of the biggest issues to consider when thinking about hackers is that hackers don't just attack your organization without reason. There is always a target and a motivation to attack a target. Otherwise, the hacker, just like anyone else, isn't going to bother. The targets that a hacker chooses are the primary means of interaction with your organization rather than a secondary consideration. The hacker needs to see that the target can be easily accessed. In some cases, the attack almost becomes a game for the hacker, a kind of puzzle to solve. Yes, there are monetary or other reasons for the attack, but if you can get money out of any of a variety of targets, then the target itself must become part of the motivation for this particular attack. The following sections discuss some hacker targets that help demonstrate an attack pattern and a reason for choosing that particular target and not some other target in an organization.

## Hosted systems

Hosted systems represent an interesting target for a hacker because if the hacker can break into the underlying hosted system, then it becomes possible to attack a number of customers at the same time rather than just one customer. So, the potential benefits of such an attack are high, but they also tend to be random in nature. However, because hosted systems have the resources to employ better and more modern security techniques, the hurdles the hacker must jump are also higher. Consequently, you often see hosted systems attacked by groups of more competent hackers rather than complete novices. An experienced hacker has already had the thrill of breaking into a difficult system, so the reward is often monetary based (even if the hacker has been hired by a competitor, the hacker is still being paid, so the hacker's reward is still monetary in nature).

## Networks

Networks often provide specific access to a specific kind of customer. Instead of hunting for a high-value reward of a random nature, the hacker has some specific reward in mind. Consequently, it often pays to ask what the network has to offer that the hacker specifically wants. This differs from other targets when the benefits are not so easily discerned. Because the hacker has something specific in mind, it's possible to put more effort into a smaller target because the benefit is known at the outset. In addition, a network setup is less likely to have advanced security and the resources to overcome the hacker attack, so the hacker has a higher probability of achieving a particular goal. Hackers of all stripes attack networks, so you can't be sure about the hacker's skill, knowledge, and experience level until you start modeling the hacker's behavior. Hackers with less skill, knowledge, and experience are more likely to fall into traps you set and to be dissuaded from continuing an attack.

## Mobile devices

Mobile devices represent groups of related individuals because they aren't actually separate: people use them to communicate with each other but without the protections that your network provides for desktop systems. In addition, people lend devices to each other, so you can't really be sure whether the malware infection came from that person's usual list of mobile devices. Public access also means that others could be listening in, adding to a group-like environment.

The hacker is often looking for a very quick return with little effort to obtain a resource such as credit card numbers. The hacker may simply want to turn the machine into a zombie or use its resources to perform some other attack. The point is that this kind of attack is against a small target that likely doesn't have the largest number of security resources. In many cases, the attack is by an individual or a small group of hackers.

## Customers

A business's customers are its second most important resource after data. Hackers are interested in customers, too. Getting a list of customers and the legitimate company representatives who interact with them is pure gold to a hacker organization because it can then perform social engineering and other attacks to achieve all sorts of goals. Normally, these hackers aren't just computer smart; they're people smart as well, which can make them particularly difficult for a security professional to deal with.

## Public venues and social media

Hackers who are really people smart don't need to attack anything. They can often get into a public venue, especially a social media site, and discover everything they need to know to perform an attack successfully at a later time. People often leave too many clues as to their social and work life in places that hackers can easily access and mine for data. When the attack comes, it's usually successful because the hacker is well prepared and knows precisely what to expect from the individual and usually knows precisely what benefit will be provided as well. From an organizational perspective, public venues and social media are extremely dangerous because they allow the hacker nearly infinite preparation time while the security professional has nearly no time to prepare. Because of the amount of data involved, this kind of hacking is usually done by well-organized hacking groups.

Not all hacker targets will affect you or your organization, so it pays to focus on those targets that affect you most. The next section of the chapter takes this topic a bit further by exploring hacker goals as they relate to your organization and its use of ML techniques. When you consider the goals that a hacker has when attacking your organization, you begin to see *why* the hacker is doing it and may be able to come up with solutions that are outside the usual defensive box. In some circles, this sort of defense is known as **thwarting hackers with behavioral science**, but first, you must understand the hacker to make it work.

## Defining hacker goals

Previous chapters have discussed various kinds of hacker goals in specific scenarios. *Chapter 4,* in particular, pays attention to the hacker as part of the threat environment. So, you already know that hackers have goals such as stealing money or credentials, causing mayhem seemingly for the sheer joy of doing so, and working for others to perform espionage, sabotage, or forms of political maneuvering. These are all effects of the hacker's personal goals; the outcome after a hacker chooses a target and an attack vector. However, they don't really look at why the hacker would perform an attack in the first place. Obviously, hackers are relatively smart, so they could earn a legitimate living doing something else, so why be a hacker? The sections that follow look at hacker goals from a more behavioral stance than previous chapters have done and assess the ability of ML to help you ascertain those behavioral goals.

---

### Is the hacker smart, or are people unobservant?

By reading online, you can find psychological profiles about hackers that say many of them aren't really all that intelligent. One article, *Hackers aren't smart – people are stupid* (`https://blog.erratasec.com/2016/02/hackers-arent-smart-people-are-stupid.html`) simplifies the issue a little too much and wants to blame the victim. These articles are correct in pointing out that people are sometimes unobservant (to use a better term than many of the articles do). However, some exploits can't be pulled off by a teenager waiting for someone to make a mistake. For example, even if security professionals don't want to admit it, social engineering attacks require people skills and people smarts. Reading *7 Ways Smart People Get Hacked by Social Engineering* (`https://www.delcor.com/resources/blog/7-ways-smart-people-get-hacked-by-social-engineering`) provides some insight into this. The article, *Optus data breach: Is your business prepared for a cyberattack?* (`https://retrac.com.au/optus-data-breach-is-your-business-prepared-for-a-cyberattack/`) takes a more moderate approach and lists common ways that businesses get hacked today. One of the more important points in this article is that the author states that hackers can buy off-the-shelf software to hack your business, but someone has to create that software, someone smart enough to do it. The most important question you can ask as you review articles online is, "*Where do these magic tools come from?*"

## Data stealing

Everyone likes secrets. If you have a secret, then you feel important, and you know that you have something valuable to trade for something else that you want. **Data stealing** is all about garnering secrets that are valuable in some way. They represent several things to the hacker in addition to feeling important and obtaining monetary gain. The idea of deciphering a puzzle that no one else has solved has a certain appeal to it. In addition, there is the concept of gaining an advantage over an entity, of having power over it. So, the hacker's goal is more than just making some money. It involves gaining intangible benefits as well. The article *Why do hackers want your personal information?* (`https://www.f-secure.com/us-en/home/articles/why-do-hackers-want-your-personal-information`) talks about personal information mostly, but it also points out the use of this information for things such as extortion or gaining illegal access to organizational resources. Modeling behavior can make these intangible benefits clearer and make thwarting the hacker easier by removing or diminishing these potential gains.

## Data modification

The act of **modifying** data can allow a hacker to subtly modify the behavior of an ML model or induce mistruths into a dataset with all sorts of potential outcomes. Reading *How hackers use AI and machine learning to target enterprises* (`https://www.techtarget.com/searchsecurity/tip/How-hackers-use-AI-and-machine-learning-to-target-enterprises`) tells you that there is actually quite a bit more than data modification at stake, but data modification is a primary means of achieving goals. Often these outcomes are monetary in nature. For example, a hacker might break into the stock market computers and feed them false data showing that a particular company's stock is suddenly worthless. The hacker's organization could then buy the stock for pennies and make a fortune once the real data is restored. Of course, this is a pretty wild sort of attack and is unlikely to happen, but it's actually possible (see `https://www.investopedia.com/news/are-your-stocks-danger-getting-hacked/` for some ideas on how this can happen). This kind of attack is usually more subtle, and, again, the idea is to gain power over some entity without the entity knowing about it. The hacker manipulates the strings behind the curtain where no one can see them. In fact, the hacker may actually think the attack is funny because the target ends up being the hacker's puppet in a real sense. Understanding the psychology of the hacker's attack may make non-security counterattacks possible, such as denying the hacker any satisfaction from performing the attack (such as one-upmanship, personal gratification, and so on) in the first place, except for the tangible elements that the hacker does get away with.

> ### Undermining trust and confidence
>
> Many hacker goals involve undermining trust and confidence in data, an organization, an individual, a group, or in just about anything else. Society operates on the dual requirements of trust and confidence. Once an entity has lost either or both of them, the entity becomes ineffective or possibly ceases to exist. The undermining process often involves half-truths, misdirected truths, and damaged truths rather than outright mistruths, which are all too easily revealed. The motivations may involve revenge but could also reflect the hacker's own sense of humor. The hacker may be making a political statement or righting a perceived wrong. To understand the potential for trust and confidence, a security professional must often go outside the organization to look for events that have fostered the particular behavior in question. There is a possibility that the reason actually lies outside any activity that your business performs.

## Algorithm modification

**Algorithm modification** refers to modifying the code used to create a model so as to produce an output that isn't what was expected. This allows the hacker to damage data, obtain a monetary benefit, exact revenge, or any number of other tangible goals. The article *How to Hack AI?* (`https://broutonlab.com/blog/how-to-hack-ai-machine-learning-vulnerabilities`) discusses some of the techniques hackers use to modify AI, which essentially means modifying or tricking the algorithms in some way. However, the behavioral goal, in this case, is often one of one-upmanship. The hacker feels smarter than the people who developed the model. In addition, the hacker may want to induce fear of the unexpected, which comes out as being just another kind of power play. Obviously, building a higher security wall is part of the solution in this case, but techniques such as downplaying the amount of knowledge needed to modify the algorithm are also helpful. Anything that denies the hacker the secondary benefits of such an attack will reduce the hacker's desire to perform the attack in the first place.

## System damage

System damage often occurs to obtain some type of monetary compensation in order to provide a fix. Someone who attacks an organization with ransomware has a tangible goal of a nice payday. However, there are other, possibly easier ways to get a payday. It often comes down to a need to damage all sorts of things, not just computers. Perhaps the hacker is simply angry at the world and not even angry at the owner of the system that has been damaged. The attack becomes a means of acting out frustration or some other source of anger, in addition to the payment it provides. It's not the security professional's job to fix the hacker's anger management problems, but determining that there is an anger management problem will help provide an alternative to some types of damaging attacks. Reading *Individuals may legitimize hacking when angry with system or authority* at `https://www.sciencedaily.com/releases/2020/10/201022125522.htm` can provide insights into this issue.

No matter what reason a hacker has for attacking your organization or what strategy you choose to thwart the attack, you still have to know that there is an actual attack. The next section focuses on the use of various monitoring and alerting techniques to predict attacks (within reason) before they occur, reduce the time that a hacker has to attack you, diminish the effectiveness of the attack, and possibly prevent it from happening at all. So, monitoring is your first line of defense, and alerting is an indicator from this line of defense that you need to do something quickly.

# Monitoring and alerting

The *Using supervised learning example* section of *Chapter 5, Keeping Your Network Clean,* shows one method for monitoring your network for unusual patterns. In this case, you monitor API calls that are coming into your network from an outside source. Previous chapters have also provided you with examples of email filtering, anomaly detection, malware detection, and fraud detection. All of these kinds of detection are helpful, but monitoring and alerting for hacker attacks, in general, is harder. The point of the sections that follow is to show that you can create a combination of detection methods to ascertain the health of your organization in general so that it becomes possible to create an alert when there is a high probability that a hacker attack is about to begin.

## Considering the importance of lag

Humans don't act instantly. Even when humans are actively engaged in something, there is a reaction time to consider. For example, try the interesting and somewhat fun driving test at `https://www.justpark.com/creative/reaction-time-test/`. You'll find that even though you're looking at the display for the required prompt, there is a lag. This lag in reaction time can be important for security issues. By tracking suspected hacker activities, you can often guess the hacker's next move through the hacker's behavior and the associated reaction time. That's part of the point of the example in the section that follows. Lag time is important to security professionals, yet the concepts of tracking behavior and anticipating the next step often go unnoticed.

The previous sections of the chapter discuss hacker targets and hacker goals. These sections tell you where to look and what to look for regarding behavior. The specific behavior that you decide to track is based on your business, security focus, and what the hacker is currently doing. In every case, when you monitor the hacker's activity and determine their behavior based on that activity, you see lags in various ways. Here are some lags to consider:

- The hacker's reaction time to new security features or changes
- A hacker's response time to current events
- The reaction time to any perceived notice that you give the hacker
- Changes in the hacker's activity based on incursions into your organization
- Modifications of behavior based on insider information; this reaction time can be especially important when you're trying to track the insider as well

- The reaction time to any changes in organizational policy

- Notice of any changes to the organization's makeup

The thing to notice about this list is that *change* leads to lag. Any change will incur a lag in the hacker's behavior, and you can use ML to find and track that lag with a reasonable level of accuracy. More importantly, the lag and the hacker's reaction to the change tell you a lot of things about the hacker. The driving test at the beginning of this section can more or less predict your age based solely on your reaction time to clicking a button. The test knows nothing about you, yet it's able to make this prediction. You have a lot more information about hackers who are trying to get into your organization, so it's time to put behaviors and lags to work in discovering more about hackers and how to modify their behavior.

## An example of detecting behavior

**Behavior detection** is relatively new in the world of ML. However, the techniques for making a prediction are based on well-established models. The example in this section anticipates human behavior based on two datasets from two completely different sources. It models a real-world scenario in that you often have to get creative and combine data from multiple sources to catch a hacker through behavioral prediction. Just knowing API call patterns, as discussed in *Chapter 5,* wouldn't be enough in the real world, you'd need to look into other patterns. Just what you need to track depends on the behavior you want to model.

This example builds a behavior detection scenario by combining data about COVID admissions into hospitals, new cases, existing hospital cases (for ongoing treatment), and its effect on retail sales and recreational activities. The example asks the question of whether it's possible to create a correlation between these two disparate sources of data and then use that correlation to correctly predict future behavior. More importantly, it also shows you that there is indeed a behavioral lag between the COVID events and the COVID effect. The following sections help you build and try this somewhat complex example. You can also find the source code for this example in the `MLSec; 09; Human Behavior Prediction.ipynb` file in the downloadable source.

### *Obtaining the data*

The two datasets come from completely different sources, which is actually a good thing because you know that there is no interaction between the two. Often, when you work through a security matter of this kind, there won't be any interaction between your data sources. The first dataset comes from the GOV.UK Coronavirus (COVID-19) UK site at `https://coronavirus.data.gov.uk/details/download`. The site allows you to set the parameters of the data you download. The example uses the following parameters:

- Data release date (must be set first): Archive: 2021-12-31

- Metrics:

- hospitalCases

- newAdmissions

- newCasesByPublishDate

- Data format: CSV

Here is a link you can use to obtain the dataset with all of the questions already answered: `https://api.coronavirus.data.gov.uk/v2/data?areaType=overview&metric=hospitalCases&metric=newAdmissions&metric=newCasesByPublishDate&format=csv`. The data used is a little older because of the second dataset that the example uses. Otherwise, you can play around with various dates and up to five metrics.

The second dataset comes from the Google COVID-19 Community Mobility Reports site at `https://www.google.com/covid19/mobility/`. Click the **Region CSVs** button to download the mobility reports for all of the regions that Google tracks. You can also use this direct link: `https://www.gstatic.com/covid19/mobility/Region_Mobility_Report_CSVs.zip`. This series of `.csv` files should appear in the `Region_Mobility_Report_CSVs` subdirectory of your code directly.

### Importing and combining the datasets

This section helps you create a combined dataset from the two datasets downloaded in the previous section. As with most real-world datasets, you need to reshape the data to make the two datasets fit together in a manner that will actually work for your analysis. There is also the issue of missing data to consider and whether the data appears in the right format. Only when the data is completely remediated can you combine the two datasets together to create a single dataset to use for developing a model. The following steps show how to perform the various tasks:

1. Import the required packages:

   ```
   import pandas as pd
   ```

2. Read the Google regional reports into a dataset:

   ```
   mobility_df = pd.read_csv(
       "Region_Mobility_Report_CSVs/" \
       "2020_GB_Region_Mobility_Report.csv")
   print(mobility_df)
   ```

*Figure 9.1* shows three major issues with this dataset. The first is that the dates begin with **2020-02-15**, so you don't have a full year to work with from 2020. The second is that the date is in text format, and you need an actual date to work with. The third is that the actual data columns that the example will use are really long. It's hard to imagine typing `retail_and_recreation_percent_change_from_baseline` every time you need the retail and recreation change data:

```
                          place_id        date  \
0         ChIJqZHHQhE7WgIReiWIMkOg-MQ  2020-02-15
1         ChIJqZHHQhE7WgIReiWIMkOg-MQ  2020-02-16
2         ChIJqZHHQhE7WgIReiWIMkOg-MQ  2020-02-17
3         ChIJqZHHQhE7WgIReiWIMkOg-MQ  2020-02-18
4         ChIJqZHHQhE7WgIReiWIMkOg-MQ  2020-02-19
...                               ...         ...
133377    ChIJh-IigLwxeUgRAKFv7Z75DAM  2020-12-27
133378    ChIJh-IigLwxeUgRAKFv7Z75DAM  2020-12-28
133379    ChIJh-IigLwxeUgRAKFv7Z75DAM  2020-12-29
133380    ChIJh-IigLwxeUgRAKFv7Z75DAM  2020-12-30
133381    ChIJh-IigLwxeUgRAKFv7Z75DAM  2020-12-31

          retail_and_recreation_percent_change_from_baseline  \
0                                                      -12.0
1                                                       -7.0
2                                                       10.0
3                                                        7.0
4                                                        6.0
```

Figure 9.1 – The Google dataset has some format issues to deal with

3.  Get rid of the data columns that you don't actually need from the dataset. Modify the date column to make it usable by the model. In addition, rename the columns so that they're easier to use (or at least easier to type). Finally, just use the first 321 rows so that the amount of data matches that found in the UK dataset:

```
Mobility_df.drop(
    ['country_region_code', 'country_region',
        'sub_region_1', 'sub_region_2', 'metro_area',
        'iso_3166_2_code', 'census_fips_code',
        'place_id'], axis='columns', inplace=True)
mobility_df['date'] = \
    pd.to_datetime(mobility_df['date'])
mobility_df.rename(columns={
'retail_and_recreation_percent_change_from_baseline':
    'retail_and_recreation'}, inplace=True)
mobility_df.rename(columns={
'grocery_and_pharmacy_percent_change_from_baseline':
```

```
            'grocery_and_pharmacy'}, inplace=True)
mobility_df.rename(columns={
    'parks_percent_change_from_baseline':
        'parks'}, inplace=True)
mobility_df.rename(columns={
    'transit_stations_percent_change_from_baseline':
        'transit'}, inplace=True)
mobility_df.rename(columns={
    'workplaces_percent_change_from_baseline':
        'workplaces'}, inplace=True)
mobility_df.rename(columns={
    'residential_percent_change_from_baseline':
        'residential'}, inplace=True)
mobility_df = mobility_df.head(321)
print(mobility_df)
```

As shown in *Figure 9.2*, the dataset is now a lot easier to use:

```
          date  retail_and_recreation  grocery_and_pharmacy  parks  transit \
0    2020-02-15                  -12.0                  -7.0  -35.0    -12.0
1    2020-02-16                   -7.0                  -6.0  -28.0     -7.0
2    2020-02-17                   10.0                   1.0   24.0     -2.0
3    2020-02-18                    7.0                  -1.0   20.0     -3.0
4    2020-02-19                    6.0                  -2.0    8.0     -4.0
..          ...                    ...                   ...    ...      ...
316  2020-12-27                  -58.0                 -24.0    3.0    -59.0
317  2020-12-28                  -51.0                 -24.0   16.0    -70.0
318  2020-12-29                  -47.0                 -11.0   14.0    -64.0
319  2020-12-30                  -42.0                  -3.0   18.0    -63.0
320  2020-12-31                  -53.0                   2.0   14.0    -67.0

     workplaces  residential
0          -4.0          2.0
1          -3.0          1.0
2         -14.0          2.0
3         -14.0          2.0
4         -14.0          3.0
..          ...          ...
316       -29.0         10.0
317       -77.0         26.0
318       -68.0         24.0
319       -66.0         23.0
320       -69.0         24.0

[321 rows x 7 columns]
```

Figure 9.2 – The Google dataset shows just the information needed to create the model

4.  Read the UK COVID data into a dataset:

```
Covid_df = pd.read_csv('overview_2022-09-22.csv')
print(covid_df)
```

*Figure 9.3* shows that the UK dataset also presents problems for the example. First, the dates aren't in the same order as the Google dataset. In addition, the dates are textual, rather than actual dates. Some of the columns are also missing data. You'll also notice that the current dataset contains 959 rows, but that's because it includes multiple years of data, whereas the Google dataset is exclusive to 2020:

```
       areaCode        areaName  areaType        date  hospitalCases  \
0     K02000001  United Kingdom  overview  2022-09-15         5729.0
1     K02000001  United Kingdom  overview  2022-09-14         5705.0
2     K02000001  United Kingdom  overview  2022-09-13         5792.0
3     K02000001  United Kingdom  overview  2022-09-12         5849.0
4     K02000001  United Kingdom  overview  2022-09-11         5911.0
..          ...             ...       ...         ...            ...
954   K02000001  United Kingdom  overview  2020-02-04            NaN
955   K02000001  United Kingdom  overview  2020-02-03            NaN
956   K02000001  United Kingdom  overview  2020-02-02            NaN
957   K02000001  United Kingdom  overview  2020-02-01            NaN
958   K02000001  United Kingdom  overview  2020-01-31            NaN

     newAdmissions  newCasesByPublishDate
0              NaN                    NaN
1              NaN                    NaN
2              NaN                    NaN
3              NaN                    NaN
4            544.0                    NaN
..             ...                    ...
954            NaN                    0.0
955            NaN                    0.0
956            NaN                    0.0
957            NaN                    0.0
958            NaN                    2.0

[959 rows x 7 columns]
```

Figure 9.3 – The UK dataset has ordering, formatting, and missing value issues to deal with

5.  Get rid of the data columns that you don't need. Change the date column to something that the model can use. Deal with the missing values in the various columns and put the data into sorted order:

```
Covid_df.drop(['areaCode', 'areaName', 'areaType'],
    axis='columns', inplace=True)
covid_df['date'] = pd.to_datetime(covid_df['date'])
covid_df['hospitalCases'] = \
    covid_df['hospitalCases'].fillna(0)
covid_df['newAdmissions'] = \
```

```
        covid_df['newAdmissions'].fillna(0)
    covid_df['newCasesByPublishDate'] = \
        covid_df['newCasesByPublishDate'].fillna(0)
    covid_df.sort_values(by=['date'], ignore_index=True,
        inplace=True)
    print(covid_df)
```

*Figure 9.4* shows that most of the issues in using the data are now addressed:

```
            date  hospitalCases  newAdmissions  newCasesByPublishDate
0     2020-01-31            0.0            0.0                    2.0
1     2020-02-01            0.0            0.0                    0.0
2     2020-02-02            0.0            0.0                    0.0
3     2020-02-03            0.0            0.0                    0.0
4     2020-02-04            0.0            0.0                    0.0
..           ...            ...            ...                    ...
954   2022-09-11         5911.0          544.0                    0.0
955   2022-09-12         5849.0            0.0                    0.0
956   2022-09-13         5792.0            0.0                    0.0
957   2022-09-14         5705.0            0.0                    0.0
958   2022-09-15         5729.0            0.0                    0.0

[959 rows x 4 columns]
```

Figure 9.4 – The UK dataset now contains just the necessary
information and most other issues are addressed

6.  Take out the data for 2021 and place it into a new dataset. You will use this data later to make a prediction, but don't worry about it for now:

```
Dates_2021 = (covid_df['date'] >= '2021-01-01') & \
    (covid_df['date'] <= '2021-12-31')
covid_df_2021 = covid_df.loc[dates_2021]
```

7.  Retain just the 2020 data for use in creating the model. Notice that the start date for the 2020 data is adjusted to match the start date of the Google dataset:

```
Dates_2020 = (covid_df['date'] >= '2020-02-15') & \
    (covid_df['date'] <= '2020-12-31')
covid_df = covid_df.loc[dates_2020]
print(covid_df)
```

As shown in *Figure 9.5*, the two datasets now have the same number of rows, which is essential before you can combine them:

```
          date  hospitalCases  newAdmissions  newCasesByPublishDate
15  2020-02-15            0.0            0.0                     0.0
16  2020-02-16            0.0            0.0                     0.0
17  2020-02-17            0.0            0.0                     0.0
18  2020-02-18            0.0            0.0                     0.0
19  2020-02-19            0.0            0.0                     0.0
..         ...            ...            ...                     ...
331 2020-12-27        22767.0         2871.0                 30501.0
332 2020-12-28        24052.0         3131.0                 41385.0
333 2020-12-29        25552.0         3250.0                 53135.0
334 2020-12-30        26554.0         3289.0                 50023.0
335 2020-12-31        26581.0         2924.0                 55892.0

[321 rows x 4 columns]
```

Figure 9.5 – The two datasets are finally compatible, so you can merge them together

8.  Merge the two datasets together now that you know they're compatible:

```
merged_df = pd.merge(covid_df,
    mobility_df,
    on='date')
print(merged_df)
```

This may seem like a lot of work to obtain a dataset to use, but this really is a simple example of the kinds of dataset manipulation you need to perform with real-world data to do anything more than simple analysis. To really become a security detective, you need to perform combinations of this sort to find the patterns that everyone else is missing and that the hacker doesn't know about. Yes, we're using COVID for this example, but the principles are the same when working with other kinds of data.

### Creating the training and testing datasets

Now that you have a dataset of merged data from multiple sources to use, it's time to create the training and testing datasets that will actually be used to develop a model. Use the following steps to perform this task:

1.  Import the required packages:

```
from sklearn.model_selection import train_test_split
```

2.  Perform the actual splitting of the merged dataset from the previous section:

```
train, test = train_test_split(merged_df,
    test_size=0.25)
```

```
print(f"Train Size: {train.shape}")
print(f"Test Size: {test.shape}")
```

*Figure 9.6* shows the shape of each of the datasets. They're a bit small but still quite usable. It's important to remember that more data is usually better when it comes to ML needs:

```
Train Size: (240, 10)
Test Size: (81, 10)
```

Figure 9.6 – The two datasets are a little small but quite usable for this example

3.  Use the split datasets to create a training and testing features matrix and a training and testing target vector. The goal is to determine the correlation, if any, between COVID statistics and retail and recreation activity levels. In other words, the datasets will help determine whether human behavior is affected by COVID:

```
train_features = train[
    ['hospitalCases', 'newAdmissions',
    'newCasesByPublishDate']].values
train_target = train['retail_and_recreation'].values
test_features = test[
    ['hospitalCases', 'newAdmissions',
    'newCasesByPublishDate']].values
test_target = test['retail_and_recreation'].values
```

Now that the data is ready for building a model, it's time to see how to create a regression model for use with this particular data. The example begins with a **k-nearest neighbor (kNN)** regressor, but you also see a random forest regressor and an XGBoost regressor used later in the chapter.

---

**kNN versus a random forest regressor versus an XGBoost regressor**

The kNN regressor is a tried and true algorithm that you see used in many examples online. In fact, you can hardly find an example that doesn't at least mention kNN. However, newer algorithms have come onto the scene that can provide better results and far greater flexibility. So, many data scientists have started using these newer algorithms in place of kNN. When it comes to random forest regressors or XGBoost regressors, the main issue is whether the result is computed in parallel or sequentially. Both algorithms are ensembles of learners, but the random forest regressor uses a voting mechanism to come up with the result, while XGBoost feeds the output of one model into the next model. This chapter shows you all three algorithms so that you can compare them. Of course, newer is not always better, and there are a lot more than just the three algorithms used here.

### Building the kNN model

The kNN algorithm (where *k* is the number of neighbors) is relatively simple, but it's also quite flexible because you can use it for both classification (determining which group a data point belongs to) and regression (the examination of one or more independent variables on a dependent variable). It's based on the principle that things that are close together tend to have a lot in common. The knowledge gained about a single point tends to translate into knowledge about the points around it within reason. You saw the kNN algorithm used for classification in the *Checking another model* section of *Chapter 8, Locating Potential Fraud*. Now you'll see it in action for regression.

One of the essentials of using kNN is the selection of how many neighbors to use for the examination process. Think about a grocery store. If you see a can of peas and then examine the cans one row up or one column over, they're likely to be peas too. However, if you start looking four rows up and four columns over, then you might see other vegetables, but you'd still see canned vegetables. Now, if you look six rows up and six columns over, you might start seeing something altogether different, perhaps juice in bottles or canned mushrooms; still something canned, but definitely not vegetables any longer. At some point, the number of neighbors becomes too large and the characteristics no longer hold as being close enough. If you use too large a value, then the model could be underfitted, while too small a value can cause overfitting. Experimentation is needed to find the right value, which is especially the case for regression problems.

In addition to problems with underfitting and overfitting, kNN can also suffer from the curse of dimensionality. It's not enough for kNN to have a close relationship between two points in one dimension; it must have a close relationship with every dimension. The example uses only three dimensions: hospital cases, new admissions, and new cases. However, when working with security data, the number of dimensions can increase very quickly, which means that you need more data to ensure that the model you build will work well. The modeling of human behavior for security needs can create an impressive number of dimensions, but very little data, so it's essential to tune your dataset (as has been done in this case) to get to the crux of the problem you're trying to resolve. With all of these issues in mind, here are the steps needed to create the kNN model for this example:

1. Import the required packages:

   ```
   from sklearn.neighbors import KNeighborsRegressor
   ```

2. Create the model and fit it to the training set. This part of the example uses 5 neighbors. Later, you will see the effects on the model of using values of both 3 and 7 neighbors. When working with `KNeighborsRegressor`, the value of k, which is provided by `n_neighbors`, is critical. The documentation for this algorithm at `https://scikit-learn.org/stable/modules/generated/sklearn.neighbors.KNeighborsRegressor.html` shows that there are other parameters you can change, and you should at least try them when fine-tuning your model. However, to start with, use the default settings and modify `n_neighbors` to optimize your results:

```
knn = KNeighborsRegressor(n_neighbors = 5)
knn.fit(train_features, train_target)
```

Now that you have a basic model to use, it's time to test it.

### Testing the model using kNN

It's time to test the model. Something important to remember is that the call to `train_test_split()` earlier actually randomizes the data. In other words, the training and testing process don't use the data in the same sequence as it appears in the original dataset, so you won't see anything resembling a smooth curve in the testing output. Rather, what you're looking for is how well the model fits its curve to the data points provided. You see the actual effect of the prediction later in the chapter, but for now, the focus is on testing using the following steps:

1.  Import the required packages:

    ```
    import numpy as np
    from sklearn.metrics import mean_squared_error
    from sklearn.metrics import r2_score
    import matplotlib.pyplot as plt
    ```

2.  Perform a prediction using the training data. Then, test the goodness of that prediction using both **root-mean-square error** (**RSME**, lower is better) and $R^2$ (higher is better) measures:

    ```
    train_pred = knn.predict(train_features)
    trainRSME = np.sqrt(mean_squared_error(train_target,
            train_pred))
    trainR2 = r2_score(train_target, train_pred) * 100
    print(f"Training RSME: {trainRSME}")
    print(f"Training R2 Score: {trainR2}")
    ```

    *Figure 9.7* shows the typical output from this step:

    ```
    Training RSME: 7.560798017846175
    Training R2 Score: 90.01955840683688
    ```

    Figure 9.7 – The output shows the result of the kNN with 5 neighbors training prediction

3. Perform a prediction using the testing data for comparison. If the statistics are close to what you achieve with the training data, then the model will likely perform well within its given error rate and variance:

```
test_pred = knn.predict(test_features)
testRSME = np.sqrt(mean_squared_error(test_target,
        test_pred))
testR2 = r2_score(test_target, test_pred) * 100
print(f"Testing RSME: {testRSME}")
print(f"Testing R2 Score: {testR2}")
```

*Figure 9.8* shows the typical output when performing the testing prediction step. Note that the RSME is higher and the $R^2$ score is lower, which indicates that the model may not perform as well as might have been anticipated:

```
Testing RSME: 10.031924350434442
Testing R2 Score: 81.88173745447082
```

Figure 9.8 – The output shows the result of the kNN with 5 neighbors testing prediction

4. Plot the prediction against the original data to see how well the model works:

```
x = np.arange(len(test_target))
plt.scatter(x, test_target, color='blue',
    label='Original')
plt.plot(x, test_pred, color='red',
    label='Prediction')
plt.legend()
plt.show()
```

*Figure 9.9* shows that the model follows acceptably in this case but could possibly follow better:

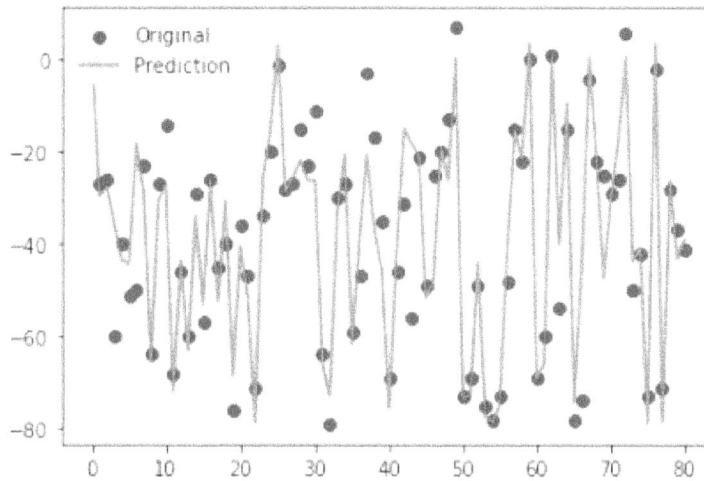

Figure 9.9 – The plot shows that kNN with 5 neighbors works acceptably but not great

Of course, this test uses just one value of *k*. You have no idea how well the model will work with other values. The next section tests two other values of *k*, 3 and 7, to see how well the model works with those values.

### Using a different-sized neighborhood with kNN

As the book progresses, you will discover more methods of ensuring that you get correct results from the ML models you create. Testing various model configurations is essential, especially in security scenarios that are already riddled with stealth and misinformation. The following steps show what happens when you use different values of *k* with kNN to perform a regression:

1. Perform the same kNN tests using a value of n_neighbors = 7:

```
knn = KNeighborsRegressor(n_neighbors = 7)
knn.fit(train_features, train_target)
train_pred = knn.predict(train_features)
trainRSME = np.sqrt(mean_squared_error(train_target,
    train_pred))
trainR2 = r2_score(train_target, train_pred) * 100
print(f"Training RSME: {trainRSME}")
print(f"Training R2 Score: {trainR2}")
test_pred = knn.predict(test_features)
testRSME = np.sqrt(mean_squared_error(test_target,
    test_pred))
```

```
testR2 = r2_score(test_target, test_pred) * 100
print(f"Testing RSME: {testRSME}")
print(f"Testing R2 Score: {testR2}")
plt.scatter(x, test_target, color='blue',
    label='Original')
plt.plot(x, test_pred, color='red',
    label='Prediction')
plt.legend()
plt.show()
```

Compare the output in *Figure 9.10* with the outputs in the previous section and you will see that changing the value to n_neighbors = 7 moves the model in the wrong direction. Every indicator shows worse performance at this level:

```
Training RSME: 8.526933120141974
Training R2 Score: 87.30595018803965
Testing RSME: 10.268025648490804
Testing R2 Score: 81.01887535277126
```

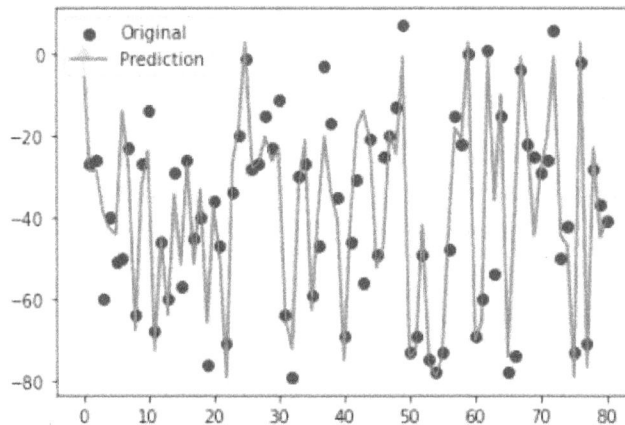

Figure 9.10 – The statistics and plot show that kNN with 7 neighbors shows worse performance

2.  Perform the same kNN tests using a value of n_neighbors = 3:

```
knn = KNeighborsRegressor(n_neighbors = 3)
knn.fit(train_features, train_target)
train_pred = knn.predict(train_features)
trainRSME = np.sqrt(mean_squared_error(train_target,
        train_pred))
```

```
trainR2 = r2_score(train_target, train_pred) * 100
print(f"Training RSME: {trainRSME}")
print(f"Training R2 Score: {trainR2}")
test_pred = knn.predict(test_features)
testRSME = np.sqrt(mean_squared_error(test_target,
        test_pred))
testR2 = r2_score(test_target, test_pred) * 100
print(f"Testing RSME: {testRSME}")
print(f"Testing R2 Score: {testR2}")
plt.scatter(x, test_target, color='blue',
    label='Original')
plt.plot(x, test_pred, color='red',
    label='Prediction')
plt.legend()
plt.show()
```

The output shown in *Figure 9.11* tells you that using a value of n_neighbors = 3 improves performance significantly:

```
Training RSME: 7.239871596978826
Training R2 Score: 90.84883865562644
Testing RSME: 9.51361741524911
Testing R2 Score: 83.70556085749895
```

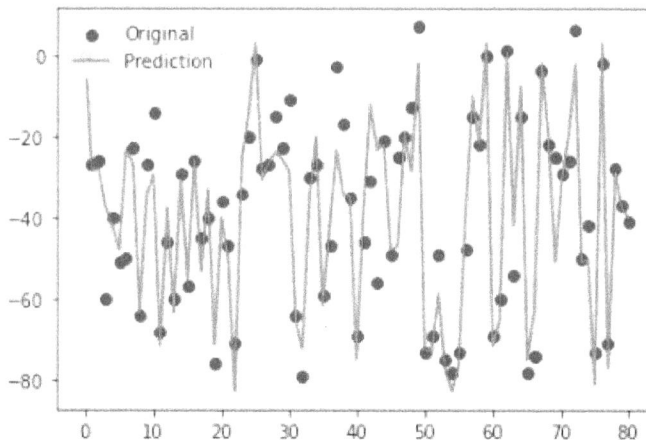

Figure 9.11 – The statistics and plot shows that kNN with 3 neighbors works the best so far

Look at the plots carefully at this point, and you'll notice that there is a tendency toward overfitting as the number of neighbors becomes smaller. If you were to use a value of `n_neighbors = 1`, you would find that performance becomes worse again. A value of `n_neighbors = 3` seems to work best for this particular dataset and model configuration.

---

**Using an odd value for *k* in kNN**

The general consensus is that using an odd value for *k* is best. The use of an odd number prevents the risk of a tie when performing the calculation.

---

### Building and testing using a random forest regressor

It usually pays to look at more than one algorithm when creating a model. You likely don't have time to work with every algorithm out there, but there are some classics that do appear in more than a few places. The random forest regressor is one of these favorites. The example in the *Setting the example up* section of *Chapter 8, Locating Potential Fraud,* uses `DecisionTreeClassifier` to perform fraud analysis. A random forest regressor is an ensemble of decision tree regressors to put the power of the crowd into the decision-making process. In this case, each decision tree works just like normal, but the output is calculated as the mean of each tree's output so that you can obtain a better result. The *Using ensemble learning* section of *Chapter 3, Mitigating Inference Risk by Avoiding Adversarial Machine Learning Attacks,* discusses some of the advantages and disadvantages of using ensemble learning. It's time to see an ensemble at work using the following steps:

1.  Import the required packages:

    ```
    from sklearn.ensemble import RandomForestRegressor
    ```

2.  Fit the model to the data. In the case of `RandomForestRegressor`, the key tuning features are the maximum tree depth, `max_depth`, and the number of estimators (trees), `n_estimators`. The example has already tuned both of these arguments for you, but feel free to play around with them to see how each of them works toward the goal of obtaining a better prediction. Note that when looking at the documentation for this algorithm, `https://scikit-learn.org/stable/modules/generated/sklearn.ensemble.RandomForestRegressor.html`, you will find that there are a lot more arguments than for kNN to help in tuning your model:

    ```
    rf = RandomForestRegressor(max_depth = 4,
        n_estimators=500)
    rf.fit(train_features, train_target)
    ```

3. Perform a prediction and measure its goodness against both the training and testing data:

```
train_pred = rf.predict(train_features)
trainRSME = np.sqrt(mean_squared_error(train_target,
        train_pred))
trainR2 = r2_score(train_target, train_pred) * 100
print(f"Training RSME: {trainRSME}")
print(f"Training R2 Score: {trainR2}")
test_pred = rf.predict(test_features)
testRSME = np.sqrt(mean_squared_error(test_target,
        test_pred))
testR2 = r2_score(test_target, test_pred) * 100
print(f"Testing RSME: {testRSME}")
print(f"Testing R2 Score: {testR2}")
```

The output in *Figure 9.12* shows that this algorithm provides better performance than kNN, but the code shows that you need to do a little more work to get it:

```
Training RSME: 7.024166548977007
Training R2 Score: 91.3860155018798
Testing RSME: 8.893716635030863
Testing R2 Score: 85.75984790693091
```

Figure 9.12 – The statistics show that the random forest regressor
doesn't perform quite as well as kNN in this case

4. Create a plot showing how the model actually works against the data:

```
plt.scatter(x, test_target, color='blue',
    label='Original')
plt.plot(x, test_pred, color='red',
    label='Prediction')
plt.legend()
plt.show()
```

*Figure 9.13* shows the results in this case:

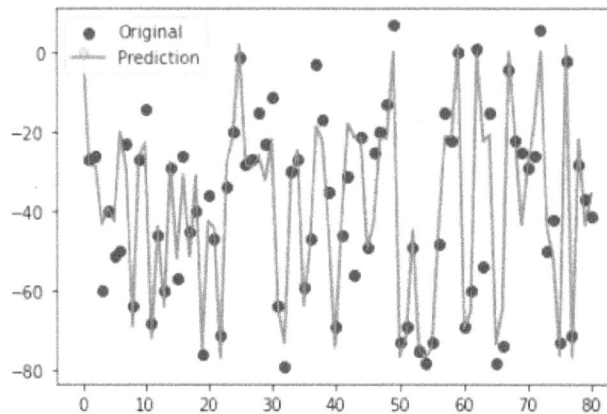

Figure 9.13 – The output of the random forest regressor prediction

It's interesting to review the differences in how the various models interact with the data as part of a plot.

## Building and testing an XGBoost regressor

The final algorithm to test against the behavior data is XGBoost, which is another ensemble of simple learners, but instead of using a voting setup to determine the result, the data from one model is fed into the next model to reduce the amount of error in the prediction. Overall, gradient boosting, the basis of XGBoost, is best for working with unbalanced data, as often happens with security scenarios. However, because of the large number of arguments that XGBoost provides (https://xgboost.readthedocs.io/en/stable/), it can prove incredibly difficult to tune. The following steps show how to work with XGBoost:

1.  Install the XGBoost package as needed. The version of XGBoost used for this example is 1.6.2. If you use a different version, you may find that you need to adjust the code slightly and that you'll obtain slightly different results:

```
modules = !pip list
installed = False
for item in modules:
    if ('xgboost' in item):
        print('XGBoost installed: ', item)
        installed = True
if not installed:
    print('Installing XGBoost...')
    !pip install xgboost
```

2.  Import the required package and fit the model. This example sets three of the arguments: max_depth to determine the depth of each tree, n_estimators to determine the number of trees, and eta to determine the learning rate. There are a great many other arguments that you can use to tune your model once you have got it to a basic level of competence:

```
from xgboost import XGBRegressor
xg = XGBRegressor(max_depth = 3, n_estimators=750,
    eta=0.3)
xg.fit(train_features, train_target)
```

The output of this step, shown in *Figure 9.14*, gives you some idea of the incredible number of tuning arguments supported by this algorithm:

```
                              XGBRegressor
XGBRegressor(base_score=0.5, booster='gbtree', callbacks=None,
            colsample_bylevel=1, colsample_bynode=1, colsample_bytree=1,
            early_stopping_rounds=None, enable_categorical=False, eta=0.3,
            eval_metric=None, gamma=0, gpu_id=-1, grow_policy='depthwise',
            importance_type=None, interaction_constraints='',
            learning_rate=0.300000012, max_bin=256, max_cat_to_onehot=4,
            max_delta_step=0, max_depth=3, max_leaves=0, min_child_weight=1,
            missing=nan, monotone_constraints='()', n_estimators=750, n_jobs=0,
            num_parallel_tree=1, predictor='auto', random_state=0, reg_alpha=0, ...)
```

Figure 9.14 – The XGBRegressor output shows an incredible number of tuning arguments

3.  Perform a prediction and measure its goodness against both the training and testing data:

```
train_pred = xg.predict(train_features)
trainRSME = np.sqrt(mean_squared_error(train_target,
        train_pred))
trainR2 = r2_score(train_target, train_pred) * 100
print(f"Training RSME: {trainRSME}")
print(f"Training R2 Score: {trainR2}")

test_pred = xg.predict(test_features)
testRSME = np.sqrt(mean_squared_error(test_target,
        test_pred))
testR2 = r2_score(test_target, test_pred) * 100
print(f"Testing RSME: {testRSME}")
print(f"Testing R2 Score: {testR2}")
```

The output, shown in *Figure 9.15*, demonstrates that this model provides the best performance of those tested for this example:

```
Training RSME: 1.2282842929226978
Training R2 Score: 99.73660211630327
Testing RSME: 8.604377627388082
Testing R2 Score: 86.67132506775764
```

Figure 9.15 – The XGBRegressor provides the best performance of all the models tested

4.  Create a plot showing how the model actually works against the data:

```
plt.scatter(x, test_target, color='blue',
    label='Original')
plt.plot(x, test_pred, color='red',
    label='Prediction')
plt.legend()
plt.show()
```

Compare the plot shown in *Figure 9.16* with the one in *Figure 9.13* and you can see the smoothing effect of the gradient boosting process:

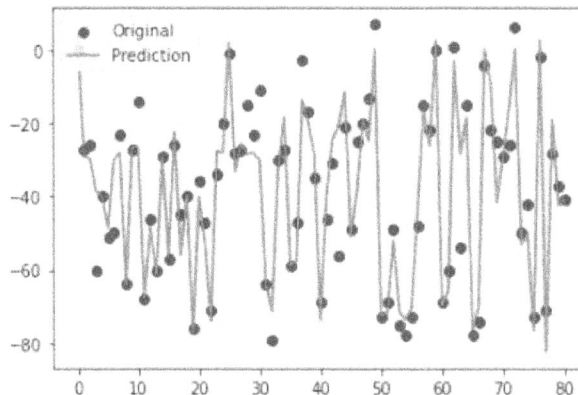

Figure 9.16 – The output of the XGBoost regressor prediction

Choosing an algorithm and then tuning it can require time and patience, but it's well worth the effort. The better the model's predictions, the greater the chance that the hacker will be caught.

## Putting the data in perspective

The plots you've seen so far show the randomized data used for training and testing but don't really tell you much about how the model actually performs using the original dataset in a real-world way. What you really need is a plot showing real-world events, which rely on the following code:

```
plt.plot(covid_df['date'],
    covid_df['hospitalCases'].values,
    color='blue', label='Hospital Cases')
plt.plot(covid_df['date'],
    covid_df['newAdmissions'].values,
    color='green', label='New Admissions')
plt.plot(covid_df['date'],
    covid_df['newCasesByPublishDate'].values,
    color='orange', label='New Cases')
plt.xticks(rotation=90)
fig = plt.gcf()
fig.set_size_inches(10, 8)
plt.legend()
plt.show()
```

The output shown in *Figure 9.17* demonstrates that there is a correlation between the three kinds of data used as features for this example:

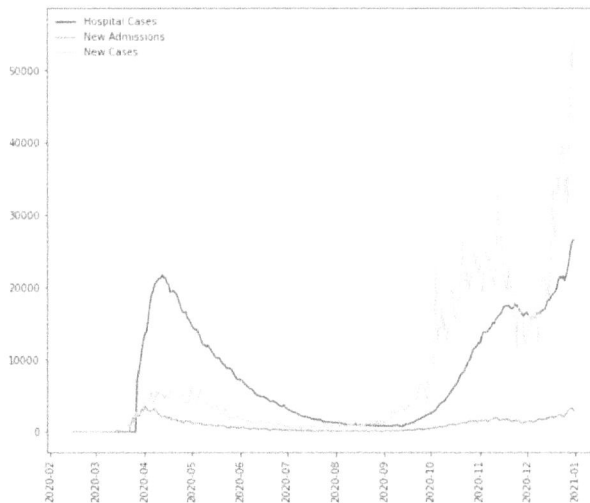

Figure 9.17 – The plot shows a correlation between the three data features

Because the XGBoost model produced the best results during testing, the example uses it to produce the prediction needed to test how the model works using the full dataset with the following code:

```
xg_full_pred = xg.predict(covid_df[[
        'hospitalCases', 'newAdmissions',
            'newCasesByPublishDate']].values)
plt.plot(covid_df['date'], xg_full_pred,
    color='red', label='Prediction')
plt.plot(covid_df['date'],
    mobility_df['retail_and_recreation'],
    color='blue', label='Actual Data')
plt.xticks(rotation=90)
plt.legend()
fig = plt.gcf()
fig.set_size_inches(10, 8)
plt.show()
```

The output shown in *Figure 9.18* demonstrates that as the number of COVID cases increases, the retail and recreation activities decrease (notice that the numbers are negative). However, also notice that there are lags. COVID increases first, then comes the decrease in activity:

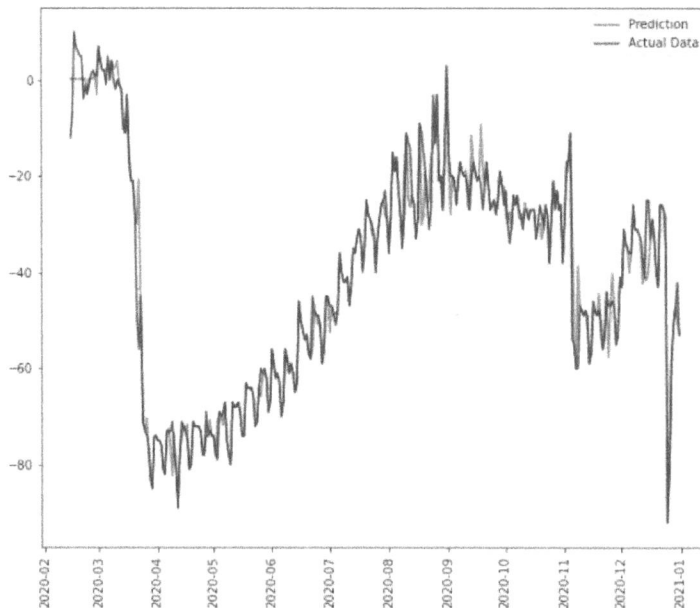

Figure 9.18 – The plot shows an inverse correlation between COVID and activity

What this example has done is demonstrated that you can model human behavior with the correct data. The problem is finding the correct data, which means becoming a detective and figuring the hackers out.

## Predicting new behavior based on the past

It's fine that the model works on data it has already worked with during training and testing. However, working with the full dataset doesn't really fit the bill if you're going to predict future human behavior. You already know what hackers have done in the past, but what are they going to do tomorrow? Now the example uses the data previously saved from 2021 that the model hasn't seen. The following code shows COVID activity for the following year:

```
plt.plot(covid_df_2021['date'],
    covid_df_2021['hospitalCases'].values,
    color='blue', label='Hospital Cases')
plt.plot(covid_df_2021['date'],
    covid_df_2021['newAdmissions'].values,
    color='green', label='New Admissions')
plt.plot(covid_df_2021['date'],
    covid_df_2021['newCasesByPublishDate'].values,
    color='orange', label='New Cases')
plt.xticks(rotation=90)
plt.legend()
fig = plt.gcf()
fig.set_size_inches(10, 8)
plt.show()
```

The results shown in *Figure 9.19* show a lull in COVID activity before it increases again:

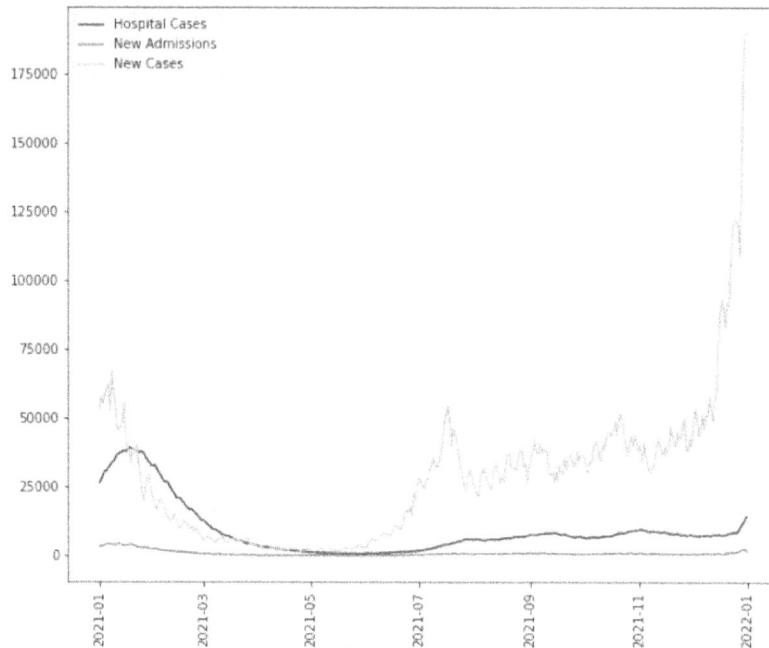

Figure 9.19 – A plot showing the 2021 COVID data

Before you can move forward, you need to obtain the actual retail and recreation data for 2021 using this code:

```
mobility_df = pd.read_csv(
    "Region_Mobility_Report_CSVs/" \
    "2021_GB_Region_Mobility_Report.csv")
mobility_df.drop(
    ['country_region_code', 'country_region',
        'sub_region_1', 'sub_region_2', 'metro_area',
        'iso_3166_2_code', 'census_fips_code',
        'place_id'], axis='columns', inplace=True)
mobility_df['date'] = pd.to_datetime(mobility_df['date'])
mobility_df.rename(columns={
        'retail_and_recreation_percent_change_from_baseline':
        'retail_and_recreation'},
    inplace=True)
```

This is essentially the same code used to obtain the 2020 data, but for 2021 instead. Now let's see what the predicted activity is like. However, in this case, you see the effects of using kNN, random forest regressor, and XGBoost regressor because there is something interesting to see. The following code performs the prediction and presents the plot for you:

```
knn_2021_pred = knn.predict(covid_df_2021[[
    'hospitalCases', 'newAdmissions',
        'newCasesByPublishDate']].values)
rf_2021_pred = rf.predict(covid_df_2021[[
        'hospitalCases', 'newAdmissions',
            'newCasesByPublishDate']].values)
xg_2021_pred = xg.predict(covid_df_2021[[
        'hospitalCases', 'newAdmissions',
            'newCasesByPublishDate']].values)
plt.plot(covid_df_2021['date'],
    knn_2021_pred,
    color='red', label='kNN Prediction')
plt.plot(covid_df_2021['date'],
    rf_2021_pred,
    x
plt.plot(covid_df_2021['date'],
    xg_2021_pred,
    color='orange', label='XG Prediction')
plt.plot(covid_df_2021['date'],
    mobility_df.head(365)['retail_and_recreation'],
    color='blue', label='Actual Data')
plt.xticks(rotation=90)
plt.legend()
fig = plt.gcf()
fig.set_size_inches(10, 8)
plt.show()
```

*Figure 9.20* shows that the correlation between predicted activity levels and COVID activity remains in the 2021 data. Now, this is data that the model hasn't seen at all, and there was little tuning done. In addition, a larger dataset would have made a difference (the smaller size was partially to keep computing costs low). So, the prediction follows the actual data reasonably closely. Notice that the random forest regressor provides a smoother prediction that follows the actual data a tiny bit closer than kNN. This is the effect of the ensemble and why you may want to choose it over other methods of making a prediction. The XGBoost regressor output is the best of the three but lacks the smoothness of the random forest regressor. With further tuning, either model would do a better job of predicting human behavior.

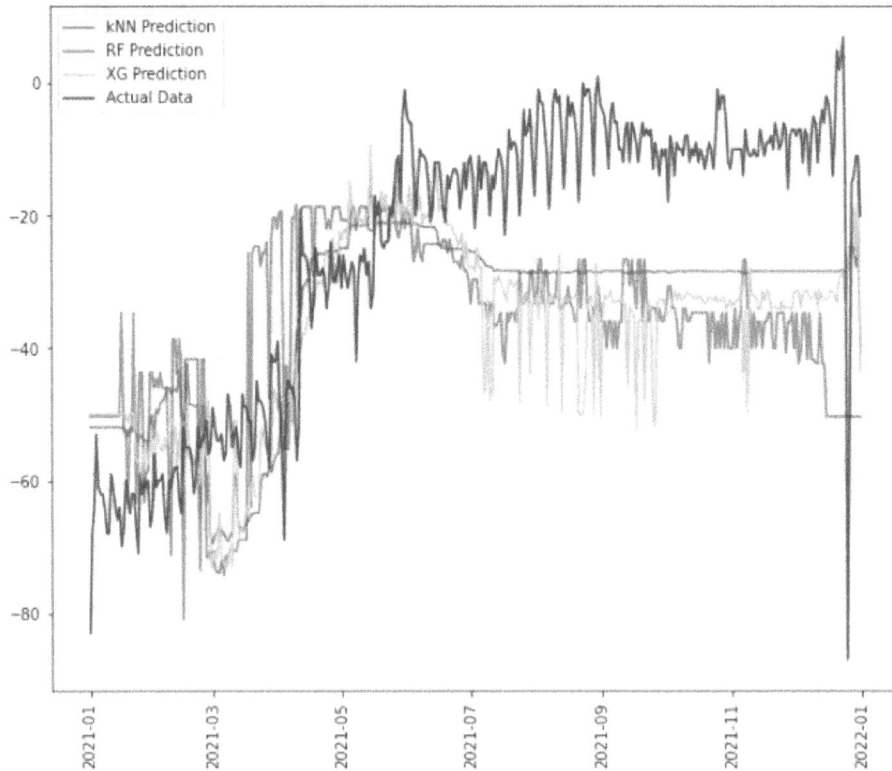

Figure 9.20 – The plot shows an inverse correlation in the 2021 data as well

What this all means is that it's possible to predict future human activity given knowledge of the past. However, the prediction is never completely accurate. Notice that in July 2021, the prediction diverges from the actual data because the prediction is modeled on what is happening with the COVID cases, while the real people in the real world were getting used to COVID and learning to live with it in favor of pursuing their normal activities. It still takes some knowledge on the part of the human to ensure that the data and predictive values are used correctly. The important takeaway from this example is that human behavior is predictable.

Using these prediction methods also work with security, but you need a lot more details to make it work (which is why this example took the direction that it did) Articles, such as *Attack Pattern Detection and Prediction* (https://towardsdatascience.com/attack-pattern-detection-and-prediction-108fc3d47f03), point out that prediction methods are now being used to anticipate hacker activity based on past attacks. It really isn't enough any longer to have a good defense in place; it's more important to know what could possibly happen in the future. Unfortunately, there isn't any sort of generalized prediction strategy that you can use now, much as you might use generalized defenses. Prediction is usually business specific.

## Locating other behavioral datasets

This chapter shows you a single example of a behavioral dataset. You may actually need something different to perform some experiments. Unfortunately, there are no datasets currently available that actually contain hacker-specific information. It may be possible to generate such a dataset for your organization using the technique in the *Building a data generator* section of *Chapter 5, Keeping Your Network Clean.*

One good place to find behavioral datasets is on GitHub at `https://github.com/topics/human-behavior`. The site contains a number of datasets to choose from, including some non-text datasets that contain images. One of the more interesting datasets is `cent-patates`, which tells you how to improve your winnings from the French lottery. Mind you, I haven't actually tested the dataset or the theories of its author.

The Mendeley Data site at `https://data.mendeley.com/research-data/?search=Human%20Behavior%20Dataset` provides a huge number of datasets. The problem is wading through all of them to find the one you want. This is the place to look for unusual datasets that rely on images, videos, or even sound formats. Note that the same site also provides access to a number of research documents and dissertations on up-and-coming technologies that really aren't ready for use today but are still interesting to read about.

In some cases, you can find very specialized datasets with strings attached, such as the *Web-Hacking Dataset for the Cyber Criminal Profiling* at `https://ocslab.hksecurity.net/Datasets/web-hacking-profiling`, which requires you to jump through some hoops to get it. For one thing, you need to fill out a form with some information about yourself that isn't required by most sites. This dataset is also only useful for research purposes, as stated by the documentation that comes with it. Given that the data is from over a period of 15 years, it also likely includes outdated techniques. However, it is one of the few datasets that could act as a jumping-off point for working with hacker data.

The takeaway from this section is that hackers really can be dangerous, that you really can use techniques to get around the attacks they perform, and that monitoring is an essential part of any security strategy. The next section discusses the use of ML to improve security and reliability. Static strategies are becoming ever less effective today, so you need to consider just how to overcome hacker attacks through the flexibility that ML can provide.

## Improving security and reliability

This last section is about your goal in studying hackers and their behavior and the improvement of security and reliability. Security doesn't work when it's too hard or too time-consuming to implement. Systems that are complex are usually unreliable. To actually improve the environment at an organization, it's essential to limit the difficulty, complexity, and time-consuming nature of the solution. If you were to implement every idea in this book, you'd do nothing but spend your days managing the security of your system. The idea is to use the hacker information in this chapter to tailor a solution that meets your organization's needs.

Behavior monitoring is the key to ending the constant struggle to build the walls of your organization's security infrastructure higher, only to see hackers cresting over the top. Building higher walls is proven not to work well, at least, not for very long, so you need a strategy that tailors your interactions with hackers based on hacker behavior so that you can do things like create more interesting honeypots that will keep hackers busy and out of your network.

Using behavior monitoring as one of the tools in your security arsenal tends to reduce complexity while improving both overall security and reliability. However, as the example demonstrates, human interpretation of the results and knowledge of actual conditions is essential. Novelties in the data (as described in *Chapter 6*) aren't compensated for in the model. Consequently, divergences in predicted output and actual output can signal when you need to rebuild a model to obtain accurate results.

It's also important to remember a prime rule in statistics that statistics apply to groups, not to individuals. The behavior modeling you perform applies to hackers as a group, the particular group plaguing your organization, but still a group. An individual hacker can still present surprises that would appear as outliers in the data and that you need to be aware of when working through the meaning of a particular prediction.

## Summary

If you take anything at all away from this chapter, it should be that hackers are real people. The thought that a hacker is some mindless entity out there, somewhere, whose only goal in life is to ruin your day is a shortcut to getting in your own way when it comes to dealing with hacker-created security issues. Understanding hacker behavior, realizing that hackers attack specific targets for a reason, considering that a form of attack is designed to emphasize hacker strengths, and then tailoring a solution that your organization will actually use are all part of a strategy to thwart hacker incursions. This chapter has reviewed the hacker in a unique way to help you create better, more flexible solutions.

*Chapter 10* is a different take on ML security, deepfakes. A **deepfake** is an output of an ML application, such as a graphic or audio file, that can fool human experts easily in many cases. You may think deepfakes are more science fiction than anything else. Yet, when you run the code, you'll find that it's real, it's in existence today, and a deepfake (correctly done) can do quite a bit of damage to your organization, its data, and the people who run it.

# Further reading

The following bullets provide you with some additional reading that you may find useful in further understanding the materials in this chapter:

- Consider how mobile devices represent a significant threat to your business: *5 Ways Your Mobile Device Can Get Malware* (`https://www.securitymetrics.com/blog/5-ways-your-mobile-device-can-get-malware`)

- Read more about psychology and hacking: *Psyber Security: Thwarting Hackers with Behavioral Science* (`https://www.psychologicalscience.org/observer/psyber-security-thwarting-hackers-with-behavioral-science`)

- Understand overfitting and underfitting issues with kNN: *K-Nearest Neighbors* (`https://www.codecademy.com/learn/introduction-to-supervised-learning-skill-path/modules/k-nearest-neighbors-skill-path/cheatsheet`)

- Read about the effect of the curse of dimensionality when working with kNN: *k-Nearest Neighbors and the Curse of Dimensionality* (`https://towardsdatascience.com/k-nearest-neighbors-and-the-curse-of-dimensionality-e39d10a6105d`)

- Consider the difference between RSME and $R^2$, and the need for both when using regression: *RMSE versus R-Squared: Which Metric Should You Use?* (`https://www.statology.org/rmse-vs-r-squared/`)

- Discover more about algorithm differences: *Comparative Study on Classic Machine learning Algorithms* (`https://towardsdatascience.com/comparative-study-on-classic-machine-learning-algorithms-24f9ff6ab222`)

- See how the different methodologies used by random forest regressors and XGBoost regressors play out in creating a model: *XGBoost versus Random Forest* (`https://medium.com/geekculture/xgboost-versus-random-forest-898e42870f30`)

# Part 3 – Protecting against ML-Driven Attacks

Up to this point in the book, someone could potentially experience security issues that rely on traditional methods that are modified to meet the demands of the ML environment. However, some advanced hackers employ ML applications specifically designed to thwart security measures through a learning process. In other words, the application has no other purpose than to somehow break into a system and achieve a particular set of hacker goals.

This section includes the following chapters:

- *Chapter 10, Considering the Ramifications of Deepfakes*
- *Chapter 11, Leveraging Machine Learning against Hacking*

# Considering the Ramifications of Deepfakes

**Deepfakes** have received more than a little press over the last few years because everyone thinks it's pretty amazing when Diep Nep becomes Morgan Freeman (`https://www.youtube.com/watch?v=oxXpB9pSETo`). Yes, deepfakes (also known as **synthetic reality**) have some fantastic uses for both work and pleasure. However, the technology also represents one of the most terrifying security issues for users of deep learning today. How can you trust anything said by anyone during a video chat when you can't even be sure you're talking with the person in question, rather than a hacker? That video security system with facial recognition that your organization just purchased is completely worthless when it comes to deepfakes because it's possible now to create fake faces (`https://www.theverge.com/tldr/2019/2/15/18226005/ai-generated-fake-people-portraits-thispersondoesnotexist-stylegan`). Any thoughts you had of using biometric authentication is a waste of time because, with the right deepfake, it's possible to create face biometrics (`https://www.wired.com/story/deepmasterprints-fake-fingerprints-machine-learning/`). It gets worse, and in this chapter, you will become aware of all the details.

There is no magic involved in deepfakes, as you will see in this chapter with the use of autoencoder and **generative adversarial network** (**GAN**) examples. There are many other ways to create a deepfake, but these two methods are quite illustrative and relatively fast at creating them (with relatively being the operative term). Running the examples and seeing how they progress over time will help you understand that deepfakes rely on stable technology that nearly anyone can employ to make security for your organization a true nightmare. The only thing that is preventing the rampant use of deepfakes today is that they're time-consuming to create and the person creating them has to have the correct skills (both of which are topics that this chapter covers as well). With these issues in mind, this chapter discusses the following topics:

- Defining a deepfake
- Understanding autoencoders
- Understanding GANs

# Technical requirements

This chapter requires that you have access to either Google Colab or Jupyter Notebook to work with the example code. The requirements to use this book is to use the section of *Chapter 1*, *Defining Machine Learning Security*, which provides additional details on how to set up and configure your programming environment.

You really do benefit from having a **graphics processing unit** (**GPU**) to run the examples in this chapter. They will run without a GPU but expect to take long coffee breaks while you wait for the code to complete running. This means choosing **Runtime | Change Runtime Type** in Google Colab, then selecting **GPU** in the **Hardware Accelerator** dropdown. Desktop users will want to review the *Checking for a GPU with a nod toward Windows* section of the chapter for desktop system instructions.

Set up your system to run TensorFlow. Google Colab users should read `https://colab.research.google.com/github/tensorflow/docs/blob/master/site/en/tutorials/quickstart/beginner.ipynb`. The *Installing TensorFlow* section of this chapter provides details on how to perform the software installation on desktop systems, where you'll generally get far superior performance than when using Google Colab. Your system must meet the following minimal requirements:

- Operating system:

  - Ubuntu 16.04 or higher (64-bit)

  - macOS 10.12.6 (Sierra) or higher (64-bit) (no GPU support)

  - Windows Native – Windows 7 or higher (64-bit)

  - Windows WSL2 – Windows 10 19044 or higher (64-bit)

- Python 3.7 through Python 3.10

- Pip version 19.0 or higher, or 20.3 or higher on macOS

- Visual Studio 2015, 2017, or 2019 on Windows systems

- NVidia GPU Support (optional):

  - GPU drivers version 450.80.02 or higher

  - **Compute Unified Device Architecture** (**CUDA**) Toolkit version 11.2

  - cuDNN SDK version 8.1.0

  - TensorRT (optional)

When testing the code, use a test site, test data, and test APIs to avoid damaging production setups and to improve the reliability of the testing process. Testing over a non-production network is highly recommended. Using the downloadable source is always highly recommended. You can find the downloadable source on the Packt GitHub site at `https://github.com/PacktPublishing/Machine-Learning-Security-Principles` or on my website at `http://www.johnmuellerbooks.com/source-code/`.

# Defining a deepfake

A deepfake (sometimes deep fake) is an application of deep learning to images, sound, video, and other forms of generally non-textual information to make one thing look or sound like something else. The idea is to deceive someone into thinking a thing is something that it's not.

This chapter doesn't mean to imply that the use of deepfakes will always deceive others in a bad way. For example, it's perfectly acceptable to take a family picture, then put it through an autoencoder or a GAN and make it look like a Renoir painting. In fact, some deepfakes are amusing or even educational. The point at which a deepfake becomes a problem is when it's used to bypass security or perform other seemingly impossible tasks. In a court of law, a deepfake video could convince a jury to convict someone who is innocent. Throughout the following sections, you will learn more about deepfakes from an ML security perspective.

> **Identifying deepfakes**
>
> It's still possible to detect deepfakes in a number of ways. Even though the **artifacts** (an anomaly that presents information that shouldn't be there) are getting smaller and harder to detect, images created using deepfake technology often do have artifacts. Voices are still very tough to reproduce, so they're often a dead giveaway that something is faked. Deepfakes usually don't do well with side views, so simply watching when someone turns their head will likely give the deepfake away (as will a person never turning their head to the side, which is unnatural). Differences in mannerisms and all sorts of other subtle clues often give a deepfake away. However, deepfakes are becoming more realistic as technology advances, so new ways of detecting them will have to be invented.

## Modifying media

Media modifications can serve all sorts of purposes. For example, someone could use media modifications to change security camera footage or to make it appear that someone has said or done something that they really haven't. Reading *Deep Fakes' Greatest Threat Is Surveillance Video* (`https://www.forbes.com/sites/kalevleetaru/2019/08/26/deep-fakes-greatest-threat-is-surveillance-video/?sh=5cf6b7c44550`) will give you a better idea of just how big a threat this sort of modification can be.

It's important to realize that deepfakes can modify any type of media. Seeing the results of deepfakes on YouTube is entertaining, but such videos are hardly the tip of the iceberg of what deepfakes can do. For example, there isn't any reason to believe that a deepfake can't modify sensor data from various kinds of inputs such as cameras, microphones, temperature sensors, and so on. A terrorist could easily modify such inputs to create an emergency or keep someone from detecting one. The problem gets worse. A field commander might end up misdirecting troops based on bad footage from a drone camera whose output was overridden and deepfake supplied in its place.

## Common deepfake types

Some deepfake types are relatively common today and it pays to spend time learning about them. These kinds of deepfakes may not directly affect your organization, but they do provide a basis for understanding the potential for misusing deepfakes. The following list provides you with some examples of commonly misused deepfakes:

- **Fake news**: People base their actions on information they receive from various sources, including the news. It's possible to use fake news, which is a form of documentation for something that never really happened, to modify human behavior and possibly use that modification to their advantage. **Fake news generators** (see an example of this at https://www.thefakenewsgenerator.com/) make it very tough to differentiate real news from fake news. Deepfakes make it possible to place a trusted news commentator on screen with a fake news story that they never filmed.

- **Hoaxes**: A *deepfake hoax* is a hoax in which the perpetrator is attempting to scam a group into believing something is true for personal gain. One of the most interesting current scams is deepfaked online interviews used to secure remote employment.

- **Voice impersonation**: A *deepfake voice impersonation* falls into a special category because people will often react to a phone call or other vocal communication that occurs remotely without even thinking about it. They hear what they think is the person's voice and then do what the voice says. So, it's possible that a deepfake could do things such as telling someone to stop payment on an important shipment of goods that manufacturing needs to complete a sale, giving a competitor an edge. Voice impersonation can also make it possible to bypass biometric security.

- **Fake people**: Creating fake people using deepfake techniques is the practice of using deep learning to make people who look real, but aren't. What you need to think about is the potential for scams. For example, fake people would make perfect candidates for government programs. With all of the right documentation (deepfaked, of course), a person could apply for just about anything and get it as long as a personal meeting isn't required.

- **Pornography**: One of the first uses of deepfake technology to gain public notoriety was the proliferation of fake pornographic videos and images. Whether of public figures or private citizens, the use of non-consensual sexual images is harmful to those depicted. Unscrupulous hackers may use deepfakes to manufacture blackmail material.

This list of the uses of deepfakes by hackers and scammers doesn't even begin to tell you about the many ways in which deepfakes are used today. It's now possible to spend days reading articles about all of the ways in which deepfakes have been used to scam people. An issue here is that the technology is still in its natal state. The technology is such that you can expect to find deepfakes everywhere all the time.

## The history of deepfakes

Deepfakes have been around for a while. The basis of what you see today about deepfakes started in a 1997 paper, *Video Rewrite: Driving Visual Speech with Audio*, written by Christoph Bregler, Michele Covell, and Malcolm Slaney (`http://chris.bregler.com/videorewrite/`). The essence of the technique is to automate some of the graphic effects that movie studios use to create movies. Of course, deepfake technology was built on previous technology that worked with audio and facial expressions in a 3D space.

Early videos had some serious problems with regard to the **uncanny valley**: people could tell something was off with the facial expressions because they weren't quite right and often felt creepy. Further work starting in 2000 looked into making faces more lifelike. One of the best examples of this focus is the *Active appearance models* whitepaper, which is available at `https://ieeexplore.ieee.org/document/927467`.

However, the first time deepfake technology gained mainstream attention was the 2018 deepfake video of performer and filmmaker Jordan Peele impersonating President Obama (`https://www.youtube.com/watch?v=cQ54GDm1eL0`). The deepfake is good enough that you don't really know it's a deepfake until Jordan Peele reveals himself near the end. Even though the video really was convincing, experts were finally able to detect it was a deepfake based on how the facial expressions were presented. However, this won't be a problem for long because the technology continues to advance. For example, current technologies address flaws such as moving eyebrows when the person stops talking.

Now that you have some idea of what deepfakes are all about, it's time to prepare to create one. The next section of this chapter discusses the specialized setup used to work with both autoencoders and GANs. This special setup is necessary because the computing power needed to create a deepfake is significant.

## Creating a deepfake computer setup

Creating a deepfake requires building serious models, using specialized software on systems that have more than a little computing horsepower. The system used for testing and in the screenshots for this chapter is more modest. It has an Intel i7 processor, 24 GB of RAM, and an NVidia GeForce GTX 1660 Super GPU. This system is used to ensure that the examples will run in a reasonable amount of time, with *reasonable* being defined as building a model in about half an hour or less. The example as a whole will require more time, likely in the hour range. The following sections will help you install a TensorFlow setup that you can use for autoencoder and GAN development without too many problems, and help you test your setup to ensure it actually works.

## Installing TensorFlow on a desktop system

Desktop developers may already have TensorFlow installed, but if you're not sure then you likely don't. The technique for creating the advanced models in this chapter relies on using TensorFlow (`https://www.tensorflow.org/`). The basic reason for going with this route is that the development process is easier and you can create a relatively simple example sooner so that you can see how such a model would work. In order to use TensorFlow, you must install the required support. You can verify that you have the required support installed by opening an `Anaconda` prompt and typing the following:

```
conda list tensorflow
```

Alternatively, you can enter the following code:

```
pip show tensorflow
```

If you have TensorFlow installed, it will show up as one of the installed packages on your system. This section assumes that you have the Conda utility available on your desktop system. If you installed the Anaconda suite on your machine, then you have Conda available by default. Otherwise, you need to install Miniconda using the instructions at `https://docs.conda.io/en/latest/miniconda.html` for your platform. If your system is too old to support the current version of Miniconda, then you can't run the examples in this chapter.

In addition to ensuring you have access to the Conda utility on Windows systems, you must also have Visual Studio 2015, Visual Studio 2017, or Visual Studio 2019 installed before you do anything else. You can obtain a free copy of Visual Studio 2019 Community Edition at `https://learn.microsoft.com/en-us/visualstudio/releases/2019/release-notes`.

Once you know you have a version of the Conda utility available, you can use these steps to set up the prerequisite tools for the examples in this chapter. The following steps will work for most platforms. However, if you encounter problems, you can also find a set of steps at `https://www.tensorflow.org/install/pip`. Select your platform at the top of the instruction list:

1.  Type `conda install nb_conda_kernels` and press *Enter*. Type `y` and press *Enter* when asked.

2.  Type `conda create --name tf python=3.9` and press *Enter* to create a clean environment for your `TensorFlow` installation. Type `y` and press *Enter* when asked.

3.  Type `conda activate tf` and press *Enter*.

> **Deactivation**
>
> Type `conda deactivate tf` and press *Enter* to deactivate the TensorFlow environment when you no longer need to keep the environment running.

4.  Type `conda install ipykernel` and press *Enter*. Type y and press *Enter* when asked.

5.  Download and install the NVIDIA driver for your platform and GPU type from `https://www.nvidia.com/Download/index.aspx`.

6.  Test the installation by typing `nvidia-smi` and pressing *Enter*. You should see a listing of device specifics, along with the processes that are currently using the GPU.

7.  Type `conda install -c conda-forge cudatoolkit=11.2 cudnn=8.1.0` and press *Enter* to install the CUDA toolkit and cuDNN SDK. Type y and press *Enter* when asked. Note that this step can take a while to complete.

8.  Depending on your platform and how you have the CUDA toolkit installed, you need to add a path statement to your platform so that it can find the CUDA toolkit. The paths commonly needed are as follows:

```
NVIDIA GPU Computing Toolkit\CUDA\v11.3\bin
NVidia GPU Computing Toolkit\CUDA\v11.3\lib\x64
NVidia GPU Computing Toolkit\CUDA\v11.3\include
```

9.  Type `pip install --upgrade pip` and press *Enter* to verify that you have the latest version of pip installed.

10. Type `pip install tensorflow` and press *Enter*. Type y and press *Enter* when asked.

11. Type `ipython kernel install --user --name=tf` and press *Enter*. This step will make the environment appear in the menus so that you can access it.

Now that you have TensorFlow installed, you can verify that your installation will work correctly using the instructions in the next section. These instructions ensure that you can access your GPU and that your GPU is of the correct type.

## Checking for a GPU

Training a model with a GPU might only require half an hour, but training it without one could easily cost you six or more hours. You want to make sure that your GPU is actually working, accessible, and of the right type before you start working with the example code. Otherwise, you have long waits and sometimes inexplicable errors to deal with when the code itself is fine.

The `MLSec; 10; Check for GPU Support.ipynb` file contains source code and some detailed instructions for configuring and checking your GPU setup on Windows. The source code works equally well for Linux and Windows systems, so Linux users can test their setup too. When using the downloadable source, you must set the kernel to use your `tf` environment by choosing **Kernel | Change Kernel | Python [conda env:tf]**. Otherwise, even if you have a successful TensorFlow installation, Jupyter Notebook won't use it. The following steps test the setup from scratch:

1. Create a new Python file by choosing the `tf` environment option from the **New** drop-down list in place of the usual Python 3 (ipykernel) option, as shown in *Figure 10.1*.

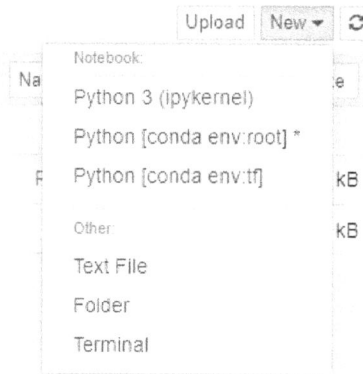

Figure 10.1 – The menu for selecting which environment to use

2. Type `!conda env list` in the first cell. Click **Run.** You will see output similar to *Figure 10.2*. The output will vary by your platform and Conda setup. Notice that the `tf` environment has an asterisk next to it, indicating that it's the one in use.

```
# conda environments:
#
base                     C:\Users\John\anaconda3
tf                  *    C:\Users\John\anaconda3\envs\tf
```

Figure 10.2 – A list of the available Conda environments

3. Type the following code to determine the selected environment using the operating system environment variables instead. If the method in *step 2* doesn't work, this one always does, but it provides less information:

```
import os
print (os.environ['CONDA_DEFAULT_ENV'])
```

The output of `tf` shows that the `tf` environment is selected as the Conda default environment for scripts you create and run in Jupyter Notebook.

4.  Type the following code to obtain a list of computing devices on your system, which will include your GPU if TensorFlow is correctly installed:

```
from tensorflow.python.client import device_lib
device_lib.list_local_devices()
```

The output shown in *Figure 10.3* is representative of the output you will see when you run this cell. However, the specifics will vary by GPU and your system may actually have multiple GPUs installed, which would mean that the list would include all of them.

```
[name: "/device:CPU:0"
device_type: "CPU"
memory_limit: 268435456
locality {
}
incarnation: 5122621535799984449
xla_global_id: -1,
name: "/device:GPU:0"
device_type: "GPU"
memory_limit: 4187553792
locality {
  bus_id: 1
  links {
  }
}
incarnation: 13742849836653392168
physical_device_desc: "device: 0, name: NVIDIA GeForce GTX 1660 SUPER, pci bus id: 0000:01:00.0, com
pute capability: 7.5"
xla_global_id: 416903419]
```

Figure 10.3 – A list of the processing devices on the local system

5.  Type the following code to verify that the GPU is recognized as a GPU:

```
import tensorflow as tf
tf.config.list_physical_devices('GPU')
```

The output shown in *Figure 10.4* is typical. It tells you the device name and the device type. Only a GPU with the correct hardware support will show up. Consequently, if you don't see your GPU, it's too old to use with TensorFlow or it lacks the appropriate drivers.

```
[PhysicalDevice(name='/physical_device:GPU:0', device_type='GPU')]
```

Figure 10.4 – A description of a GPU on the local system

6.  Verify that the GPU will actually perform the required math when interacting with TensorFlow using the following code. Note that this requires the creation of tensors using `tf.constant()`:

```
tf.debugging.set_log_device_placement(True)
a = tf.constant([[1.0, 2.0, 3.0],
    [4.0, 5.0, 6.0]])
b = tf.constant([[1.0, 2.0],
    [3.0, 4.0],
    [5.0, 6.0]])
c = tf.matmul(a, b)
print(c)
```

The output shown in *Figure 10.5* is precisely what you should see. This code creates two matrixes that it then multiplies together using `matmul()`.

```
Executing op _EagerConst in device /job:localhost/replica:0/task:0/device:GPU:0
Executing op _EagerConst in device /job:localhost/replica:0/task:0/device:GPU:0
Executing op MatMul in device /job:localhost/replica:0/task:0/device:GPU:0
tf.Tensor(
[[22. 28.]
 [49. 64.]], shape=(2, 2), dtype=float32)
```

Figure 10.5 – The result of matrix multiplication using TensorFlow

At this point, you know that your setup will work with the code in the chapter, so you can proceed to the next section about autoencoders with confidence. When creating the examples in this chapter, always remember to use the `tf` environment to start a new code file or modify the kernel using the **Kernel | Change Kernel** menu command.

Now that you have a deepfake setup to use, it's time to employ it by creating an example. The autoencoder is the simplest of the deepfake technologies to understand, so we will cover it in the next section.

# Understanding autoencoders

An **autoencoder** encodes data and compresses it, then decodes data and decompresses it, which doesn't seem like a very helpful thing to do. However, it's what happens during the encoding and decoding process that makes autoencoders useful. For example, during this process, the autoencoder can remove noise from a picture, sound, or video, thus cleaning it up. Autoencoders are simpler than GANs and they're commonly used today for the following important tasks (in order of relevance):

- Data de-noising
- Data dimensionality reduction

- Teaching how more complex techniques work

- Detail context matching (where the autoencoder receives a small high-resolution piece of an image as input and is able to find it in a lower-resolution target image)

- Toy tasks, such as jigsaw puzzle solving

- Simple image generation

The third use means that anyone taking a class on more advanced machine learning techniques will likely encounter autoencoders before GANs because autoencoders are definitely simpler. However, from a hacker's perspective, the data de-noising use is probably more important because a model doesn't actually know what de-noising is. Instead, it's simply an algorithm removing unwanted patterns from data and replacing those elements with wanted patterns. So, a hacker could de-noise data in a manner that benefits them and would likely be unnoticeable to the end user, even though it's quite noticeable to any software encountering the data. In the following sections, you will see more details about autoencoders and how to create one.

---

**Not just videos**

It's essential to remember that deepfakes can be images, sounds, videos, and any other sort of media you can think of. For example, it might be possible to deepfake smells or sensations (not that I've personally encountered either, but it's possible). Sites such as `https://www.tamikothiel.com/cgi-bin/LendMeYourFace.cgi` make it possible to create a deepfake of your face into an image. If you need a fake face, you can create one at `https://generated.photos/`. For a security professional, thinking of deepfakes only as videos will get you into trouble because deepfakes can take all sorts of forms. Consider this, what happens when someone deepfakes the badges used to gain access to your organization? This chapter uses images for deepfakes because they provide a good starting point for infinitely more complex videos. Think about a video as simply being a series of images (which it is) and you'll appreciate why starting with images is a good idea.

---

## Defining the autoencoder

An autoencoder is categorized as self-supervised learning. In other words, there are still labels involved, but the targets are generated by the model itself. The **target function**, the part of the model that is involved in selecting and creating labels, must be designed in such a way that it recognizes useful features in the training data. Part of this process is to provide rules that prioritize functionality; for instance, it's critical to consider **visual macro structure** (elements needed to provide the minimum connectivity required to tell a coherent world story) as being more important than **pixel-level details** (providing a precise match between elements). The concepts vary between autoencoders depending on the purpose the autoencoder serves.

The decoding process is also critical. It can generate output that modifies the input in a specific manner, such as removing noise. A hacker can modify the decoding process by sending bad inputs that modify the manner in which the decoder does things, such as removing noise. If a hacker has direct access to the application, it's possible to modify the decoder in specific ways that modify the output at the pixel level so that humans can't see the changes, but the software does. Consequently, the training of the decoder often involves some trial and error to obtain the specific effect required in the output. You must test decoding periodically to ensure that the decoder continues to work as specified. There are a number of characteristics, besides being self-supervised, that make autoencoders a specific kind of deep learning technology:

- They are data specific, which means that they don't work well on every kind of data in a particular category, only the data that they're trained to work with. As a consequence, you can't use them in a general way, such as compressing sound in the same manner that **MPEG-2 Audio Layer III (MP3)** does.

- The process is lossy, which means that even in the best situation, the output will show some loss of detail.

- An autoencoder is trained with data, not engineered to perform encoding and decoding in a particular way.

Creating an autoencoder requires these three parts:

- **Encoder**: A neural network designed to compress the data using specific algorithms, but the compression relies on the data used to train the neural network.

- **Decoder**: A neural network designed to decompress the data using algorithms that match the compression process provided by the encoder.

- **Loss function**: The method of determining the amount of compression versus the amount of loss that occurs. The loss function defines a balance between how much compression can occur for a given amount of loss, the two characteristics have an inverse relationship.

## Working with an autoencoder example

This section looks at a simple autoencoder that has multiple layers, but this autoencoder simply compresses and then decompresses a series of images. The idea is that you will see that the compression and decompression process is learned by a neural network, rather than hardwired like a **coder-decoder (CODEC)** is. *Figure 10.6* gives you an idea of what this autoencoder looks like graphically.

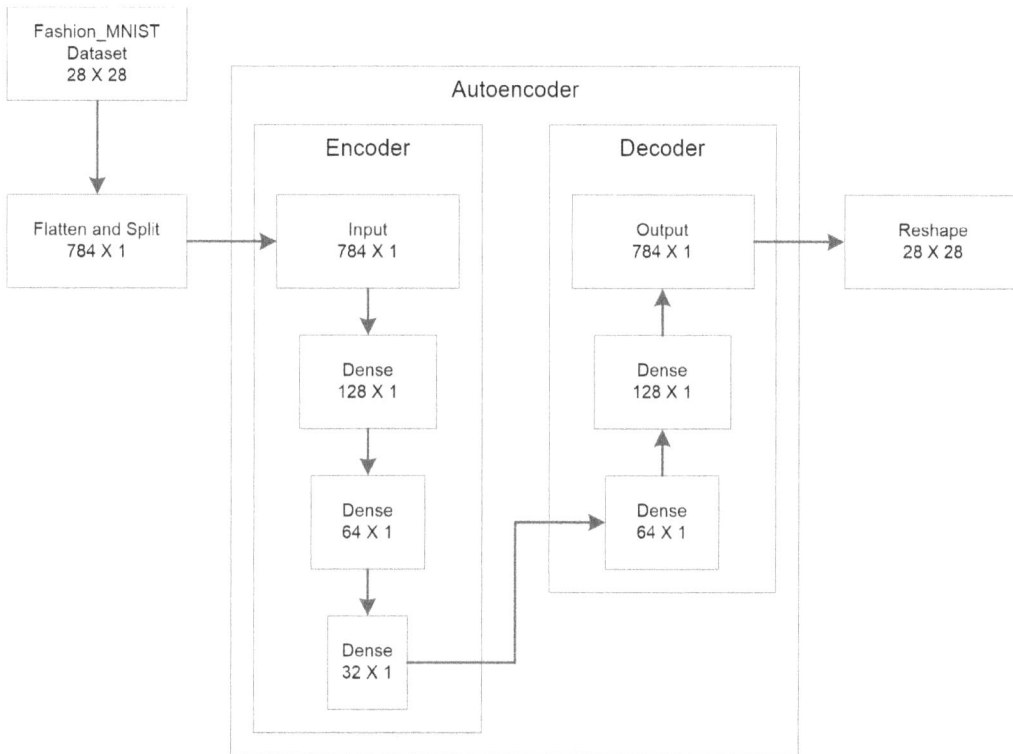

Figure 10.6 – A block diagram of the example autoencoder

The autoencoder consists of two separate programming entities, the encoder and the decoder. Later, you will see in the code that the loss function is included as part of the autoencoder. You can find the source code for this example in the `MLSec; 10; Autoencoder.ipynb` file in the downloadable source.

Now that you have some idea of what the code will do for this example, it's time to look at the various pieces shown in *Figure 10.6*. The following sections detail the items in the block diagram so you can see how they work in Python.

### Obtain the Fashion-MNIST dataset

This example works with a number of different image datasets. However, in this case, you see it used with the common `Fashion-MNIST` dataset because it offers a level of detail that some other datasets don't provide, and it works well in shades of gray. The deepfake aspect of this particular dataset is that you can use it to see how images react to compression and decompression using a trained model, rather than other process. What you need to think about is what sorts of modifications could be made at each layer to make the image look like something it isn't. The following steps show how to obtain and configure the dataset in the example:

1.  Import the required packages:

```
from tensorflow.keras.datasets import fashion_mnist
import matplotlib.pyplot as plt
```

2.  Divide the dataset into training and testing datasets:

```
(x_train, _), (x_test, _) = fashion_mnist.load_data()
x_train = x_train.astype('float32') / 255.
X_test = x_test.astype('float32') / 255.
Print (x_train.shape)
print (x_test.shape)
```

When you perform this step, you will see the output shown in *Figure 10.7*. Unlike other splitting methods shown so far in the book, this one is automatic. However, you could also split it manually if desired. Notice that the shape shows the size of each image is 28x28 pixels.

```
(60000, 28, 28)
(10000, 28, 28)
```

Figure 10.7 – The output showing the split between training and testing datasets

3.  Create the showFigures() function to provide a means of displaying any number of the dataset images:

```
def showFigures(dataset, n_items=10, title="Test"):
    fig, axs = plt.subplots(1, n_items,
        constrained_layout=True)
    fig.suptitle(title, y=0.65, fontsize=16,)
    for i in range(n_items):
        plt.gray()
        axs[i].imshow(dataset[i])
        axs[i].get_xaxis().set_visible(False)
        axs[i].get_yaxis().set_visible(False)
    plt.show()
```

4.  Test the showFigures() function:

```
showFigures(x_train, title="Training Data")
```

The output shown in *Figure 10.8* demonstrates that this is a good dataset to choose for looking at the level of detail. In addition, the images do look good in shades of gray.

Training Data

Figure 10.8 – Ten of the images from the training dataset

Now that you have a dataset to use, it's time to start working with it using an autoencoder. The next section addresses the encoder part of the autoencoder.

### Build an encoder

The first part of an autoencoder is the encoder. This is the code that does something to compress or otherwise manipulate the data as part of the input stream. This example exclusively uses the Dense layers, but there are a great many more layer types listed at https://keras.io/api/layers/ that perform tasks other than simple compression. For example, instead of flattening the image outside of the model, you could use the Flatten layer to flatten it inside the model. The following steps show how to create the encoder for this example:

1.  Import the required packages:

    ```
    from tensorflow.keras import layers
    from tensorflow import keras
    import tensorflow as tf
    import numpy as np
    ```

2.  Reshape the training and testing datasets for use with the model:

    ```
    x_train = x_train.reshape((len(x_train),
        np.prod(x_train.shape[1:])))
    x_test = x_test.reshape((len(x_test),
        np.prod(x_test.shape[1:])))
    print(x_train.shape)
    print(x_test.shape)
    ```

The output from this step, shown in *Figure 10.9*, indicates that each image is now a 784-bit vector, rather than a 28x28-bit matrix.

```
(60000, 784)
(10000, 784)
```

Figure 10.9 – Each image is now a 784-bit vector

3.  Create the encoder starting with an `Input` layer and then adding three `Dense` layers to progressively compress the image. All three of the `Dense` layers use the **rectified linear activation unit (ReLU)** activation function, which outputs the value directly when it is positive, otherwise, it outputs a value of 0 (`https://machinelearningmastery.com/rectified-linear-activation-function-for-deep-learning-neural-networks/`).

```
encoder_inputs = keras.Input(shape=(784,))
encoded = layers.Dense(
    128, activation="relu")(encoder_inputs)
encoded = layers.Dense(
    64, activation="relu")(encoded)
encoded = layers.Dense(
    32, activation="relu")(encoded)
```

Notice that the code describes the shape of each layer and the activation function. A `Dense` layer offers a lot more in the way of configuration options, as explained at `https://keras.io/api/layers/core_layers/dense/`. The value in parenthesis after the call to `Dense()` is a linking value. It links the new layer to the previous layers. So, the first `Dense` layer is linked to `encoder_inputs`, while the second `Dense` layer is linked to the first `Dense` layer. The next step is to build the decoder, which of course links to the encoder.

### Build a decoder

The decoder in this example simply consists of more `Dense` layers. However, instead of making each succeeding layer denser, it makes each layer less dense, as shown in the following code:

```
decoded = layers.Dense(64, activation="relu")(encoded)
decoded = layers.Dense(128, activation="relu")(decoded)
decoded = layers.Dense(784, activation="sigmoid")(decoded)
```

The decoder follows the same pattern as the encoder, but the layers go up in size, as shown in *Figure 10.6*. Notice that the first `Dense` layer is linked to the final encoded layer. The final `Dense` layer uses a sigmoid activation function, which has a tendency to smooth the output (`https://machinelearningmastery.com/a-gentle-introduction-to-sigmoid-function/`)

and ensures that the output value is always between 0 and 1. The encoder and decoder are essentially separate right now as shown in *Figure 10.6*, so it's time to put them into the autoencoder box so that they can work together as described in the next section.

## Build the autoencoder

What this section is really about is putting things into boxes, starting with the autoencoder box. These boxes are models that are used to allow the autoencoder to perform useful work. You'll see that they work together to create a specific type of neural network that can manipulate data in various ways, depending on the layers you use. The following steps take you through the process of building the autoencoder, which involves creating several models:

1.  Create the autoencoder model and display its structure:

    ```
    autoencoder = keras.Model(encoder_inputs, decoded)
    autoencoder.summary()
    ```

    As shown in *Figure 10.10*, the autoencoder model incorporates all of the layers of the encoder and decoder we constructed in earlier sections.

```
Model: "model"
_____
 Layer (type)              Output Shape             Param #
=================================================================
 input_1 (InputLayer)      [(None, 784)]            0

 dense (Dense)             (None, 128)              100480

 dense_1 (Dense)           (None, 64)               8256

 dense_2 (Dense)           (None, 32)               2080

 dense_3 (Dense)           (None, 64)               2112

 dense_4 (Dense)           (None, 128)              8320

 dense_5 (Dense)           (None, 784)              101136

=================================================================
Total params: 222,384
Trainable params: 222,384
Non-trainable params: 0
_____
```

Figure 10.10 – The structure of the autoencoder as a whole

2. Compile the autoencoder, which includes adding a loss function and an optimizer so that the autoencoder works efficiently:

```
autoencoder.compile(optimizer='adam',
    loss='binary_crossentropy')
```

3. Create the encoder model and display its structure:

```
encoder = keras.Model(encoder_inputs,
    encoded, name="encoder")
print(encoder.summary())
```

The output shown in *Figure 10.11* demonstrates that this model is the first half of the autoencoder.

```
Model: "encoder"
_____
Layer (type)                 Output Shape              Param #
=================================================================
input_1 (InputLayer)         [(None, 784)]             0

dense (Dense)                (None, 128)               100480

dense_1 (Dense)              (None, 64)                8256

dense_2 (Dense)              (None, 32)                2080

=================================================================
Total params: 110,816
Trainable params: 110,816
Non-trainable params: 0
_____
None
```

Figure 10.11 – The structure of the encoder model

4. Create the decoder model and display its structure:

```
encoded_input = keras.Input(shape=(32,))
decoder_layer_1 = \
    autoencoder.layers[-3](encoded_input)
decoder_layer_2 = \
    autoencoder.layers[-2](decoder_layer_1)
decoder_layer_3 = \
    autoencoder.layers[-1](decoder_layer_2)
decoder = keras.Model(encoded_input, decoder_layer_3,
```

```
        name="decoder")
    print(decoder.summary())
```

This code requires a little more explanation than the encoder model. First, the encoded_input has a shape of 32, now because it's compressed. Second, each of the decoded layers of the model comes from the compiled autoencoder, which is why they're referred to as autoencoder.layers[-3], autoencoder.layers[-2], and autoencoder.layers[-1]. If you count up the autoencoder model layers shown in *Figure 10.10*, you will see that this arrangement basically begins with the first level of the decoder and works down from there. As with building the decoder, you must also connect the layers together as shown. *Figure 10.12* shows the structure of the decoder model.

```
Model: "decoder"
_____
 Layer (type)                Output Shape              Param #
=================================================================
 input_2 (InputLayer)        [(None, 32)]              0

 dense_3 (Dense)             (None, 64)                2112

 dense_4 (Dense)             (None, 128)               8320

 dense_5 (Dense)             (None, 784)               101136

=================================================================
Total params: 111,568
Trainable params: 111,568
Non-trainable params: 0
_____
None
```

Figure 10.12 – The structure of the decoder model

It's time to create and train the model as a whole by fitting it to the data. This process tracks the learning curve of the neural network so you can see how it works.

## Create and train a model from the encoder and decoder

You still have to fit the autoencoder to the data. Part of this process is optional. The following example uses TensorBoard (https://www.tensorflow.org/tensorboard) to track how the learning process goes and to provide other information covered in the next section:

1.  Import the required packages:

```
from keras.callbacks import TensorBoard
```

2.  Use magics to load TensorBoard into memory. It isn't included by default:

    ```
    %load_ext tensorboard
    ```

3.  Fit the autoencoder model to the data:

    ```
    autoencoder.fit(
        x_train, x_train, epochs=50, batch_size=256,
        shuffle=True, validation_data=(x_test, x_test),
        callbacks=[TensorBoard(log_dir='autoencoder')])
    ```

Defining the number of epochs determines how long to train the model. batch_size determines how many of the samples are used for training during any given epoch. Setting shuffle to True means that the dataset is constantly shuffled to promote better training. Unlike other models that you've worked with, this one automatically validates the data against the test set specified by validation_data. Finally, the callbacks argument tells the fit() function to send learning data to TensorBoard and also tells TensorBoard where to store the data. When you run this cell, you will see the epoch data is similar to that shown in *Figure 10.13* (shortened for the book).

```
Epoch 1/50
235/235 [==============================] - 4s 10ms/step - loss: 0.3793 - val_loss: 0.3205
Epoch 2/50
235/235 [==============================] - 2s 9ms/step - loss: 0.3110 - val_loss: 0.3073
Epoch 3/50
235/235 [==============================] - 2s 9ms/step - loss: 0.3021 - val_loss: 0.3014
```

Figure 10.13 – Epoch data for each epoch of model training

At this point, you can start to see how the model learned by examining the statistics shown in the next section.

### Obtaining and graphics model statistics

The process for viewing the statistics is relatively easy. All you need to do is start TensorBoard using the following code:

```
%tensorboard --logdir 'autoencoder'
```

The output is a relatively complex-looking display containing all sorts of interesting graphs, as shown in *Figure 10.14*. To get the full benefit from them, consult the guide at https://www.tensorflow.org/tensorboard/get_started.

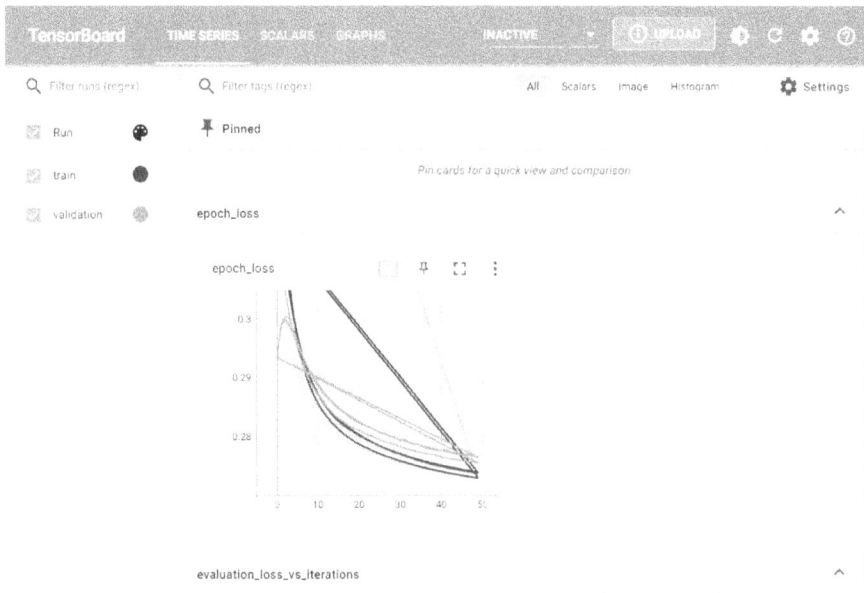

Figure 10.14 – One of many TensorBoard statistical outputs

In this case, what you see is the effect of the model steadily learning. The loss becomes less during each epoch so overall it ends up being quite small. Hovering your mouse over the graph shows data point information with precise values.

> **TensorBoard! Why won't you just die?**
>
> This particular note is designed to save you from pulling out your hair given the advice that you'll likely receive online if you're using a Windows system. Restarting Windows won't do anything for you, so don't waste your time. Windows doesn't support the `kill` command, so adding `!kill <PID>` to your code isn't going to produce anything but an error message if you try it after seeing the `Reusing TensorBoard on port 6006...` message. You may get lucky and find that using the `tasklist` command (`https://learn.microsoft.com/en-us/windows-server/administration/windows-commands/tasklist`) will display the `TensorBoard.exe` application, but it's unlikely in many cases because it isn't actually running. If you do see `TaskBoard.exe`, you can use the `taskkill` command to stop it. However, what you'll end up doing in most cases is to halt and close your Notebook, then stop the Jupyter Notebook server as you normally do at the end of a session. To fix the problem, start by deleting the TensorBoard log files that were created during the fitting process. Next, locate the `\Users\<Username>\AppData\Local\Temp\.tensorboard-info` directory on your system and delete it. At this point, you can restart Jupyter Notebook and go on your merry way, hair intact.

## Testing the model

This section answers the question of whether the model will compress and decompress data with minimal loss. The following steps show you how to test this:

1.  Perform the required prediction:

    ```
    encoded_imgs = encoder.predict(x_test)
    decoded_imgs = decoder.predict(encoded_imgs)
    ```

    The prediction process isn't a single step in this case. So, what you see is the output shown in Figure 10.15, which helps you track the prediction process.

    ```
    313/313 [==============================] - 1s 2ms/step
    313/313 [==============================] - 1s 2ms/step
    ```

    Figure 10.15 – Tracking the prediction process

2.  Display the input and output figures:

    ```
    showFigures(x_test.reshape(10000,28,28),
        title="Original Testing Data")
    showFigures(decoded_imgs.reshape(10000,28,28),
        title="Modified Testing Data")
    ```

    Part of this process reshapes the data so that you can display it on screen. The output shows ten of the images for side-by-side comparison, as shown in *Figure 10.16*.

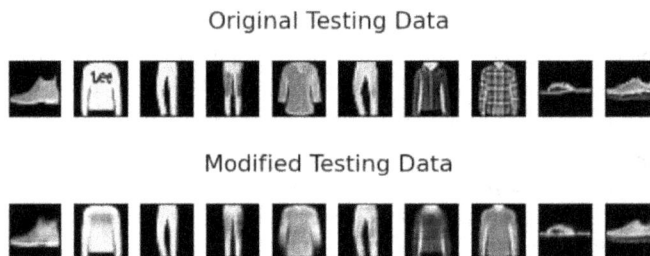

### Original Testing Data

### Modified Testing Data

Figure 10.16 – A side-by-side comparison of input to output

The loss of detail should tell you something about the potential security issues with autoencoders. Because any data manipulation you perform is likely to cause some type of degradation, it pays to choose your models and configuration carefully. Otherwise, it becomes very hard to determine whether a particular issue is the result of hacker activity or simply due to a bad model.

### *Seeing the effect of bad data*

At the outset of this example, you discovered that autoencoders learn how to transform specific data. That is, if you feed the autoencoder what amounts to bad data, even if that data isn't from a hacker, then the results are going to be less useful. This section puts that theory to the test using the following steps. What is important to note is that the model isn't trained again; you're using the same model as before to simulate the introduction of unwanted data:

1. Import the required packages:

```
from tensorflow.keras.datasets import mnist
```

2. Split the data into training and testing sets:

```
(x_train, _), (x_test, _) = mnist.load_data()
x_train = x_train.astype('float32') / 255.
x_test = x_test.astype('float32') / 255.
print (x_train.shape)
print (x_test.shape)
```

3. Reshape the training and testing data:

```
x_train = x_train.reshape((
    len(x_train), np.prod(x_train.shape[1:])))
x_test = x_test.reshape((
    len(x_test), np.prod(x_test.shape[1:])))
print(x_train.shape)
print(x_test.shape)
```

4. Perform a prediction:

```
encoded_imgs = encoder.predict(x_test)
decoded_imgs = decoder.predict(encoded_imgs)
```

5. Compare the input and output:

```
showFigures(x_test.reshape(10000,28,28),
    title="Original Testing Data")
showFigures(decoded_imgs.reshape(10000,28,28),
    title="Modified Testing Data")
```

*Figure 10.17* shows the results of the comparison. The results, needless to say, are disappointing.

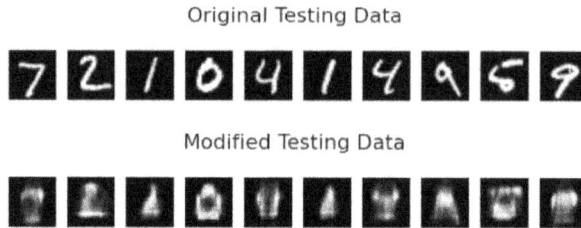

Original Testing Data

Modified Testing Data

Figure 10.17 – A side-by-side comparison of using the model with the wrong data

Remember that this is a basic example where you're in full control of everything that happens. From a security perspective, you need to consider what would happen if a hacker fed your autoencoder bad data without you knowing it. Suddenly, you might start seeing unexpected results and may not be able to track them down very easily.

## Understanding CNNs and implementing GANs

**Convolutional neural networks (CNNs)** are great for computer vision tasks. For example, you might partly depend on facial recognition techniques to secure your computing devices, buildings, or other infrastructure. By adding facial recognition to names and passwords (or other biometrics), you provide a second level of protection. However, as shown in the *Seeing adversarial attacks in action* section of *Chapter 3, Mitigating Inference Risk by Avoiding Adversarial Machine Learning Attacks*, it's somewhat easy to fool the facial recognition application.

The problem isn't the facial recognition application but rather the underlying model, which has been trained with good pictures of the various employees. The way around this problem is to create a dataset that contains both real and fake images of the employees so that the CNN learns to recognize the difference. *Figure 10.18* shows a potential setup for training purposes.

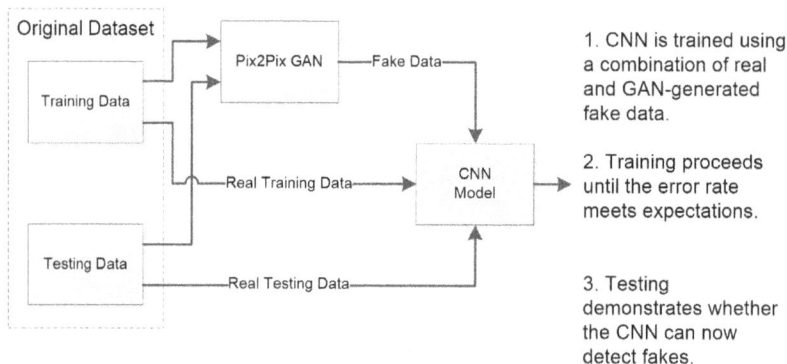

Figure 10.18 – A Pix2Pix GAN use to supplement model training

Using a Pix2Pix GAN is the approach suggested in *Developing a Robust Defensive System against Adversarial Examples Using Generative Adversarial Networks* (https://www.mdpi.com/2504-2289/4/2/11/pdf). Of course, now the issue is finding a way to generate fake images of real employees. That's where a **Pix2Pix** GAN comes into play. You can feed it pictures of the real employee and then have the Pix2Pix GAN create any number of fake images that are tampered with in specific ways. The CNN will learn to differentiate between real and fake images based on the Pix2Pix output. From a hacker's perspective, trying to fool security cameras (as an example) now becomes a lot harder because the security camera software is trained to recognize fake faces.

## An overview of a Pix2Pix GAN

Phillip Isola (et. al) originally presented the idea of a Pix2Pix GAN in the *Image-to-Image Translation with Conditional Adversarial Networks* whitepaper (https://arxiv.org/abs/1611.07004), in 2016. You'd follow essentially the same process to perform the task for your ML application:

1.  Choose an employee picture to modify.

2.  Choose a model.

3.  Perform the translation.

4.  Add the result to a dataset containing both real pictures and fake pictures.

This Pix2Pix GAN example relies on a combination of a U-Net generator and a PatchGAN discriminator, as shown in *Figure 10.19*.

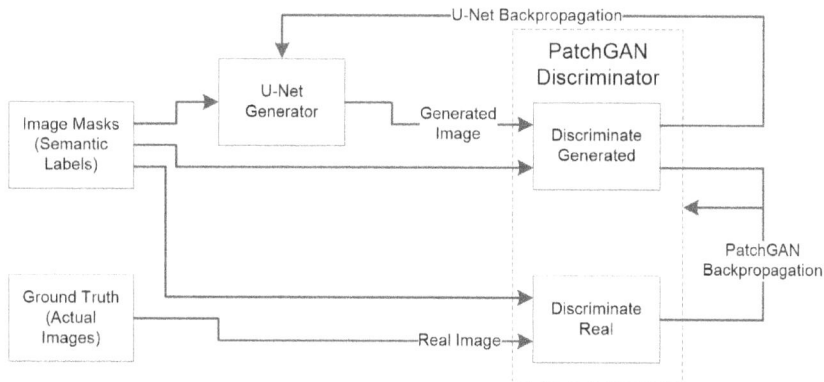

Figure 10.19 – Diagram of a Pix2Pix GAN with U-Net generator and PatchGAN discriminator

Each time the generator creates a new image, the discriminator tests it. If the discriminator can accurately determine whether the image is fake or real, then the generator is updated with new weights so it can produce better output. When the discriminator is unable to determine the fake images from the real ones, the discriminator weights are updated instead. In this way, the two models keep working against each under until the generator can produce a usable output.

## Obtaining and viewing the images

Finding a Pix2Pix GAN example that's specific enough for use in a security setup is hard, which is why this chapter goes into more detail about creating one. This process begins by obtaining a dataset containing the images needed to train the GAN from https://www.kaggle.com/datasets/ balraj98/facades-dataset. Once you download the dataset, unarchive it in the facades subdirectory of the example code. The following steps will get you started on manipulating the images:

1.  Import the required packages:

    ```
    import tensorflow as tf
    from matplotlib import pyplot as plt
    ```

2.  Define the image size and location:

    ```
    IMG_WIDTH = 256
    IMG_HEIGHT = 256
    PATH = 'facades/'
    ```

3.  Create a function for displaying the images:

    ```
    def load(image_file):
        raw_image = tf.io.read_file(image_file)
        decode_image = tf.image.decode_jpeg(raw_image)
        image = tf.cast(decode_image, tf.float32)
        return image
    ```

4.  Separate two images from the rest:

    ```
    real_image = load(PATH+'trainA/40_A.jpg')
    input_image = load(PATH+'trainB/40_B.jpg')
    ```

5.  View the images:

    ```
    fig, axes = plt.subplots(nrows=1, ncols=2)
    axes[0].imshow(input_image/255.0)
    axes[0].set_title("Input Image")
    ```

```
axes[1].imshow(real_image/255.0)
axes[1].set_title("Real Image")
```

*Figure 10.20* shows the **Input image** (semantic labels) on the left and the **Real image** (ground truth) on the right.

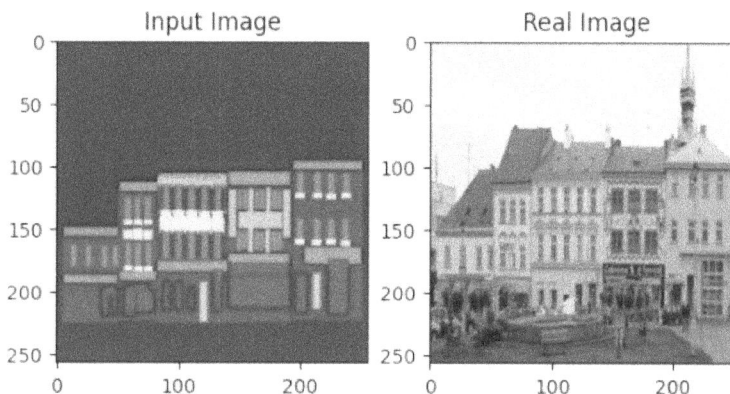

Figure 10.20 – Input image (semantic labels) and real image (ground truth) used as GAN input

Notice how the input image mimics the real image, using colors to label the input image so that the generator has an easier time creating a believable output. The colors relate to features in the real image. There is no actual guide on which colors to use, the colors simply serve to delineate various features. Usually, a human hand creates the input image using techniques such as those discussed at `https://ml4a.github.io/guides/Pix2Pix/`. Distortions in the input image will modify the output, as you see later in the example. A hacker could contaminate the input image database in a manner that modifies the output in specific ways that are to the hacker's advantage. The next section tells you about the image manipulation requirements.

## Manipulating the images

The images aren't very useful in their original form, so it's important to know how to make the images appear in the way that you need them to appear, without losing any data. The following steps show you how to achieve this:

1.  Create a function to resize the images to 286x286x3 to allow for random cropping, resulting in a final size of 256x256x3:

    ```
    def resize(input_image, real_image, height, width):
        input_image = tf.image.resize(
            input_image, [height, width],
    ```

```
        method=tf.image.ResizeMethod.NEAREST_NEIGHBOR)
    real_image = tf.image.resize(
        real_image, [height, width],
        method=tf.image.ResizeMethod.NEAREST_NEIGHBOR)
    return input_image, real_image
```

2.  Create a function to crop the images to a 256x256x3 size in a random manner:

```
@tf.autograph.experimental.do_not_convert
def random_crop(input_image, real_image):
    stacked_image = tf.stack(
        [input_image, real_image], axis=0)
    cropped_image = tf.image.random_crop(
        stacked_image,
        size=[2, IMG_HEIGHT, IMG_WIDTH, 3])
    return cropped_image[0], cropped_image[1]
```

3.  Define a function to add jitter to the image:

```
@tf.function()
def random_jitter(input_image, real_image):
    input_image, real_image = resize(
        input_image, real_image, 286, 286)
    input_image, real_image = random_crop(
        input_image, real_image)
    if tf.random.uniform(()) > 0.5:
        input_image = \
            tf.image.flip_left_right(input_image)
        real_image = \
            tf.image.flip_left_right(real_image)
    return input_image, real_image
```

4.  Create a plot to show four image pairs consisting of an input image and a real image:

```
fig, axes = plt.subplots(nrows=2, ncols=4,
    figsize=(12, 6))
fig.tight_layout(pad=2)
for i in range(2):
    for j in range(0, 4):
```

```
if j%2 == 0:
changed_input_image, changed_real_image = \
random_jitter(input_image, real_image)
axes[i, j].imshow(changed_input_image/255.0)
axes[i, j].set_title("Input Image")
axes[i, j +1].imshow(changed_real_image/255.0)
axes[i, j + 1].set_title("Real Image")
```

*Figure 10.21* shows the typical output at this step, with the various modifications labeled.

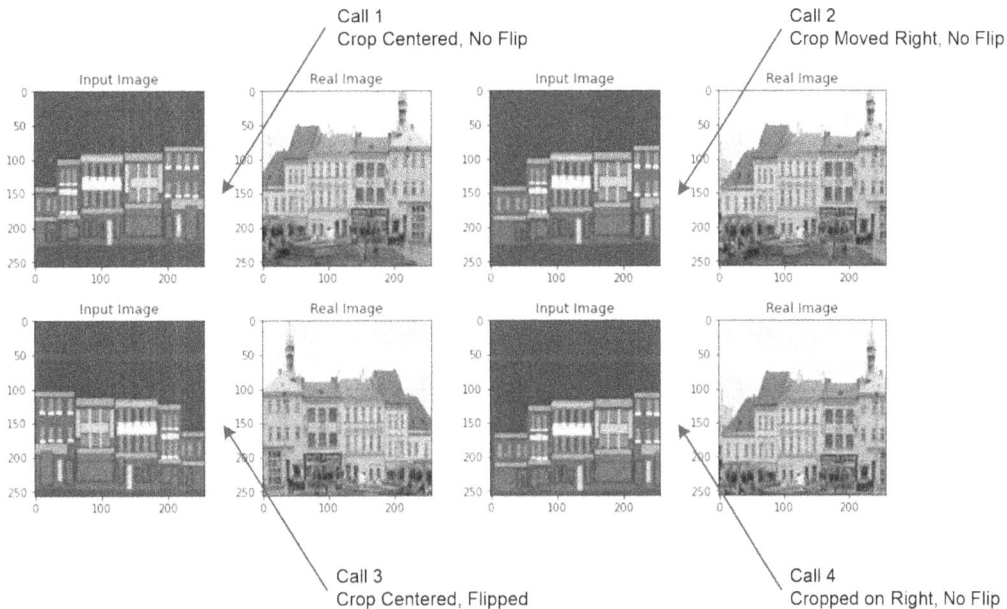

Figure 10.21 – The output of the image modification tests

5.  Define a normalizing function. The act of **normalization** prevents training problems that can occur due to differences in the various images:

```
def normalize(input_image, real_image):
    input_image = (input_image / 127.5) - 1
    real_image = (real_image / 127.5) - 1
    return input_image, real_image
```

6. Perform the image normalization:

```
normal_input_image, normal_real_image = \
    normalize(input_image, real_image)
print(normal_input_image)
```

This final step tests the normalization process. *Figure 10.22* shows the typical output. The actual output is significantly longer than shown.

```
tf.Tensor(
[[[-0.9764706  -0.9764706   0.5294118 ]
  [-0.99215686 -0.9764706   0.5686275 ]
  [-0.99215686 -1.          0.6862745 ]
  ...
  [-1.         -1.          0.7411765 ]
  [-1.         -1.          0.7411765 ]
  [-1.         -1.          0.7411765 ]]
```

Figure 10.22 – The normalized façade image output

Note that while it's easy to determine that the data is acceptable in form, it's not possible to tell that it's the right data. Hacker modifications would be impossible to detect at this point, which is why you need to perform testing and verification of data when it's in a form that you can detect modifications. Now that everything is in place to manipulate the images, it's time to create the actual datasets.

## Developing datasets from the modified images

As with all of the machine learning examples in the book so far, you need a training dataset and a testing dataset to use with the model. The following steps show how to create the required image datasets:

1. Import the required packages:

```
import os
```

2. Create a function to load a training image. The training images have to be randomized for the model to work well. However, the testing images should appear as normal to truly test the model in a real-world setting:

```
@tf.autograph.experimental.do_not_convert
def load_image_train(files):
    input_image = load(files[0])
    real_image = load(files[1])
```

```
input_image, real_image = \
    random_jitter(input_image, real_image)
input_image, real_image = \
    normalize(input_image, real_image)
return input_image, real_image
```

3. Create a function to load a testing image. Note that the images must be resized to 256x256 so that they appear the same as in the original dataset:

```
@tf.autograph.experimental.do_not_convert
def load_image_test(files):
    input_image = load(files[0])
    real_image = load(files[1])
    input_image, real_image = resize(input_image,
        real_image, IMG_HEIGHT, IMG_WIDTH)
    input_image, real_image = normalize(input_image,
        real_image)
    return input_image, real_image
```

4. Specify the dataset parameters. This is the same as the batch_size setting used when fitting the autoencoder in the previous example:

```
BUFFER_SIZE = 200
BATCH_SIZE = 1
```

5. Create a list of files to process for the training dataset:

```
real_files = os.listdir(PATH+'trainA')
real_files = \
    [PATH+'trainA/'+file for file in real_files]
input_files = os.listdir(PATH+'trainB')
input_files = \
    [PATH+'trainB/'+file for file in input_files]
file_list = \
    list(map(list, zip(input_files, real_files)))
for file in file_list:
    print(file)
```

The input and real files appear in two different directories. Yet, these files are actually paired with each other. Consequently, what this code does is create a list of lists, where each pair appears in its own list. *Figure 10.23* shows the output from this step.

```
['facades/trainB/100_B.jpg', 'facades/trainA/100_A.jpg']
['facades/trainB/101_B.jpg', 'facades/trainA/101_A.jpg']
['facades/trainB/102_B.jpg', 'facades/trainA/102_A.jpg']
['facades/trainB/103_B.jpg', 'facades/trainA/103_A.jpg']
['facades/trainB/104_B.jpg', 'facades/trainA/104_A.jpg']
['facades/trainB/105_B.jpg', 'facades/trainA/105_A.jpg']
['facades/trainB/106_B.jpg', 'facades/trainA/106_A.jpg']
['facades/trainB/107_B.jpg', 'facades/trainA/107_A.jpg']
['facades/trainB/108_B.jpg', 'facades/trainA/108_A.jpg']
['facades/trainB/109_B.jpg', 'facades/trainA/109_A.jpg']
['facades/trainB/10_B.jpg', 'facades/trainA/10_A.jpg']
['facades/trainB/110_B.jpg', 'facades/trainA/110_A.jpg']
['facades/trainB/111_B.jpg', 'facades/trainA/111_A.jpg']
['facades/trainB/112_B.jpg', 'facades/trainA/112_A.jpg']
['facades/trainB/113_B.jpg', 'facades/trainA/113_A.jpg']
['facades/trainB/114_B.jpg', 'facades/trainA/114_A.jpg']
['facades/trainB/115_B.jpg', 'facades/trainA/115_A.jpg']
['facades/trainB/116_B.jpg', 'facades/trainA/116_A.jpg']
['facades/trainB/117_B.jpg', 'facades/trainA/117_A.jpg']
```

Figure 10.23 – The pairings of input and real files used to create the training dataset

6.  Load the training dataset. The first step actually creates the dataset from the `file_list` prepared in the previous step. This series of filenames is used to load the images using the `load_image_train()` function defined in *step 2*. Because the dataset is currently in a specific order, the `shuffle()` function randomizes the image order. Finally, the number of batches to perform is set:

```
train_dataset = \
    tf.data.Dataset.from_tensor_slices(file_list)
train_dataset = \
    train_dataset.map(load_image_train,
        num_parallel_calls=4)
train_dataset = train_dataset.shuffle(BUFFER_SIZE)
train_dataset = train_dataset.batch(BATCH_SIZE)
```

7.  View the result. The dataset includes both features and labels:

```
features, label = iter(train_dataset).next()
print("Example features:", features[0])
print("Example label:", label[0])
```

*Figure 10.24* shows a very short example of what you'll see as output.

```
Example features: tf.Tensor(
[[[-0.8901961  -0.60784316  0.47450984]
  [-0.92941177 -0.6392157   0.39607847]
  [-0.9137255  -0.60784316  0.4039216 ]
  ...
  [-0.84313726 -0.6627451   0.5529412 ]
  [-0.8509804  -0.6392157   0.6862745 ]
  [-1.         -0.8352941   0.5529412 ]]
```

Figure 10.24 – A list of tensors based on the input and real images

8.  Create a list of files to process for the testing dataset. This is basically a repetition of the process for the training dataset:

```
real_files = os.listdir(PATH+'testA')
real_files = \
    [PATH+'testA/'+file for file in real_files]
input_files = os.listdir(PATH+'testB')
input_files = \
    [PATH+'testB/'+file for file in input_files]
file_list = list(map(list, zip(input_files,
        real_files)))
```

9.  Load the testing dataset:

```
test_dataset = \
    tf.data.Dataset.from_tensor_slices(file_list)
test_dataset = test_dataset.map(load_image_test)
test_dataset = test_dataset.batch(BATCH_SIZE)
```

The datasets are finally ready to use. Now it's time to create the generator part of the Pix2Pix GAN.

## Creating the generator

A U-Net generator is known as such because it actually forms a kind of U in the method it uses for processing data. The process consists of **downsampling** (encoding), which compresses the data, and **upsampling** (decoding), which decompresses the data. *Figure 10.25* shows the U-Net for this example.

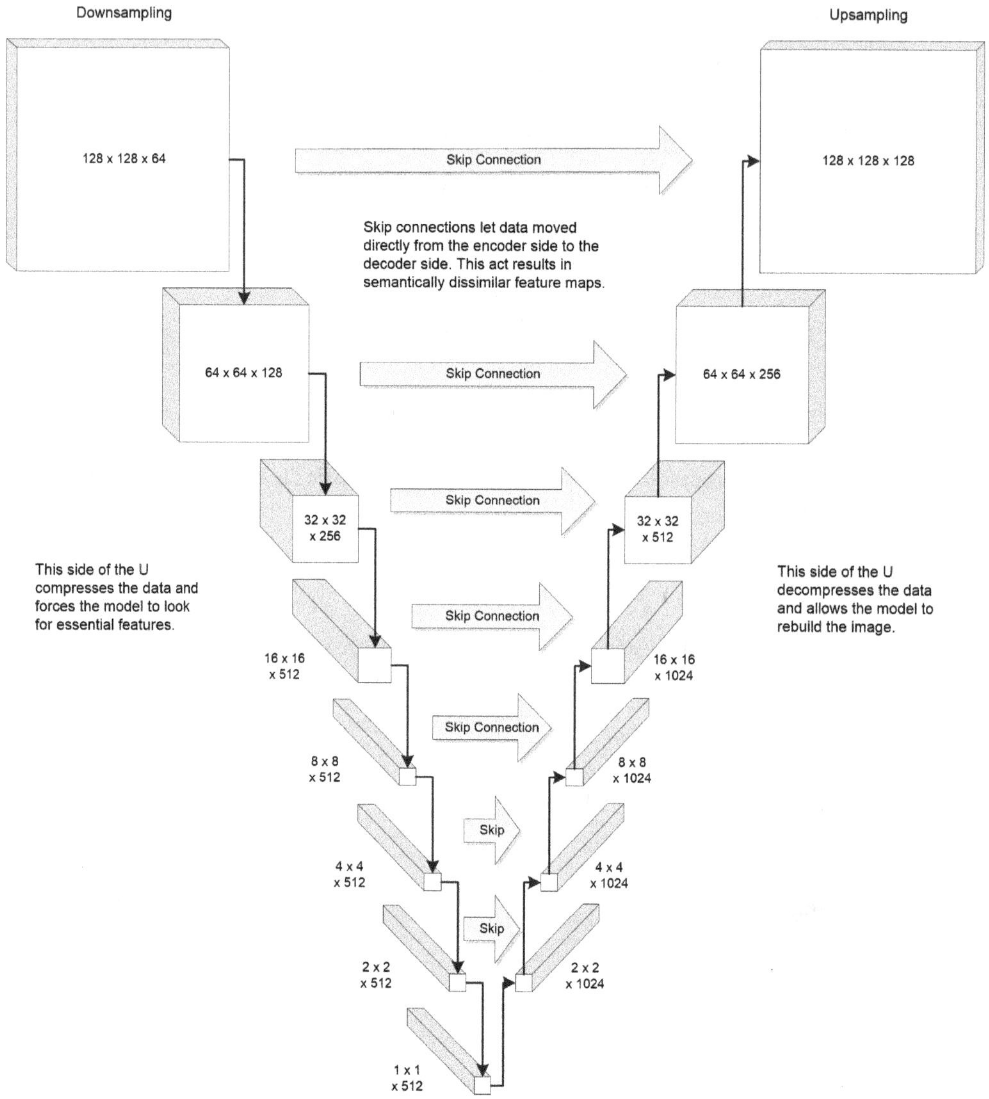

Downsampling

Upsampling

128 x 128 x 64

Skip Connection

128 x 128 x 128

Skip connections let data moved
directly from the encoder side to the
decoder side. This act results in
semantically dissimilar feature maps.

64 x 64 x 128

Skip Connection

64 x 64 x 256

32 x 32 x 256

Skip Connection

32 x 32 x 512

This side of the U
compresses the data and
forces the model to look
for essential features.

This side of the U
decompresses the data
and allows the model to
rebuild the image.

16 x 16 x 512

Skip Connection

16 x 16 x 1024

8 x 8 x 512

Skip Connection

8 x 8 x 1024

4 x 4 x 512

Skip

4 x 4 x 1024

2 x 2 x 512

Skip

2 x 2 x 1024

1 x 1 x 512

Figure 10.25 – A diagram of a U-Net generator

If you think the graphic in *Figure 10.25* looks sort of like a fancy version of the autoencoder in *Figure 10.6*, you'd be right in a way. The U-Net generator does compress and decompress data like the autoencoder but it does so in a smarter way so that it can generate a new image from the existing one. Unfortunately, the model is susceptible to the same forms of hacking as an autoencoder is. Sending bad inputs will affect this model just as much as affects an autoencoder, so you need to exercise care in keeping your model free from hacker activity. Of course, there is a lot more going on than just compression and decompression. The following sections build each element of the U-Net generator in turn.

### Defining the downsampling code

Downsampling relies on a number of layers to compress the data. These layers accomplish the following purposes:

- `Conv2D`: Convolutes the layer inputs to produce a tensor of output values. Essentially, this is the part that compresses the data.

- `BatchNormalization`: Performs a transformation to keep the output mean close to 0 and the standard deviation close to 1.

- `LeakyReLU`: Provides a leaky version of a ReLU that provides activation for the layer. The term **Leaky ReLU** means that there is a small gradient applied when the input is negative.

Now that you have a better idea of what the downsampling layers mean, it's time to look at the required code. The following steps show how to build this part of the U-Net generator:

1. Create a downsample() function:

```
OUTPUT_CHANNELS = 3

def downsample(filters, size):
    initializer = tf.random_normal_initializer(
        0., 0.02)
    result = tf.keras.Sequential()
    result.add(
        tf.keras.layers.Conv2D(filters, size,
            strides=2, padding='same',
            kernel_initializer=initializer,
            use_bias=False))
    result.add(
        tf.keras.layers.BatchNormalization())
    result.add(tf.keras.layers.LeakyReLU())
    return result
```

2. View the downsample() function results. This code tests the downsample() function using just one image:

```
down_model = downsample(3, 4)
down_result = \
    down_model(tf.expand_dims(input_image, 0))
print (down_result.shape)
```

The output is the shape of the tested image, as shown in *Figure 10.26*.

$$(1, \ 128, \ 128, \ 3)$$

Figure 10.26 – Output of the initial downsample() function test

When you compare this test to *Figure 10.26*, you will see that the batch size is reflected in the first return value, the size of the image is reflected in the second and third values, and the number of filters in the fourth value. Because this is the first step of compression, the original 256x256 size of the input image is reduced in half.

### *Defining the upsampling code*

Upsampling relies on a number of layers to decompress the data. These layers accomplish the following purposes:

- `Conv2DTranspose`: Deconvolutes the layer inputs to produce a tensor of output values. Essentially, this is the part that decompresses the data.

- `BatchNormalization`: Performs a transformation to keep the output mean close to 0 and the standard deviation close to 1.

- ReLU: Provides activation for the layer.

As you can see, the upsampler follows a process similar to the downsampler, just in reverse. The following steps describe how to create the upsampler and test it:

1.  Create the upsample() function:

```
def upsample(filters, size):
    initializer = tf.random_normal_initializer(
        0., 0.02)
    result = tf.keras.Sequential()
    result.add(
        tf.keras.layers.Conv2DTranspose(filters,
            size, strides=2, padding='same',
            kernel_initializer=initializer,
            use_bias=False))
    result.add(
        tf.keras.layers.BatchNormalization())
    result.add(tf.keras.layers.ReLU())
    return result
```

2. View the `upsample()` function results:

```
up_model = upsample(3, 4)
up_result = up_model(down_result)
print (up_result.shape)
```

This code is upsampling the downsampled image, so *Figure 10.27* shows that the image is now 256x256 again.

$$(1, 256, 256, 3)$$

Figure 10.27 – The upsampled result of the downsampled image

As with the autoencoder example, you now have a downsampler (which is akin to the encoder) and an upsampler (which is akin to the decoder). However, you don't have an entire U-Net generator yet. The next section takes these two pieces and puts them together to create the generator depicted in *Figure 10.25*.

## Putting the generator together

You now have everything needed to create a U-Net like the one depicted in *Figure 10.25* (referencing the figure helps explain the code). The following code puts everything together. The comments tell you about the size changes that occur as the images are downsampled, then upsampled:

```
def Generator():
    inputs = tf.keras.layers.Input(shape=[256, 256, 3])
    down_stack = [
        downsample(64, 4),    # 128, 128, 64
        downsample(128, 4),   # 64, 64, 128
        downsample(256, 4),   # 32, 32, 256
        downsample(512, 4),   # 16, 16, 512
        downsample(512, 4),   # 8, 8, 512
        downsample(512, 4),   # 4, 4, 512
        downsample(512, 4),   # 2, 2, 512
        downsample(512, 4),   # 1, 1, 512
    ]
    up_stack = [
        upsample(512, 4),     # 2, 2, 1024
        upsample(512, 4),     # 4, 4, 1024
        upsample(512, 4),     # 8, 8, 1024
        upsample(512, 4),     # 16, 16, 1024
```

```
        upsample(256, 4),      # 32, 32, 512
        upsample(128, 4),      # 64, 64, 256
        upsample(64, 4),       # 128, 128, 128
    ]
    initializer = tf.random_normal_initializer(0., 0.02)
    # 256, 256, 3
    last = tf.keras.layers.Conv2DTranspose(
        OUTPUT_CHANNELS, 4, strides=2, padding='same',
        kernel_initializer=initializer, activation='tanh')
    x = inputs
    skips = []
    for down in down_stack:
        x = down(x)
        skips.append(x)
    skips = reversed(skips[:-1])
    for up, skip in zip(up_stack, skips):
        x = up(x)
        x = tf.keras.layers.Concatenate()([x, skip])
    x = last(x)
    return tf.keras.Model(inputs=inputs, outputs=x)

generator = Generator()
```

The code begins, just as the autoencoder did, with the creation of an Input layer. Next, comes a series of downsampling and upsampling steps to perform. The process ends with a final call to Conv2DTranspose(), defined as function last(), which returns the image to its former size of 256x256x3. So, just as with the autoencoder, you have an input, compression stages, decompression states, and an output layer. The main difference, in this case, is that the model uses the tanh activation, which is similar to the sigmoid activation used for the autoencoder, except that tanh works with both positive and negative values.

As shown in *Figure 10.25*, this model allows for the use of skips, where the data goes from some level of the downsample directly to the corresponding layer of the upsample without traversing all of the layers. This approach allows the generator to create a better model because not every image is processed to precisely the same level. The skips are completely random. The final step is to actually create the generator with a call to Generator().

## *Defining the generator loss function*

The loss function helps optimize the generator weights. As the generator produces images, the weights help determine changes in generator output so that the generator output better matches the original image. The following steps show how to create the generator loss function for this example:

1. Create the required `loss_object` function, which uses cross-entropy to determine the difference between true labels and predicted labels:

   ```
   loss_object = \
       tf.keras.losses.BinaryCrossentropy(
       from_logits=True)
   ```

2. Specify the LAMBDA value, which controls the amount of regularization applied to the model:

   ```
   LAMBDA = 100
   ```

3. Define the loss function, which includes calculating the initial loss (`gan_loss`) and the L1 loss (`l1_loss`). Then, use them to create a `total_gen_loss` for the generator as a whole:

   ```
   def generator_loss(disc_generated_output,
       gen_output, target):
       gan_loss = loss_object(
           tf.ones_like(disc_generated_output),
           disc_generated_output)
       l1_loss = tf.reduce_mean(
           tf.abs(target - gen_output))
       total_gen_loss = gan_loss + (LAMBDA * l1_loss)
       return total_gen_loss, gan_loss, l1_loss
   ```

The example uses the `BinaryCrossentropy()` loss function as a starting point for creating a hand-tuned loss function for the U-Net. It then calculates an additional L1 loss value, which minimizes the error from the sum of all the absolute differences between the true values and the predicted values. The total loss is then calculated by adding the cross-entropy loss to the L1 loss (after equalizing the two values) after having multiplied it by a constant. Empirically it has been demonstrated that this combination of losses helps the network to converge faster and in a more stable way.

## Creating the discriminator

The PatchGAN is a type of discriminator where a patch of a specific size (30x30x1 in this case) is run across images to determine whether they're fake or real. This example begins with two 256x256x3 images: the first is of the input image from the generator, while the second is the target image from the dataset.

To create the patch, the images are downsampled (compressed). After that, the compressed images are processed in various ways, as determined by the model. *Figure 10.28* shows the process graphically.

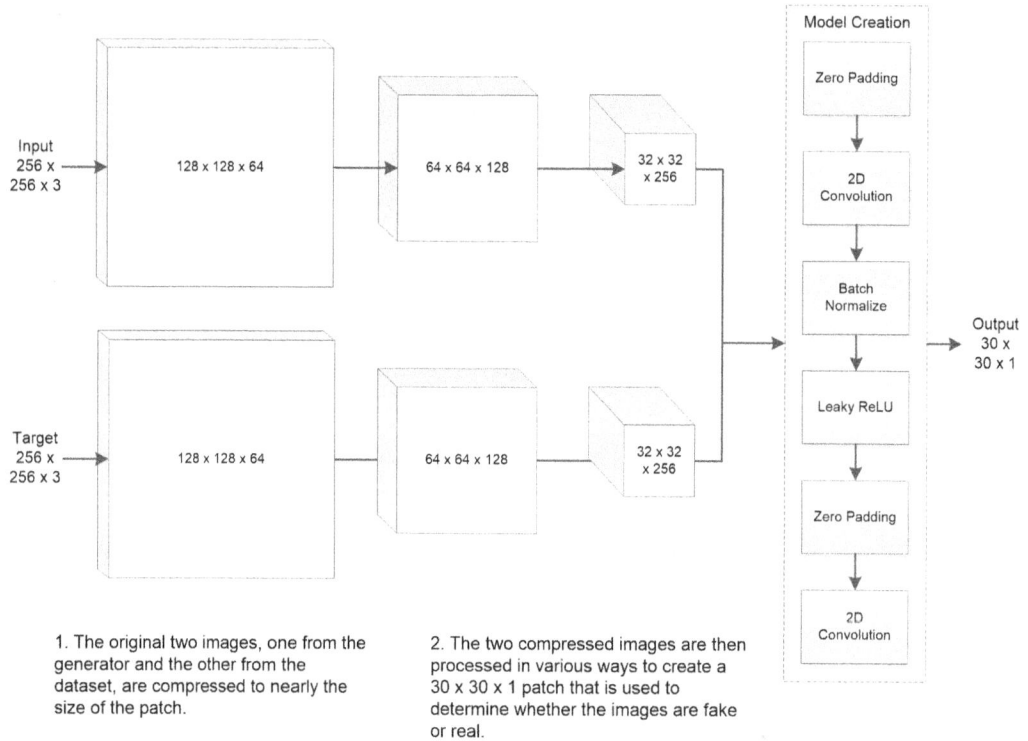

Figure 10.28 – A diagram of the PatchGAN discriminator

As with the generator, the discriminator consists of a number of pieces that are best understood when discussed individually. The following sections tell you about them.

### Putting the discriminator together

Even though *Figure 10.28* may look a little complex, the actual coding doesn't require much effort after working through the intricacies of the U-Net generator. In fact, the code is rather short, as shown here. The comments tell you about the downsampling process so that you can compare it with *Figure 10.28*. There are similarities between this model and the autoencoder from earlier, but in this case, you can see there are two inputs instead of one, so you need to concatenate them:

```
Def Discriminator():
    initializer = tf.random_normal_initializer(0., 0.02)
    inp = tf.keras.layers.Input(
```

```
        shape=[256, 256, 3], name='input_image')
    tar = tf.keras.layers.Input(
        shape=[256, 256, 3], name='target_image')
    # 256, 256, channels*2)
    x = tf.keras.layers.concatenate([inp, tar])

    down1 = downsample(64, 4)(x)         # 128, 128, 64
    down2 = downsample(128, 4)(down1) # 64, 64, 128
    down3 = downsample(256, 4)(down2) # 32, 32, 256

    # 34, 34, 256
    zero_pad1 = tf.keras.layers.ZeroPadding2D()(down3)

    # 31, 31, 512
    conv = tf.keras.layers.Conv2D(
        512, 4, strides=1,
        kernel_initializer=initializer,
        use_bias=False)(zero_pad1)
    batchnorm1 = \
        tf.keras.layers.BatchNormalization()(conv)
    leaky_relu = tf.keras.layers.LeakyReLU()(batchnorm1)

    # 33, 33, 512
    zero_pad2 = \
        tf.keras.layers.ZeroPadding2D()(leaky_relu)

    # 30, 30, 1
    last = tf.keras.layers.Conv2D(
        1, 4, strides=1,
        kernel_initializer=initializer)(zero_pad2)
    return tf.keras.Model(inputs=[inp, tar], outputs=last)

discriminator = Discriminator()
```

The goal of this code is to create a useful model—one that takes the two image sets as input and provides a 30x30 patch as output. Images are processed in patches. It uses these steps to make the determination between fake and real images:

1. Obtain the input and target images sized at 256x256x3.

2. Concatenate the images for processing. Each image should be a separate channel.

3. Downsample the images contained in x (the two concatenated channels) to 32x32x256.

4. Add zeros around the entire image so that each image is now 34x34.

5. Compress the data to 31x31x512.

6. Normalize the image data.

7. Perform the activation function.

8. Add zeros around the entire image so that each image is now 33x33.

9. Create a patch of 30x30x1.

10. Return the model consisting of the original input data and the patch.

The patch is run convolutionally across the image and the results are averaged to determine whether the image as a whole is fake or real.

### Defining the discriminator loss

As with the generator loss function, the discriminator loss function helps optimize weights but the function works with the discriminator in this case, rather than the generator. So, there are two loss functions, one for the generator and another for the discriminator. Here is the code used for the discriminator loss:

```
def discriminator_loss(disc_real_output,
    disc_generated_output):
        real_loss = loss_object(
        tf.ones_like(disc_real_output), disc_real_output)
    generated_loss = loss_object(
        tf.zeros_like(disc_generated_output),
        disc_generated_output)
    total_disc_loss = real_loss + generated_loss

    return total_disc_loss
```

There are a number of steps in calculating the loss:

1. Create a real image loss object using 1s to show this is the real image.

2. Create a generated image loss object using 0s to show this is the fake image.

3. Calculate the loss based on whether the discriminator sees the real image as real and the fake image as fake.

As you can see, the discriminator loss is different from the generator loss in that the discriminator loss determines whether the image is fake or real.

## Performing optimization of both generator and discriminator

To make the process of generating and discriminating images faster and better, the code relies on the Adam() optimization function for both the generator and the discriminator. The Adam() optimization function relies on the stochastic gradient descent method. The primary reason to use it in this case is that it's computationally efficient and doesn't require a lot of memory. A secondary reason is that it works well with noisy data, which is likely going to happen with images. The following code shows the calls for using the Adam() optimizer:

```
generator_optimizer = \
    tf.keras.optimizers.Adam(2e-4, beta_1=0.5)
discriminator_optimizer = \
    tf.keras.optimizers.Adam(2e-4, beta_1=0.5)
```

The first argument determines the learning rate for the generator and discriminator. You normally want these values to be the same or the model may not work as intended. The beta_1 value determines the exponential decay rate for the first moment (the mean, rather than the uncentered variance, which is determined by beta_2). You can read more about this function at https://www.tensorflow.org/api_docs/python/tf/keras/optimizers/Adam.

The next step is to perform the required training. However, because this process takes so long, you want to monitor it so that you can stop training and make adjustments as needed.

## Monitoring the training process

As you train your model, you will want to see how the results change over time. The following code outputs three images: the input image (semantic labels), the ground truth (original image), and the predicted (generated) image:

```
def generate_images(model, test_input, tar):
    prediction = model(test_input, training=True)
    plt.figure(figsize=(15, 15))
```

```
display_list = [test_input[0], tar[0], prediction[0]]
title = ['Input Image', 'Ground Truth',
    'Predicted Image']

for i in range(3):
    plt.subplot(1, 3, i+1)
    plt.title(title[i])
    plt.imshow(display_list[i] * 0.5 + 0.5)
    plt.axis('off')
plt.show()
```

The following code provides a quick check of three images:

```
for example_input, example_target in test_dataset.take(1):
    generate_images(generator, example_input,
        example_target)
```

*Figure 10.29* shows the typical results for an untrained model.

Figure 10.29 – The output of the generate_images() function using an untrained model

Now that you have the means for monitoring the training, it's time to do some actual training of the model, as described in the next section. One thing to consider is that monitoring does provide a method of detecting potential hacker activity. If you have set everything up correctly and your model still isn't producing the predicted image, then you might have problems with the training data, especially the input images, because small modifications would be tough to locate.

## Training the model

Each step of the training occurs in what is termed an **epoch**. It helps to review how the training will occur by viewing *Figure 10.19* as an overview. *Figure 10.25* provides details of the generator and *Figure 10.28* provides details of the discriminator. The following code finally puts together what these figures have been showing you. It outlines a single training step:

```
EPOCHS = 24


@tf.function
@tf.autograph.experimental.do_not_convert
def train_step(input_image, target, epoch):
    with tf.GradientTape() as gen_tape, \
        tf.GradientTape() as disc_tape:
            gen_output = generator(
                input_image, training=True)
        disc_real_output = discriminator(
            [input_image, target], training=True)
        disc_generated_output = discriminator(
            [input_image, gen_output], training=True)
        gen_total_loss, gen_gan_loss, gen_l1_loss = \
            generator_loss(disc_generated_output,
                gen_output, target)
        disc_loss = discriminator_loss(
            disc_real_output, disc_generated_output)
    generator_gradients = gen_tape.gradient(
        gen_total_loss, generator.trainable_variables)
    discriminator_gradients = disc_tape.gradient(
        disc_loss, discriminator.trainable_variables)
    generator_refittings = 3
    for _ in range(generator_refittings):
        generator_optimizer.apply_gradients(zip(
            generator_gradients,
            generator.trainable_variables))
    discriminator_optimizer.apply_gradients(zip(
        discriminator_gradients,
        discriminator.trainable_variables))
```

The code is following these steps during each epoch:

1.  Create two GradientTape() objects (see https://www.tensorflow.org/api_docs/python/tf/GradientTape to record operations for automatic differentiation): one for the generator (gen_tape) and one for the discriminator (disc_tape). Think of a tape used for making backups or to record other kinds of information, except that this one is recording operations.

2.  Generate an image using the generator.

3.  Generate the real output using the discriminator.

4.  Generate the predicted output using the discriminator.

5.  Calculate the generator loss.

6.  Calculate the discriminator loss.

7.  Determine how much to change each model after each training cycle and place this result in generator_gradients for the generator and discriminator_gradients for the discriminator.

8.  Apply the changes to each model, optimizing the result in each case, using generator_optimizer.apply_gradients() for the generator and discriminator_optimizer.apply_gradients() for the discriminator.

Note that you can change the EPOCHS setting as needed for your system. The more epochs you use, the better the model, but each epoch takes a significant amount of time. The next step is to define the fitting function.

> **Specifying how to train the model**
>
> It's important to remember that this is just one training step or epoch. The example performs 24 epochs to obtain a reasonable result. However, many Pix2Pix GANs go through 150 or more epochs to obtain a production-level result. During the testing process of the example, it became evident that the generator wasn't being worked hard enough. So, the example also puts the generator through three application cycles to one for the discriminator, producing a better result in a shorter time. As you work through your code, you'll likely find that you need to make tweaks like this to obtain a better result with an eye toward efficiency.

### Defining the fitting function

Now that the steps to perform for each epoch are defined, it's time to perform the fitting process. Fitting trains the generator and discriminator to produce the desired output. The following steps show the fitting process for this example:

1.  Import the required packages:

    ```
    from IPython import display
    ```

2.  Create the `fit()` function to perform the fitting:

    ```
    def fit(train_ds, epochs, test_ds):
        for epoch in range(epochs):
            display.clear_output(wait=True)
            for example_input, example_target in \
                test_ds.take(1):
                    generate_images(generator,
                        example_input,
                        example_target)
            print("Epoch: ", epoch)
            for n, (input_image, target) in \
                train_ds.enumerate():
                    print('.', end='')
                    train_step(input_image, target, epoch)
            print()
    ```

The `fit()` function is straightforward. All it does is fit the two models (generator and discriminator) to the images one at a time, make adjustments, and then move on to the next epoch. The next section performs the actual fitting process.

## Performing the fitting

So far, no one has really hit the run button. Everything is in place, but now it's time to actually run it, which is the purpose of the code in the following steps:

1.  Perform the actual fitting task:

    ```
    fit(train_dataset, EPOCHS, test_dataset)
    ```

*Figure 10.30* shows the output for a single epoch.

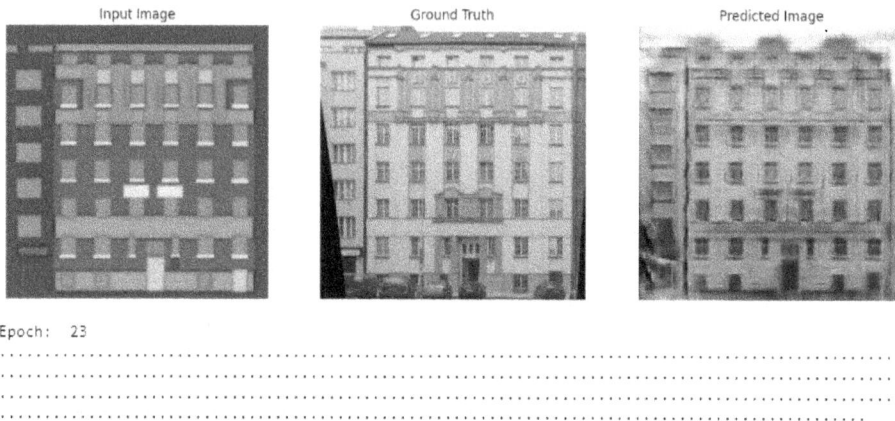

Figure 10.30 – The output from a single training epoch

2.  Check the results:

```
for example_input, example_target in \
    test_dataset.take(5):
        generate_images(generator, example_input,
            example_target)
```

The output will show five random images from the dataset. In looking at the output shown in *Figure 10.31* for a single image from the output, you can see that the input image has affected the original (ground truth) image and produced the predicted output, which isn't perfect at this point because the model requires more training.

Figure 10.31 – Sample output from the trained Pix2Pix GAN

The essential thing to remember about a Pix2Pix GAN is that it's a complex model that requires a large dataset, which gives hackers plenty of opportunity to skew your model. As shown in *Figure 10.31* (and any other Pix2Pix GAN example you want to review), it would be very hard if not impossible for a human to detect that the output has been skewed. What a human would see is that the image has been modified, hopefully in the right direction. A smart hacker could modify the model using any of a number of methods, with incorrect data being the easiest and most probable, to output just about anything.

## Summary

This chapter has provided you with the barest of overviews of deepfakes and the technologies used to create them: autoencoders and GANs. What you should take away from this chapter is the knowledge that these technologies are simply tools that someone can use for good or evil intent. From a security perspective, using deepfakes can help harden your surveillance technologies and help you implement better facial recognition strategies. Of course, you also have to be wary of hackers who modify your models, damage your data, or try to sway the output of your models in a way that is beneficial to them using other methods.

*Chapter 11* is going to move further into the security realm of GANs by looking at ways in which they are used by hackers to gain entry into your systems or by you to thwart hacker advances. The fact that GANs can learn from each experience means that the wall-building security strategies of the past have taken on a new aspect. The machines are not merely hosts for attacks any longer, they have become generators of automated attacks with the parameters of those attacks being controlled by one side or the other.

## Further reading

The following bullets provide you with some additional reading that you may find useful for understanding the materials in this chapter better:

- Enjoy some of the fabulous paintings created by Renoir: *Pierre Auguste Renoir:* `https://www.pierre-auguste-renoir.org/`

- Discover examples of deepfakes that are both interesting and somewhat terrifying: *18 deepfake examples that terrified and amused the internet:* `https://www.creativeblog.com/features/deepfake-examples`

- Read about how deepfakes can change the voices of support people on the phone: *Sanas, the buzzy Bay Area startup that wants to make the world sound whiter:* `https://www.sfgate.com/news/article/sanas-startup-creates-american-voice-17382771.php`

- Gain an understanding of how the legal system has failed with its revenge porn laws in some respects: *Are You Sure You Know What Revenge Porn Is?:* `https://www.wired.com/story/revenge-porn-platforms-grindr-queerness/`

- See how deepfakes are used to scam others: *Top 5 Deepfake Scams that Stormed the Internet this Year*: `https://www.analyticsinsight.net/top-5-deepfake-scams-that-stormed-the-internet-this-year/`

- Review examples of fake people pictures: *New York Times*: `https://www.nytimes.com/interactive/2020/11/21/science/artificial-intelligence-fake-people-faces.html`

- Consider the fact that people actually trust deepfake faces more than the real thing: *People Trust Deepfake Faces More Than Real Faces*: `https://www.vice.com/en/article/7kb7ge/people-trust-deepfake-faces-more-than-real-faces`

- Obtain a list of the best AI art generators today: *Here's how the best AI art generators compare*: `https://www.creativebloq.com/news/ai-art-generator-comparison`

- Get more information about detecting deepfakes in video: *Detect DeepFakes: How to counteract misinformation created by AI*: `https://www.media.mit.edu/projects/detect-fakes/overview/`

- Consider the ease of creating a really good deepfake and then detecting it: *How Easy Is It to Make and Detect a Deepfake?*: `https://insights.sei.cmu.edu/blog/how-easy-is-it-to-make-and-detect-a-deepfake/`

- Gain an understanding of just how terrible deepfake porn can be for women (and most likely men as well): *Deepfakes: The Latest Anti-Woman Weapon*: `https://womensenews.org/2022/05/deep-fakes-the-latest-anti-woman-weapon/`

- Discover how autoencoder approaches are improving unsupervised learning techniques: *Unsupervised Learning of Visual Representations by Solving Jigsaw Puzzles*: `https://arxiv.org/abs/1603.09246`

- Get an idea of how Pix2Pix works without coding it first: *Doodles to Pictures*: `https://mitmedialab.github.io/GAN-play/`

- Read more about the use of color for the facade dataset: *CMP Facade Database*: `https://cmp.felk.cvut.cz/~tylecr1/facade/`

- Gain insights into ReLU functionality: *A Gentle Introduction to the Rectified Linear Unit (ReLU)*: `https://machinelearningmastery.com/rectified-linear-activation-function-for-deep-learning-neural-networks//`

- Discover more about activation function differences, especially the difference between sigmoid and tanh: *Activation Functions in Neural Networks*: `https://towardsdatascience.com/activation-functions-neural-networks-1cbd9f8d91d6`

- Get a better idea of how GAN loss functions work: *A Gentle Introduction to Generative Adversarial Network Loss Functions:* `https://machinelearningmastery.com/generative-adversarial-network-loss-functions/`

- Learn more about how the `Adam()` optimization function works: *Gentle Introduction to the Adam Optimization Algorithm for Deep Learning:* `https://machinelearningmastery.com/adam-optimization-algorithm-for-deep-learning/`

# Leveraging Machine Learning for Hacking

When it comes to any sort of enforcement or security concern, it often helps to take the adversary's point of view. That's what this chapter does, to an extent. You won't see any actual exploit code (which would be unethical, this isn't a junior guide to a hacker's paradise after all), but you will encounter methods that hackers use to employ **machine learning** (**ML**) to do things such as bypass Captcha and harvest information. Discovering the techniques used can greatly aid in your own security efforts.

The chapter also reviews some of the methods used to mitigate ML attacks by hackers by taking the hacker's eye-view of things. This approach differs from previous chapters in that you're no longer looking at building a higher wall or considering the hacker's behavior based on their needs and wants, but rather looking at the world of computing from the perspective of how the hacker. In some regards, the best metaphor to use in this chapter would be a **generational adversarial network** (**GAN**) such as the Pix2Pix GAN encountered in the previous chapter. The scenario is one in which the hacker's ML code acts as a generator and your code is more like the discriminator. As the generator improves, the discriminator improves as well. Of course, the output of this security dance isn't a pretty picture or an interesting video. Rather, it's a safe system with data and users fully protected from an attack. With these issues in mind, this chapter discusses the following topics:

- Making attacks automatic and personalized
- Enhancing existing capabilities

## Making attacks automatic and personalized

It's in the hacker's best interest to automate and personalize attacks so that it becomes possible to attack multiple targets at once, but in a manner that's designed to improve the chances of success. Security professionals keep putting up new walls to keep hackers at bay, such as the use of **Completely Automated Public Turing tests to tell Computers and Humans Apart** (**CAPTCHA**) (`http://www.captcha.net/`) to make it harder to access a website through automation. Hackers keep coming up with new ways to bypass these protections by relying on automation to do it.

Once a protection is bypassed, the hacker then needs a method of interacting with users or being seen as a perfectly normal member of the computing community, which requires specific information. Harvesting information discretely helps a hacker appear invisible and makes attacks more successful. The following sections describe both of these strategies in a specific way that you can later apply generically to groups of security problems where automation and personalization are a problem.

## Gaining unauthorized access bypassing CAPTCHA

According to some sources (see `https://pub.towardsai.net/breaking-captcha-using-machine-learning-in-0-05-seconds-9feefb997694`), it's possible to break CAPTCHA solutions in as little as 0.05 seconds. Of course, the hacker needs a good ML model to perform this feat because there is little point in the hacker addressing CAPTCHA at each site without automation. In times past, hackers would train a model using a huge dataset of collected CAPTCHA images, but this is no longer the case. Research shows now that it's possible to use the deepfake strategies discussed in *Chapter 10, Considering the Ramifications of Deepfakes*, to create a dataset of synthetic images. This is one of the reasons that it's important to view deepfake strategies as more than just video or audio. A deepfake strategy makes it possible to create fakes (synthetic reality) of just about anything. CAPTCHA comes in these forms today, but other forms may appear in the future:

- **Checkmark**: The screen presents a simple box containing a checkbox with text similar to "I am not a robot" to the user. The user checks the checkbox to indicate their non-robot status. Unfortunately, this form is extremely easy to overcome and is often used to download malware to the user's machine (check out the article at `https://malwaretips.com/blogs/remove-click-allow-to-verify-that-you-are-not-a-robot/` for details). A hacker wouldn't bother to create a model to overcome this form of CAPTCHA, it really is just too easy to beat as it is.

- **Text**: A graphic displaying words, letters, numbers, or special characters appear on screen that are distorted in some manner to make it hard to use screen scraping techniques to cheat the CAPTCHA setup. The amount of distortion on some sites is severe enough to cause user frustration. Using a deepfake strategy is extremely effective in this case to bypass the CAPTCHA. Unless paired with a sound setup, this form of CAPTCHA also presents accessibility issues (see `https://www.w3.org/TR/turingtest/` for details).

- **Picture**: A series of graphic boxes depicting a common item are presented. The CAPTCHA software may ask the user to choose all of the boxes containing school buses, for example. This form of CAPTCHA does keep hackers at bay a lot better than other forms of CAPTCHA because it's a lot harder to train a good model to bypass it and it's possible to add little defects to the images that thwart even the best models. However, this form of CAPTCHA is even less accessible than any other form because it can't use the `ALT` text to describe the image to people with special needs (as that would defeat the purpose of using the CAPTCHA in the first place).

- **Sound**: Audio is provided by the system and after listening, the user types whatever the voice says into a box. The audio is machine-generated, which makes it hard for people with hearing-related disabilities to hear it (assuming they can hear it at all). Given the state of deepfakes today, this form of CAPTCHA falls between text and pictures with regard to the difficulty to overcome using automation.

- **Logic puzzles**: The person is asked to solve a puzzle, math problem, trivia, spatial problem, or something similar. This form of CAPTCHA is less common than other types because it keeps legitimate users out as well. Theoretically, some type of ML solution could overcome this form of CAPTCHA, but the solution would be specific to a particular site.

- **Biometrics**: Some type of physical characteristic is used to ascertain whether the user is a robot. Unfortunately, this form of CAPTCHA can also keep legitimate users away if they lack a particular characteristic. In addition, biometric deepfakes are available today that make overcoming this form of CAPTCHA relatively easy for the hacker (see `https://www.netspi.com/blog/technical/adversary-simulation/using-deep-fakes-to-bypass-voice-biometrics/` as an example).

To make CAPTCHA accessible for people with disabilities, some sites combine more than one form of CAPTCHA. The most common combination is text and sound, where the sound spells out or says what the text is presenting on screen. The user must specifically select the sound version by clicking a button or checking a box, with the text being the default.

---

**Liveness detection – a better alternative**

It isn't always possible to replace CAPTCHA with another technology, but there are other useful technologies that work better than CAPTCHA and are definitely outside the realm of deepfake strategies today. One of those technologies is liveness detection, which looks for natural human behaviors and the little imperfections that only humans generate. For example, the technique determines whether it's possible to detect sweat pores in facial or fingerprint images, which obviously won't appear in today's deepfakes. Reading *Liveness in biometrics: spoofing attacks and detection* (`https://www.thalesgroup.com/en/markets/digital-identity-and-security/government/inspired/liveness-detection`) will give you a better idea of how this technology, which is based on deep learning techniques, works. Interestingly enough, China is the only country in the world where deepfakes are currently regulated according to `https://recfaces.com/articles/biometric-technology-vs-deepfake`.

---

Of course, the ability to synthesize any security data, such as CAPTCHA images, doesn't answer the whole question. It would be theoretically possible to use the Pix2Pix GAN example in *Chapter 10* to take care of this part of the process. The completion of bypassing CAPTCHA or any other security measure using automation means doing more than just training a model to approximate the image part of the security measure. *Figure 11.1* shows a block diagram of what a complete solution might look like, including some items that may not seem important at first.

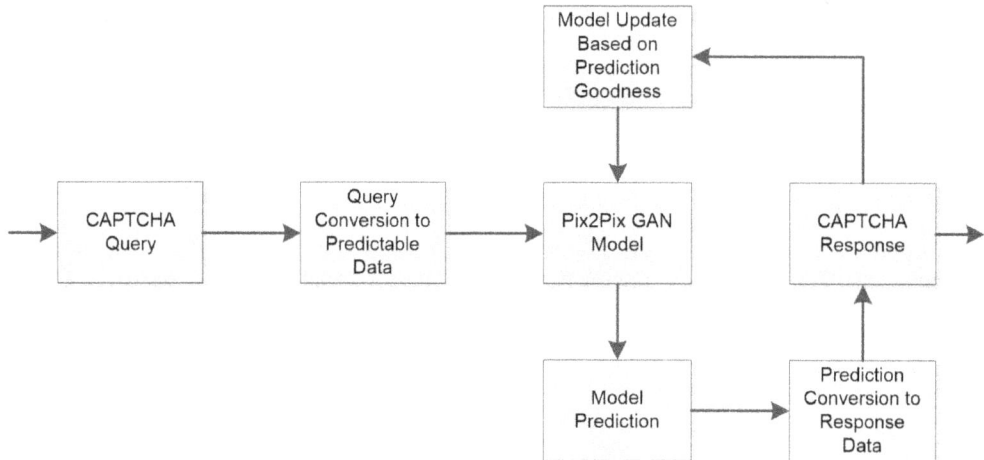

| 1. Hacker's automation accesses site that uses CAPTCHA. | 2. CAPTCHA query is converted to a form the model can understand. | 3. Model predicts the correct response. | 4. If the response is correct, the hacker gains access to the site. Otherwise, the model receives feedback information. |

Figure 11.1 – A block diagram of an automated security measure bypass scheme

The model is only as good as its ability to create a correct prediction, one that will fool CAPTCHA. As improvements are made to CAPTCHA or based on the vagaries of specific site handling of CAPTCHA, the model will receive negative responses from the CAPTCHA application. These negative responses, along with additional data (either real or synthetic) are then used to update the model. As you can see, the process is very much like the Pix2Pix GAN in *Chapter 10*, with the CAPTCHA response mechanism acting as the discriminator.

Most automatic sequences will work in a similar way. The hacker's automation must constantly update to allow access to sites or perform other tasks based on improvements in your security. You can be sure that the hacker is positively motivated to put such augmentation into place, so it's important that your security solution automatically adjusts to the hacker's automation.

Slowing the hacker down

Many security professionals think only in terms of stopping a hacker, but this is a limited view of what can be done. Setting your application up to allow so many tries from a particular IP address and then stopping access to that IP address may not seem like much of a deterrent, but it does serve to slow the hacker down. Even if the hacker has an entire army of IP addresses pounding away at your defenses, shutting down IP addresses that make four or five tries for a period of time reduces the effectiveness of that army. So, why not just shut the IP address off forever? Legitimate users make mistakes too, so unless you want to burden your administrator with endless requests to reopen an IP address, it's important to have a timed response, say 20 minutes for the first offense of four tries and moving up to 24 hours for 12 tries. Making the hacker inefficient is part of a strategy in force to keep the hacker off balance. More importantly, these deterrents can be tracked by an ML application to look for patterns of access that help you better identify which access attempts are hacker-based and which are not. For example, the user who needs several tries to gain access to the site every Monday morning is not a hacker, but an ML application can track that pattern and determine that this is a user and not a hacker.

Once a hacker gains access using some sort of automated method, it's not a very long leap toward automating the collection of data available on the compromised site. The next section deals with this next step in the fully automated attack.

## Automatically harvesting information

According to *35 Outrageous Hacking Statistics & Predictions [2022 Update]* (`https://review42.com/resources/hacking-statistics/`), hacking has become big business. Like any big business, hacking relies heavily on data: your data. It's predicted that hackers will steal 33 billion records in 2023. That's a lot of data, which means that hackers aren't taking time to manually extract these records from the victims of their exploits; like everyone else, they use automation. *Chapter 3* of the book has already told you about the various kinds of ML attacks that hackers use to automate certain kinds of data retrieval. *Chapter 4* looks at user-focused attacks that hackers use to obtain data. *Chapter 5* discusses network-based attacks. However, that's not all of the methods that hackers use, even though it may seem as if there wouldn't be any other methods. This section of the chapter focuses on automation, but it also looks at seemingly impossible methods of retrieving data that hackers use.

### Considering the sneaky approaches hackers employ in search of data

If their goal is to steal data, hackers want people to feel comfortable and unaware (hackers do have other goals as explained in *Chapter 9*). Consequently, a hacker won't hold up a huge red sign with bold yellow letters telling you that data stealing is taking place. Rather, the hacker will gain access to a system using some form of automation as described in the previous section, and then start to look around at what is available on the network drive without looking conspicuous. However, the hacker will have a limited time to look, so automation is key in making the search for valuable information efficient. You can be certain that the hacker will employ various kinds of **bots** (a software agent that is usually very small and extremely efficient in performing a task, usually data-related, requiring limited functionality).

In some cases, the hacker may actually combine methodologies, such as relying on a deepfake to gain employment as a remote employee. According to *FBI Warns Deepfakes Might Be Used in Remote Job Interviews* (`https://petapixel.com/2022/07/05/fbi-warns-deepfakes-might-be-used-in-remote-job-interviews/`), the number of hackers applying for remote work is on the rise. The hacker isn't actually interested in working for the organization, but rather uses the access provided, along with access gained through automation, to obtain significant, seemingly legitimate access to a network and other organizational resources. Unless an administrator is reviewing precisely what the hacker is doing at all times, the access may go unnoticed for a long time. Some statistics say that a data breach can go undetected for 196 days or more. The legitimacy of the hacker being engaged in lawful employment while working away in the background stealing information is something that administrators need to consider.

The ML approach to this problem is to build a model that tracks user activities in an efficient and targeted manner. For example, user activities create log entries (and if they aren't, they should). An ML application can review the logs, along with things such as access times, to build a user behavioral profile that suggests that hacking is taking place. An employee hired to work the support line really has no business looking at finances, for example, and one has to wonder how the employee gained access to such information in the first place. An ML application can learn what support line employees do during the day, even the out-of-the-ordinary sorts of things, and then compare this profile with what a specific employee is doing to discover potential hackers.

## *Defining problems with security through air-gapped computers*

Some organizations try to combat information vulnerability and loss using **air-gapped computers**, which are computers that aren't connected to any unsecured networks. This means that an air-gapped computer has no connection to the internet and may not connect to anything else at all. Typically, you can use air-gapped computers to hold the most valuable organizational secrets, which makes them a prime target for hackers. Of course, gaining access to an air-gapped computer isn't likely to follow any of the techniques found in the book so far, which is where hackers seeking employment to appear legitimate comes into play. There are four common methods of accessing air-gapped computers:

- **Electromagnetic**: Computers put out a tremendous amount of electrical energy. Even if a computer isn't directly connected to a network line, it's possible to monitor these signals in various ways. In fact, the problem goes further. Any metal can act as a conduit for electromagnetic energy. Computers that are actually meant for use in air-gapped situations undergo **TEMPEST** (see `https://apps.nsa.gov/iaarchive/programs/iad-initiatives/tempest.cfm` for details) testing to ensure they don't leak electromagnetic energy. To make certain that the computer doesn't leak information, you'd also need to use it within a **Faraday cage** (a special construct that blocks electromagnetic energy).

- **Acoustic**: With the advent of smartphones, it's possible for a hacker to hack a phone and pick up computer emanations through it. Sounds that are inaudible to humans are quite readable by the hacker. Of course, the hacker would need inside information to discover which smartphones are best to use for this purpose. The only way to ensure your air-gapped computer remains secure is to disallow the presence of smartphones in the vicinity.

- **Optical**: Cameras are everywhere today. It may not seem possible, but the status LEDs on various devices can inform a hacker as to what is happening inside a computer in these common ways: device state, device activity levels, and data content processing. Even though the human eye can't detect what is happening with the modulated LEDs used for data content processing, a computer program can make sense of the pulses, potentially compromising data. Consequently, keeping cameras out of the room is the best idea.

- **Thermal**: Theoretically, it's possible to discover at least some information about an air-gapped computer by monitoring its thermal emanations. However, the specifics of actually carrying out this kind of eavesdropping are limited. White papers such as *Compromising emanations: eavesdropping risks of computer displays* (`https://www.cl.cam.ac.uk/techreports/UCAM-CL-TR-577.pdf`) allude to this type of monitoring, but aren't very specific about how it's done.

Most organizations don't actually have air-gapped computers, even if they were sold a system that is theoretically suitable for use in an air-gapped scenario. Unless your computer is TEMPEST certified and used in a room that lacks any sort of outside cabling, it may as well be connected to the internet because some smart hacker will be able to access it. This list of air-gapped computer security issues really isn't complete, but it does show that complacency is likely to cause more problems than simply assuming that any sort of computer system is vulnerable to attack. Hackers tend to find ways of overcoming any security measure you put into place, so simply continuing to build the walls higher won't work. This is why other chapters in this book, such as *Chapter 9*, look at alternative methods of overcoming hacker issues.

Now that you have a better idea of how hackers approach the problem of overcoming your security, it's time to discover methodologies for dealing with it. The next section discusses ML approaches that don't necessarily stop hacker activity. These approaches don't build the wall higher. Instead, they make the hacker work harder, provide some monitoring capability, help you determine behavioral strategies to combat the hackers' seeming ability to attack from anywhere, and help you create a more informed environment based on the hacker's view of security.

# Enhancing existing capabilities

Enhancing security measures means not building the wall higher, but rather making the existing wall more effective and efficient. For the hacker, this means that existing strategies continue to work, but with a lower probability of success. It becomes a matter of not creating a new strategy, but of making the existing strategy work better when it comes to overcoming security solutions. This change in direction can be difficult for the hacker to overcome because it means working through a particular security strategy and any new methodologies learned may not work everywhere. In short, it means a return to some level of manual hacking in many cases because automation is no longer effective.

The following sections provide an overview of strategies for augmenting existing security strategies presented in previous chapters in a manner that doesn't build a higher wall. The goal is to slow the hacker down, create distractions, make automation less effective, and hide data in new ways. A way to look at these sections is that they provide a broader base of protections, whereas previous chapters provided a taller base of protections.

## Rendering malware less effective using GANs

The methods used to train a GAN to detect malware rely heavily on the **adversarial malware examples (AMEs)** provided. Consequently, obtaining better examples would necessarily improve GAN effectiveness without actually relying on a new technology or model creation methodology. The use of generated AMEs would allow security professionals to anticipate malware trends, rather than operate in reaction mode. The following sections describe AMEs in more detail.

### *Understanding GAN problems mitigated by AMEs*

GANs don't typically provide perfect detection or protection because their data sources are tainted or are less than optimal. In addition, model settings can be hard to tweak due to the lack of usable data. Using generated examples would combat the current issues that occur when working exclusively with real-world examples:

- **Optimization issues**: The inability to optimize means that the GAN is inefficient and can't be improved. There are a number of optimization issues that can plague a GAN:

  - **Vanishing gradient**: When the discriminator part of a GAN works too well, the generator can fail to train correctly due to the vanishing gradient problem, which is an inability of the weights within the generator to change due to the small output of the error function with respect to the current weight. One way to combat this problem in this particular scenario is using a modified min-max loss function (`https://towardsdatascience.com/introduction-to-generative-networks-e33c18a660dd`).

  - **Failure to converge**: A convergence failure occurs when the generator and discriminator don't achieve a balance during training. In fact, the generator gets near zero and the discriminator score gets near one, so the two will never converge. The *Specifying how to train the model* note in *Chapter 10* explores why the generator for that example worked three times harder than the discriminator in order to provide a good result that converges properly. You can also find some other ideas on how to correct this problem at `https://www.mathworks.com/help//deeplearning/ug/monitor-gan-training-progress-and-identify-common-failure-modes.html`.

- **Boundary sample optimization**: A problem occurs when a model is trained in such a way that it becomes impossible to make an optimal decision about data that lies on the boundary between one class and another. These issues occur because of outliners, imbalanced training sizes, or an irregular distribution in most cases. Many of the techniques found in *Chapter 6, Detecting and Analyzing Anomalies*, can help remedy the problems of outliers. The examples you've seen so far in the book show how to properly split datasets into training and testing datasets, which helps with the imbalanced training sizes issue. The remedy for the third problem, irregular distribution, is to rely on a parameterized decision boundary, such as that used by a *support vector machine (SVM)* (`https://towardsdatascience.com/support-vector-machine-simply-explained-fee28eba5496`).

- **Model collapse**: It's essential when combatting malware that the GAN be flexible and adaptive. In other words, it should provide a wide variety of outputs that reflect the inputs it receives. Model collapse occurs when a GAN learns to produce just one or two outputs because these outputs are so successful. When this problem happens, a hacker can make subtle changes to a malware attack and the GAN will continue to try the same strategies every time, which means that the hacker will be successful more often than not. Using generated AMEs provides a much broader range of inputs during training, which reduces the likelihood of model collapse. In addition to using generated AMEs, it's also possible to reduce model collapse by causing the discriminator to reject solutions that the generator stabilizes on using techniques such as Wasserstein loss (`https://developers.google.com/machine-learning/gan/loss`) and unrolled GANs (`https://arxiv.org/pdf/1611.02163.pdf`).

- **Training instability**: Training instability can result from overtraining a model in an effort to obtain a higher-quality result. As training progresses, the model can degenerate into periods of high-variance loss and corresponding lower-quality output. The discriminator loss function should sit at 0.5 upon the completion of training, assuming a near-perfect model. Likewise, the generator loss should sit at either 0.5 or 2.0 upon the completion of training. In addition, the discriminator accuracy should hover around 80%. However, depending on your settings, more training may not result in better values, which is why the *Monitoring the training process* section of *Chapter 10* describes one method for monitoring training progress and emphasizes the need to perform this task. When monitoring shows that the training process is becoming stagnant or unstable, then stopping the training process and modifying the model settings is necessary.

The use of generated examples would overcome these issues by ensuring that the input data is as close as possible to real-world data, but is organized so as to ensure the goodness of the model. When working with real-world data, issues such as data distribution crop up because it simply isn't possible to obtain a perfect dataset from the real world. The term *perfect* means perfect for model training purposes; not that any dataset is ever without flaws (using such a dataset for training wouldn't result in a reliable or useful model). By optimizing the training dataset and also optimizing the model training settings as described in the list, you can create a highly efficient model that will produce better results without actually changing any part of your strategy. The hacker will find that a new strategy for overcoming the model may not be successful due to the model's improved performance.

### Defining a mitigation technique based on AMEs

The Pix2Pix GAN example in *Chapter 10* may not seem relevant to this particular problem, but it could potentially be adapted to generate malware examples in place of pictures. As a security professional, it's important to think without a box, that is, so far outside of the box that the box no longer exists. A computer doesn't care whether the data it manipulates represents pictures or malware to a human; the computer only sees numbers. So, from the *Chapter 10* example, you would need to change the human meaning of the data and the presentation of that data in a human-understandable form. Obviously, the use of the data would also change, but again, the computer doesn't care. In creating a GAN to generate AMEs, you'd still need a generator and a discriminator. The setup of the two pieces would be different, but the process would be the same.

The most critical part of this process is obtaining example malware datasets from which to generate examples artificially. As shown in *Chapter 10*, the generation of examples usually begins with a real-world source. The *Obtaining malware samples and labels* section of *Chapter 7, Dealing with Malware*, tells you where you can obtain the required examples. What's important, in this case, is to focus on getting disabled malware examples because otherwise, you have no idea of just what will come out of the example generation process. The GAN used to detect the malware won't look for the features that actually cause the application to become active, but rather features that define the application as a whole.

The takeaway from this section is that you can make GANs used for any purpose more efficient. For example, you could just as easily use the techniques in this section to tackle fraud. Any detection task will benefit from improved GAN efficiency without having to recreate the model from scratch and without giving the hacker some new strategy to overcome.

### Hackers use AMEs too!

It's possible to find some whitepapers online that discuss the hacker eye view of AMEs. For example, *Generating Adversarial Malware Examples for Black-Box Attacks Based on GAN* (`https://arxiv.org/abs/1702.05983`) discusses the issues that hackers face when trying to overcome what appears to them as a black-box malware detector. It details that feature sets used by ML applications, most especially GANs, using the same approach found in the *Generating malware detection features* section of *Chapter 7, Dealing with Malware*. As part of the strategy to defeat the malware detector, the authors recommend feeding the black box detector carefully crafted examples that modify specific features that a detection GAN normally relies upon for detection, such as API call names, to see how the black box reacts using a substitute detector.

The name of this particular malware detection GAN defeater is **MalGAN,** and it has attracted quite a lot of attention, which means you need to know about it to devise appropriate mitigation strategies for your organization. Page 3 of this white paper even includes a block diagram of the setup used by MalGAN to defeat black box detectors that rely on too few features or static feature sets. As a consequence of this process, the hacker now has AMEs to use to train a GAN to overcome the malware detector GAN. It sounds convoluted, but the process is actually straightforward enough for an experienced hacker to follow.

This sort of example is one of the best reasons to use generated AMEs to train a GAN in the first place, whether it's a malware detector GAN or not. Just in case you're wondering, you can find proof of concept code for MalGAN at `https://github.com/lzylucy/Malware-GAN-attack`. There is also an improved version of MalGAN described in *Improved MalGAN: Avoiding Malware Detector by Leaning Cleanware Features* at `https://ieeexplore.ieee.org/document/8669079`. In the next section of the chapter, you will see a different sort of hacker favorite, spear-phishing attacks, and how ML can help mitigate this particular threat.

## Putting artificial intelligence in spear-phishing

**Spear-phishing** is a targeted attack against a particular individual or small group typically relying on emails purportedly from trusted friends or colleagues as the delivery method. When attacking using telephone calls, the tactic is called **vishing**, or voice phishing. Likewise, when attacking using text messages, the tactic is called **smishing**, for SMS-phishing. No matter how the attack is delivered, the goal of the attack is to obtain sensitive information that the hacker can use to better penetrate a network or other organizational resource. In order to make such an attack work, the hacker already needs to know a considerable amount of information about the target.

To obtain the required information, a hacker might look at public sources, as described in *Chapter 4*. However, some modern users (especially the high-profile targets that hackers want to interact with) have become a little savvier about the kinds of information made available on public sources, so more in-depth research is needed for a successful attack. A hacker would rely on using automation to gain access to organizational resources and then automate a search method for the required data, potentially relying on bots to do the job.

> **Understanding the shotgun alternative**
>
> Specially crafted spear-phishing can take several different routes, one of which could be viewed as the shotgun alternative. Instead of specifically focusing on one individual, the method focuses on a particular group. An email is crafted that should appeal to the group as a whole; the email is then sent out to a large number of individuals in that group, with the hacker hoping for a percentage of hits from the attack. It's important to realize that phishing targets range from an individual, to targeted groups, to people with particular tastes, and to those who are in the general public. However, spear-phishing tactics differ from standard phishing in that spear-phishing attacks are crafted to provide a targeted attack—a spear attack.

Whether the attack is a case of spear-phishing, vishing, smishing, or possibly a combination of all three, the idea that the attack is targeted should produce immediate action on the part of the security professional because this is the kind of attack that can't be addressed in a generic manner. Yes, you can use ML to look for patterns and raise an alarm when those patterns are spotted, but then the security professional must take a course of action that is as unique as the spear-phishing attack. The following sections provide some ideas of how a hacker views this kind of attack, how you can defend against it in the form of an ML application, and possible mitigation strategies.

### Understanding the key differences between phishing and spear-phishing

Spear-phishing is generally a bigger problem for the security professional because the hacker has taken the time to know more than usual about the target. The idea is that one high-value target is likely to garner better information than blanketing the internet as a whole and that the information gained will ultimately net a higher profit. Because of the focus, trying to get rid of the hacker is going to be more difficult—simply ignoring the email may not be enough. Here are some characteristics that differentiate spear-phishing from phishing in general (presented in the order of the gravity of the threat they represent):

- **Group affiliation**: Many spear-phishing attacks rely on ML techniques to discover the most common group affiliations of a particular target. The group affiliation is a lot more specific than whether or not the target uses Facebook or Twitter. A smart hacker will look for unexpected target affiliations, such as the American Association of Bank Directors for financial officers of banks. The email will use the correct heading for the affiliation and will do so without spelling errors in many cases. Your only indicator that the email is spear-phishing is that the URL provided doesn't go to the right place (often the URL differs by just one letter). An ML application to battle such a targeted attack would look at the URLs provided throughout the entire email to find patterns that indicate spear-phishing, rather than a legitimate email.

- **Unexpected or irregular requests from an organizational group**: One hacker's favorite is to send an email that's supposedly from HR. The target user clicks on the link provided by the email and goes to a site that looks like it belongs to HR. This kind of spear-phishing attack focuses on an organization as a whole, with the hacker hoping that a high-profile user will take the bait. It's usually the first step in a multi-step **advanced persistent threat** (APT). Once the hacker has a foot in the door, additional attacks occur to garner more information. An ML application can learn details about HR practices and methods of communication. It can look for errors in the email presentation and any incorrect URLs in the message. This particular kind of attack is generally easy to thwart if captured in its nascent stage.

- **Title specified**: A spear-phishing attack that is targeted at a specific group will have the correct title specified, such as financial officer. However, you can tell that it's targeted at a group because it likely doesn't include a name or other details. A hacker could simply obtain a list of financial officers from corporate annual reports and send the email to the entire group. An ML application to battle this kind of attack would look for the correct title and the correct corporate name (rather than a nickname). It would also look for obvious misspellings that occur because many people's names are not spelled in the traditional manner. Other indicators that an ML application could learn would include odd URLs in the email, misspellings of various terms, missing group information, and other anomalies of the type discussed in *Chapter 6*.

- **Non-monetary target**: Nation states often perpetrate spear-phishing attacks that have nothing to do with monetary gain. These probing attacks may only look for the kinds of information that someone could neglectfully pass along at the cooler. Unfortunately, it's hard to train an ML application to recognize this kind of attack with a high probability of success, so employee training is required. If someone seems too friendly in an email, then perhaps it's time to look at it as potentially hostile in nature.

- **Name included**: If the suspicious email has a name included, it's definitely a spear-phishing attack. A phishing attack has a header such as "*Dear Sir or Madam.*" However, the format of the name is an indicator of spear-phishing in some cases. Only spear-phishing attacks would use the person's full name. You need to consider the use of nicknames (did the hacker research the individual enough to know what other people actually call them). If so, the hacker may have access to a great deal of uncomfortable information about the target and special care must be taken. An ML application could look for anomalies in such an email. For example, if a nickname is used, then the person sending the email should appear in the targeted user's address book.

- **Activities included**: It's a bad idea to include the itinerary for your upcoming trip on social media for a lot of reasons. Local thieves read Facebook too. However, for the spear-phishing hacker, knowing details like this make it easier to craft an authentic-looking email that mentions enough specifics that even a security professional can be taken off-guard. Unfortunately, unless the hacker slips up in other ways, this particular form of attack is very hard to create a model to detect because there really isn't anything to detect. The activity is likely real, so the model looks for other potential ways in which the hacker could slip up within the email's content.

- **Professional acquaintance or service**: Hackers will use any resource needed to gain access to their target. In some cases, it has been reported that a hacker will resort to hacking a professional's email first, such as a lawyer, to gain the trust of the target. Hackers may also rely on organizations that provide services, such as real estate agents. One of the biggest sources of spear-phishing attacks is technical support. A hacker will pose as a representative of some type of technical support for an organization. The obvious kinds of technical support would be hardware, software, or devices. However, hackers also pose as technical support for financial organizations and utility companies (among others). An ML application could help in this case by looking for reception patterns, such as the same email being sent to a number of people in the organization or URLs that don't match the organization sending the message.

- **Friend and relative knowledge**: Hackers will often rely on knowledge of a person's associations to craft a spear-phishing email. For example, an email may say that a friend or relative recommended the person to the company or the target might receive a message providing a certain amount off a product based on the association's recommendation. Using connections between people makes the message seem a lot more believable. Of course, the easiest way to combat this particular attack vector is to call the friend or relative in question and ask if they made the recommendation.

This list isn't complete. Hackers will use any sort of foothold to gain a person's trust in order for a spear-phishing attack to succeed. According to the FBI's Internet Crime Report, the **Internet Crime Complaint Center (IC3)** received 19,369 **Business Email Compromise (BEC)** and **Email Account Compromise** (EAC) complaints in 2020 (see the *Internet Crime Report 2020*, `https://www.ic3.gov/Media/PDF/AnnualReport/2020_IC3Report.pdf`, for details). The Internet Crime Report is an important resource because it helps you understand the ramifications of spear-phishing for your organization in particular, and may offer ideas on things to look for when creating a feature set for your model.

> **Getting law enforcement involved**
>
> I am not a legal professional, so I can't advise you on your legal responsibilities. There, I had to get that out of the way. Even so, it's important to realize that the very personal sorts of attacks described in this section may eventually require the intervention of legal professionals. This is where you make HR and the legal department of your organization aware of the various kinds of attacks taking place and inquire about the direction that mitigation should take, especially when interactions with a nation state are concerned. Many security professionals aren't truly aware of the legal requirements for certain attacks, so it's important to ask rather than simply assume you can take care of the problem yourself. For example, there is now a legal requirement to report certain kinds of attacks in some countries (see *Congress Approves Cyber Attack Reporting Requirement for U.S. Companies*, `https://www.insurancejournal.com/news/national/2022/03/14/657926.htm`, as an example). Otherwise, you may find yourself in an untenable position later.

### Understanding the special focus of spear-phishing on security professionals

If you are in IT of any type, especially if a hacker can determine that you're a security professional, hackers will have a special interest in you as a target. According to research done by Ivanti (`https://www.ivanti.com/resources/v/doc/infographics/ivi-2594_9-must-know-phishing-attack-trends-ig`), here are the statistics regarding spear-phishing attacks against IT professionals:

- 85% of security professionals agree that the attacks have become more sophisticated
- 80% of IT professionals have noticed an increase in attacks in the past year
- 74% of organizations have been a victim of phishing attacks within the past year (40% of them within the last month as of the time of the report)
- 48% of C-level managers have personally fallen prey to spear-phishing attacks
- 47% of security professionals have personally fallen prey to spear-phishing attacks

The most interesting statistic of all is that 96% of organizations report that they provide cybersecurity training. The problem is severe enough that many vendors now offer phishing attack testing for employees and professional training based on test results. Obviously, training alone won't reduce the number of spear-phishing attacks, which is why ML measures are important as well. Unfortunately, as outlined so far in this chapter, ML strategies only work when the model you create is robust and flexible enough to meet the current challenges created by hackers.

## Reducing spear-phishing attacks

A useful strategy when creating a model to detect spear-phishing attacks is to look for small problems in the email. For example, if the organization's name is slightly different in various locations, there are spelling errors, the use of capitalization differs from one location to another, there are differences between the email content and the content on the legitimate site, and so on.

One of the more interesting ways to detect spear-phishing emails today is to check for differences in graphics between the legitimate site and the spear-phishing email. Some organizations focus on logos because it's easy to create a database of hash values for legitimate logos and compare it to the hash values for logos in an email. However, it may not be enough to detect logo differences if the hacker is good at copying the original and ensuring the hash values match. An ML model can learn more advanced techniques for discovering a fake, such as the layout and organization of both text and graphics.

When receiving an email that *appears* to be text, but is actually an image, it's important to think about it as a potential spear-phishing attack for the simple reason that a legitimate sender wouldn't normally have a good reason to send an image. From the user's perspective, it would be hard to see that one email contains text and another email an image of text. An ML application could learn this pattern and classify those containing text images as spear-phishing candidates. Two of the more popular algorithms for performing this work are **Visual Geometry Group-16 layer (VGG-16)** (`https://www.geeksforgeeks.org/vgg-16-cnn-model/`) and **Residual Network (ResNet)** (`https://www.geeksforgeeks.org/residual-networks-resnet-deep-learning/`). When using these models, it's important to introduce perturbations to the training set as described in the *Manipulating the images* section of *Chapter 10,* including the following:

- Cropping
- Downsampling
- Blurring
- Placing essential graphics in unexpected locations and within other graphics
- Modifying the **Hue, Saturation, Value (HSV)** color space

The kinds of perturbations that you introduce into a training dataset depend on the effect that you want to achieve in the model, with cropping, downsampling, and blurring being standard fare. In this case, the model has to be flexible enough to recognize a target graphic, such as a logo, wherever it might appear in the email, or a **text image** (a graphic that contains text, as in a screenshot of a display. It looks like text, but it's actually an image of the text). Consequently, adding the logo in places that the model doesn't expect is a required part of the training. The image may also have differences in HSV from the legitimate site, so the model has to recognize these differences as well.

The next section looks at smart bots. More and more smart bots now appear in all sorts of places online and on local machines. This section does spend a little time with legitimate smart bots but looks at them mostly from the hacker's perspective of being helpful assistants.

## Generating smart bots for fake news and reviews

There are many legitimate uses of bots online. **Smart bots** are simply bots that have an ML basis so that they can adapt to various situations and become more useful to their user. Companies such as SmartBots (https://www.smartbots.ai/) are currently engaged in making common bots, such as chatbots, into smarter assistants. For example, they offer a method of detecting when it's time to transfer a customer to a human representative, which is something a standard chatbot can't easily do. However, hackers have other ideas when it comes to smart bots, using them to improve their chances of success in various kinds of exploits, such as generating fake news and reviews. The following sections discuss the hacker-eye view of fake news and reviews.

### Understanding the uses for fake news

Fake news refers to attempts to pass partially or completely false reports of news off as fully legitimate. There are a number of reasons that fake news can be successful:

- It plays on people's fears
- It provides a sense of urgency
- The stories are often well-written
- There is just enough truth to make the story seem valid
- The stories don't make any outlandish claims
- The content is unexpected

Oddly enough, it's possible to generate fake news using ML techniques as described in *New AI Generates Horrifyingly Plausible Fake News* at https://futurism.com/ai-generates-fake-news. What isn't clear from most articles on this topic is that the model needs sources of real news to create convincing fake news. Using smart bots makes the task of gathering appropriate real news considerably easier. By tweaking the model, the resulting output can become more convincing than real news because the model creators follow a pattern that draws people's attention and then convinces them that the news is real.

Whatever hook the creators of fake news use to draw people in consists of in terms of information, hackers generally use fake news to do something other than present news. The most common form of attack is to download malware to the victim's machine. However, hackers also use fake news to get people to buy products, send money, and perform all sorts of other tasks on the hacker's behalf. These requests are often successful because the person honestly believes that what they're doing is worthwhile and because they haven't been asked to participate in that sort of task before. In short, the hacker makes the person feel important, all the while robbing them of something.

A unique form of fake news attack is the use of a legitimate news outlet to broadcast fake news. Reading *Ukrainian Radio Stations Hacked to Broadcast Fake News About Zelenskyy's Health* (`https://thehackernews.com/2022/07/ukrainian-radio-stations-hacked-to.html`) can tell you quite a bit about this form of attack. As you can imagine, the IT staff for the radio station had to make a serious effort to regain control of the radio station. Besides the delivery method, the target of this attack was the Ukrainian military and people.

Another form of fake news actually does provide real news, but the site serving it is fake. The *Chinese Hackers Deploy Fake News Site To Infect Government, Energy Targets* story at `https://www.technewsworld.com/story/chinese-hackers-deploy-fake-news-site-to-infect-government-energy-targets-177036.html` discusses how a fake news site can serve up real stories from legitimate sources such as the BBC and Sky News. So, the news isn't fake, but the site serving the news takes credit for someone else's stories. When a victim goes to the site to read the story, the ScanBox malware is downloaded to their system. The interesting aspect of this particular attack is that the target is government agencies and the energy industry, not the general public. It's an example of a targeted attack that's successful because it relies on redirection. The victim expects news (and gets it) but receives malware as well. The benefit of using real news is that the victim can discuss the site with others and have totally legitimate news to talk about, so the attack remains hidden from view. The hackers also gain the benefit that they don't have to penetrate the fortifications offered by more secure networks; the attack occurs outside the network.

Training is the main method for handling this particular threat. Employees should be trained to be skeptical. It's a cliché, but one of those that actually works: "*If it sounds too good to be true, then it probably is.*" However, ML techniques can also help pinpoint fake news sites by looking at the page content without actually downloading it. Blacklisting sites that are suspected of serving up fake news will help employees steer clear of it. In addition, models such as Grover (`https://grover.allenai.org/`) make it possible to detect AI-generated fake news.

### Understanding the focus of fake reviews

Fake reviews, unlike fake news, are usually targeted at generating money in some way. A company may decide that a flood of positive reviews on Amazon will help product sales and hire someone to generate them. Likewise, a competitor may devise to generate a wealth of negative reviews to reduce the sales of an adversary. The problem is so severe on sites such as Amazon that you can find tools that measure the real product rating using ML application techniques, such as those described in *These 3 Tools Will Help You Spot Fake Amazon Reviews* (`https://www.makeuseof.com/fake-reviews-amazon/`). Of course, the best way to generate fake reviews is by using smart bots.

Another form of fake reviews makes the apps in online stores seem better than they really are. Stories such as *Fake reviews and ratings manipulation continue to plague the App Store charts* (`https://9to5mac.com/2021/04/12/fake-reviews-and-ratings-manipulation-app-store/`) make it clear that what you think an app will do isn't necessarily what it does do. Efforts to control the effects of these fake reviews often fall short with the developer of the inferior app raking in quite a lot of money before the app is taken out of the store.

Interestingly enough, you don't even have to build a custom smart bot to upload reviews for a product. Sites such as MobileMonkey (`https://mobilemonkey.com/blog/how-to-get-more-reviews-with-a-bot`) tell you precisely how to get started with smart bots in a way that will generate a lot of reviews in a short time and produce verifiable results. If you scroll down the page, you will find that building a smart bot to do the work for you is simply a matter of filling in forms and letting the host software do the rest.

There are methods of telling real reviews from fake reviews that you can build into an ML model. Here are the most common features of real reviews:

- It contains details that don't appear on the product site
- Pros and cons often appear together
- Pictures appear with the reviews (although, pictures can be faked too)
- The reviewer has a history of providing reviews that contain both pro and con reviews
- The text doesn't just mimic the text of other reviews

As you can see, many of these methods fall into the realms of critical thinking and looking for details. Verifying facts is an essential part of tracking down real from fake reviews, and this is an area in which ML applications excel.

**Considering the non-bot form of fake reviews**

Companies such as Amazon have developed better software to detect fake reviews over the years, so some companies have resorted to other means for generating fake (or at least misleading) reviews. One technique creates reviews by selling a product to a potential reviewer and then offering to refund the cost of the product for a five-star review on specific sites. In fact, some companies offer up to a 15% commission for such reviews. The online review won't mention that the reviewer has been paid and is certainly less than objective in providing the review. It's still possible to create ML models to detect such reviews by looking for two, three, or four-star reviews, comparing ratings between online sellers, checking if the reviewer also answers product questions from other customers, and using a timestamp filter to block out review clusters. Moderate reviews that reflect a general consensus across vendor sites from people who also answer questions are usually genuine. In addition, companies often hire reviewers at the same time, so you will see clusters of potentially fake reviews.

## Reducing internal smart bot attacks

Smart bots are successful because they're small, flexible, and targeted. It's possible to have a smart bot on a network and not even know it's there. External smart bots are best avoided through a combination of training, skepticism, and site detection tools. Detecting internal smart bots (those on your network or associated systems) would begin by building a model to note any unusual network activity. For example, examining activity logs could point to clues about the presence of a smart bot.

The techniques shown in *Chapter 7* to detect and classify malware also apply to smart bots. You can look for smart bot features that make it possible to tell malware from benign software. The problem with smart bots is that they are small and targeted, so the probability of identifying a smart bot is smaller than that of detecting other kinds of malware, but it can be done.

One interesting way to possibly locate smart bots is to track various performance indicators, especially network statistics. A sudden upsurge in network traffic that you can't account for might be an indication that a smart bot is at work (once you have eliminated malware as the cause). ML applications are adept at locating patterns in performance data that a human might not see. Purposely looking for these patterns can help locate myriad network issues, including smart bots that communicate with the home base at odd times or use system resources in unexpected ways.

# Summary

This chapter is about taking the hacker-eye view of security, which is extremely porous and profoundly easy to break. The idea is to help security professionals to remain informed that being comfortable and taking half-measures won't keep a hacker at bay. Building the walls higher isn't particularly effective either. Years of one-upmanship have demonstrated that building higher walls simply means that the hacker must devise a new strategy, which often comes even before the walls are built. Creating an effective defense requires that the security professional deal with the hacker from the perspective that the hacker thinks it's possible to overcome any obstacle. This is where ML comes into play because using ML techniques makes it possible to look for hacker patterns that a security professional can exploit to enhance security without building a higher wall. It's time that the hacker becomes more comfortable with an existing strategy and lowers their guard, making it possible for the security professional to gain an advantage.

The next chapter is all about ethics. It's likely not a topic that you've seen covered much, but it's an extremely important security topic nonetheless. The ethical treatment of data, users, and resources tend to enhance security because the security professional becomes more aware of the ramifications of each action. It's no longer about just dealing with hackers, but viewing the targets of the hacker's affection with new eyes. Ethical management techniques help keep a focus on what's important with regard to security.

# Further reading

The following bullets provide you with some additional reading that you may find useful for understanding the materials in this chapter in greater depth:

- Read about how the use of skipping (see *Figure 10.25*) helps improve the CNN used to break CAPTCHA: *A novel CAPTCHA solver framework using deep skipping Convolutional Neural Networks* (https://peerj.com/articles/cs-879.pdf)

- Discover the myriad of bots used for legitimate purposes: *WhatIs.com bot* (https://www.techtarget.com/whatis/definition/bot-robot)

- See a list of commonly offered commercial bots used for various tasks: *Bot Platforms Software* (https://www.saasworthy.com/list/bot-platforms-software) and *Best Bot Platforms Software* (https://www.g2.com/categories/bot-platforms)

- Learn more about air-gapped computers: *What is an Air Gapped Computer?* (https://www.thesslstore.com/blog/air-gapped-computer/)

- Read about the details of hacking an air-gapped computer: *Four methods hackers use to steal data from air-gapped computers* (https://www.zdnet.com/article/stealing-data-from-air-gapped-computers/)

- Invest some time into reading more about the potential for smartphone security issues: *An experimental new attack can steal data from air-gapped computers using a phone's gyroscope* (`https://techcrunch.com/2022/08/24/gairoscope-air-gap-attack/`)

- Consider the ramifications of using a TEMPEST certified system: *5 Things Everyone Should Know About Tempest And Information Security* (`https://www.fiberplex.com/blog/5-things-everyone-should-know-about-tempest-and-information-security.html`)

- Discover how using a **least squares generative adversarial network (LSGAN)** can improve the creation of AME: *LSGAN-AT: enhancing malware detector robustness against adversarial examples* (`https://cybersecurity.springeropen.com/articles/10.1186/s42400-021-00102-9`)

- Find an n-gram version of MalGAN: *N-gram MalGAN: Evading machine learning detection via feature n-gram* (`https://www.sciencedirect.com/science/article/pii/S2352864821000973`)

- Learn more about using logos to detect spear-phishing emails: *Color Schemes: Detecting Logos in Phishing Attacks* (`https://www.vadesecure.com/en/blog/detecting-logos-in-phishing-attacks`)

- Consider the use of smart bots for human augmentation: *Using Smart Bots to Augment Humans* (`https://www.spiceworks.com/tech/innovation/guest-article/using-smart-bots-to-augment-humans/`)

- Read about some political motivations behind fake news: *"Hacker X"—the American who built a pro-Trump fake news empire—unmasks himself* (`https://arstechnica.com/information-technology/2021/10/hacker-x-the-american-who-built-a-pro-trump-fake-news-empire-unmasks-himself/`)

# Part 4 – Performing ML Tasks in an Ethical Manner

Some ML developers somehow feel that it's OK to use any accessible data to train models, perform testing, and create an analysis, no matter what its source might be. However, recent events show that developers of all stripes must now ensure that data is collected ethically, cleaned properly, and used correctly in a transparent manner. Not only will approaching the task in this manner keep lawsuits and regulatory implications at bay, but it also makes ML inherently more secure.

This section includes the following chapter:

- *Chapter 12, Embracing and Incorporating Ethical Behavior*

# 12

# Embracing and Incorporating Ethical Behavior

Ethical behavior with regard to AI, ML, and deep learning is becoming a more significant issue now that people have started to gain experience with these technologies and see their potential limitations. From a security perspective, ethical behavior is important for a number of reasons. For one thing, you don't want your ML model directing you to look for hackers in the wrong place because the model is biased in some way. Another issue for security professionals is the fair treatment of everyone they encounter, whether this person only exists online from your organization's perspective or not. In order to create useful models for security purposes, the data you use must not only meet the legal requirements but also produce something that is both fair and unbiased, which is quite an undertaking because bias hides everywhere.

Besides the issues of fairness and bias, it's also essential to keep privacy in mind when creating a model using any sort of data with **personally identifiable information** (**PII**) in it. You don't want your application to be the source of someone's identity theft nightmare. This chapter also provides some thoughts about how the incorporation of ethical behavior into ML applications of all sorts naturally includes data hiding and data protection in the form of federated learning and other strategies. With these issues in mind, this chapter discusses the following topics:

- Sanitizing data correctly
- Defining data source awareness
- Understanding ML fairness
- Addressing fairness concerns
- Mitigating privacy risks using federated learning and differential privacy

# Technical requirements

This chapter requires that you have access to either Google Colab or Jupyter Notebook to work with the example code. The *Requirements to use this book* section of *Chapter 1*, *Defining Machine Learning Security*, provides additional details on how to set up and configure your programming environment. When testing the code, use a test site, test data, and test APIs to avoid damaging production setups and to improve the reliability of the testing process. Testing over a non-production network is highly recommended, but not absolutely necessary. Using the downloadable source code is always highly recommended. You can find the downloadable source on the Packt GitHub site at `https://github.com/PacktPublishing/Machine-Learning-Security-Principles` or on my website at `http://www.johnmuellerbooks.com/source-code/`.

> **Dataset used in this chapter**
>
> The dataset used in this chapter is custom to this chapter because it demonstrates constructions that you shouldn't use in a real-world application. All of the values used are fictitious and any resemblance to any entity, living or dead, real or not real, or otherwise, is purely coincidence. I'm including this note because some people may find the data offensive in some way. I assure you that no offense is intended and that the data is designed to show you what to avoid to keep from offending someone else.

# Sanitizing data correctly

The act of sanitizing data is to clean it up before using it so that it doesn't contain things such as PII or unneeded features. In addition, sanitization provides benefits to ML models that shouldn't be ignored.

The example in this section relies on a database that is typical of information obtained from a corporate customer database, combined with an opinion poll. The *Importing and combining the datasets* section of *Chapter 9*, *Defending against Hackers*, shows a similar process where you combine mobility data with COVID statistics. This data combination is a common scenario today where businesses ask people's opinions about everything, but the combined form of the database is completely inappropriate as it performs an analysis of how customers feel about product characteristics. Here are the goals for this analysis (which are likely simplified from what you will encounter in the real world, but work fine here):

- Improve sales by making the produce more visually appealing
- Reduce costs by removing less popular product options
- Add new product options that should sell better than those that have been removed

Now that you have some idea of where the data came from and the goals for using it, it's time to look at the sanitization process in more detail. The following sections talk about the sanitization process, its benefits, and its goals.

# Obtaining benefits from data sanitization

It may not seem at first that a developer could gain anything from the data sanitization process and instead would spend a lot of extra time in the office manipulating data to achieve dubious results. However, the stakeholders involved in managing the data and using it for analysis purposes do gain quite a bit from the process:

- Models are faster to train because there is less useless data
- Performing analysis requires fewer resources because the models and datasets are both smaller
- Developers gain better insights into the data
- It's easier to understand the basis for decisions that the resulting model makes
- Models perform better, saving tweaking time
- Avoiding outputs that tend to personalize the information makes it easier for all stakeholders to make better decisions

As you can see, it's not just about making the data clean or avoiding bias; sanitization is about obtaining results that actually reflect the real world in a non-judgmental manner. The ability to obtain a result that you're certain reflects reality is critical to validating ML as a useful methodology to perform analysis and to provide actionable output. When an ML model learns from clean data, the result is a more precise presentation of facts that no one else may even know exists.

# Considering the current dataset

The sample dataset consists of a somewhat reduced copy of the customer data from the customer dataset and the actual survey data. The number of records is very small because they're targeted at sanitization, rather than analysis. The customer data includes the following information:

- First name
- Last name
- Primary address
- Age
- Income
- Gender

There is an assumption that the customer hasn't lied about any of the data, but this usually isn't the case. The age field is a number between 0 and 125, with 0 representing a customer who prefers not to provide an age. The income field is one of seven selected ranges:

- 0 to 19,999

- 20,000 to 39,999

- 40,000 to 59,999

- 60,000 to 79,999

- 80,000 to 99,999

- 100,000 to 119,999

- 120,000 and above

A value of 0 indicates that the person is unemployed or wishes not to provide an income value. The company is somewhat aware of gender issues so this particular field can contain M for male, F for female, X for intersex or non-binary, and N for not willing to say. The survey data consists of the following information:

- Date of survey

- Preferred product color

- Price the customer is willing to pay

- Desire to have Gadget 1 added

- Desire to have Gadget 2 added

The survey allows a limited number of colors: red, yellow, green, blue, or purple. The price value is one of six selections: 159, 169, 179, 189, 199, or 209. By not allowing any fill-in-the-blank entries, performing survey analysis should be easier, but may not reflect a customer's absolute preferences. For example, a customer might feel that 159 is too low, but 169 is too high. With all this in mind, the first step to working with the dataset is to import it. The following code shows this process. You can also find the source code for this example in MLSec; 12; Sanitizing Data.ipynb in the downloadable source:

```
import pandas as pd

sanitize_df = pd.read_csv("PollData.csv")
print(sanitize_df)
```

All this code does is read the dataset into memory and then print it so you can see what it looks like. *Figure 12.1* shows the output of this code.

```
    Date FirstName    LastName                    PrimaryAddress  Age \
0  01/15/23    George       Smith     123 Anywhere, Anywhere, WI 59999   44
1  01/15/23     Sally       Jones     123 Somewhere, Somewhere, NV 89503   32
2  01/15/23     Renee      Walker       123 Nowhere, Nowhere, CA 90011   49
3  01/16/23   Saniago   Dominguez     123 Downthere, Downthere, MN 55144   50
4  01/16/23  Abdullah       Brown         123 Upthere, Upthere, FL 33052    0
5  01/16/23    Fenhua        Yang  123 Aroundthere, Aroundthere, NY 10008   89

   Income Gender ProductColor  Price Gadget1 Gadget2
0   40000      M        Blue    199    True   False
1   80000      F         Red    169   False   False
2   60000      X      Purple    179    True    True
3  100000      N       Green    209    True    True
4   60000      M      Yellow    179   False    True
5   80000      N         Red    199    True    True
```

Figure 12.1 – The unmodified PollData.csv dataset file

No, there aren't many records, but these few records will serve just fine in showing how to sanitize the data. The next section starts the sanitization process.

## Removing PII

Bias can creep into an analysis based on name alone, not to mention where a person lives, their financial status, and so on. In fact, all sorts of data can become an issue when performing analysis, not to mention presenting ethical issues outside of analysis, such as the perception of the people in the dataset. So, the first question you need to ask is whether PII is even needed in your dataset. In many cases, you can simply drop the offending columns. PII is actually categorized into three levels:

- **Personally identifiable**: Enough of the personal information is retained that anyone can identify the person involved. However, at least some of the information is removed to ensure privacy. This is typically not a good solution even for datasets where identification is necessary because the names alone can bias the analysis and the person viewing the result.

- **Re-identifiable**: All of the personal information is removed, but an identifier is added in its place. The identifier is generally a simple number that has no value in and of itself. However, the identifier can be looked up in a separate database later should it become necessary to contact the person. This is the solution normally relied upon for medical research or other datasets where contacting a person later is absolutely necessary. It isn't a good solution for any other purpose.

- **Non-identifiable**: All of the personal information is removed so that the record has no connection whatsoever to the person to whom it applies. This is the solution used for most datasets today because it presents the lowest probability of causing analysis bias or influencing the dataset user. In addition, the removal of PII simplifies the dataset, making it more understandable, and reduces the resources needed for processing and storage.

When removing PII or anything else from an original dataset, store the modified data in a new dataset file and document the process used for modification. The original dataset should be encrypted and stored in a safe location in case the raw data is needed sometime in the future, such as to answer a legal requirement, to provide additional detail, or to recover from an error in manipulating the data. Never modify a raw dataset. If you need to add something to the original data, place it in a new dataset designed for this purpose.

The example dataset has three PII columns: `FirstName`, `LastName`, and `PrimaryAddress`. The company does want to contact the individuals later if there is a question about why they made a certain selection, so it's not possible to eliminate the data, but it also isn't needed for the analysis. The solution is to rely on a re-identifiable setup. The following steps show one technique for making the `sanitize_df` dataset re-identifiable:

1. Create an index for the existing dataset. Nothing fancy is needed; just a simple range will suffice:

```
ids = range(0, len(sanitize_df))
sanitize_df["Id"] = ids
```

2. Copy the PII data to a new dataset and save it to disk. The example assumes that this file is later encrypted and moved to a safe location:

```
saved_df = sanitize_df[[
    "Id", "FirstName",
    "LastName", "PrimaryAddress"]]
print(saved_df)


saved_df.to_csv("SavedData.csv")
```

The output shown in *Figure 12.2* demonstrates that `saved_df` contains all of the PII, plus the `Id` value used to connect `saved_df` with `sanitize_df`.

```
   Id FirstName    LastName                        PrimaryAddress
0   0    George       Smith       123 Anywhere, Anywhere, WI 59999
1   1     Sally       Jones      123 Somewhere, Somewhere, NV 89503
2   2     Renee      Walker         123 Nowhere, Nowhere, CA 90011
3   3   Saniago   Dominguez       123 Downthere, Downthere, MN 55144
4   4  Abdullah       Brown          123 Upthere, Upthere, FL 33052
5   5    Fenhua        Yang  123 Aroundthere, Aroundthere, NY 10008
```

Figure 12.2 – The contents of the saved_df dataset.

3. Drop the unneeded data from the `sanitize_df` dataset:

```
sanitize_df.drop([
    "FirstName", "LastName", "PrimaryAddress"],
    axis='columns', inplace=True)
print(sanitize_df)
```

*Figure 12.3* shows the result of removing PII from this dataset. As you can see, the records now contain no PII, but it's still possible to re-identify the original creator of the survey data.

```
    Date  Age  Income Gender ProductColor  Price  Gadget1  Gadget2  Id
0  01/15/23   44   40000      M         Blue    199     True    False   0
1  01/15/23   32   80000      F          Red    169    False    False   1
2  01/15/23   49   60000      X       Purple    179     True     True   2
3  01/16/23   50  100000      N        Green    209     True     True   3
4  01/16/23    0   60000      M       Yellow    179    False     True   4
5  01/16/23   89   80000      N          Red    199     True     True   5
```

Figure 12.3 – The contents of the modified sanitize_df dataset

The result of this process is a smaller dataset with less potential for bias. Developers often worry about data loss in a process such as this, but the example shows that there hasn't been any data loss in this case. It would be possible to completely reconstruct the original data from what is presented now.

---

**Avoiding collecting PII in the first place**

Whenever possible, it's better not to collect PII unless you absolutely have to have it for some reason. This also means avoiding things such as collecting date of birth. If you need to know how old someone is to perform your analysis, then collect their age as a numeric value in years. In addition, it's always best practice to provide a means for the person to opt out of providing a certain kind of data, whether you need the data or not. In practice, you can fill the data that hasn't been supplied using the missingness techniques found in the *Mitigating dataset corruption* section of *Chapter 2, Mitigating Risk at Training by Validating and Maintaining Datasets*. For example, if an age isn't supplied, you can always use the average of the ages that are supplied to keep the missing value from affecting your analysis. This practice will actually make your dataset easier to use in the long run. Instead of having to calculate the age value using the date of birth, the dataset will contain it directly, saving time, resources, and potential errors.

## Adding traits together to make them less identifiable

The essential goal of this section is to make data less identifiable by combining traits. However, this section also talks about other ways in which combining traits can create a better dataset. Adding traits together helps you achieve the following goals:

- Reduces dimensionality, which reduces model complexity
- Makes the dataset less susceptible to bias
- Improves the opacity of data sources
- Decreases training time
- Moderates resource usage
- Enhances security

You could possibly add a few more items to the list, but this is a very good starting point. As with many data manipulation tasks, this one serves multiple purposes in helping you create better datasets, which in turn creates better models. The example dataset has two (possibly three if you include Gender) columns that you could combine: Age and Income.

Depending on your goals, you could combine these two columns in a lot of ways, but the example takes a simple approach. If you place the ages into decades, you could create the following age groups: 1 to 19, 20 to 29, 30 to 39, 40 to 49, 50 to 59, 60 to 69, and 70 and above. There are seven age groups. If you also use the seven income levels, you could combine the two columns into a value between 1 and 49 by assigning a number to each level and viewing them by using Age as a starting point:

- 1 to 19: Groups 1 through 7
- 20 to 29: Groups 8 through 14
- 30 to 39: Groups 15 through 21
- 40 to 49: Groups 22 through 28
- 50 to 59: Groups 29 through 35
- 60 to 69: Groups 36 through 42
- 70 and above: Groups 43 through 49

You could also add the groups together or multiply them. The idea is to obtain an amalgamation that describes the groups in a particular way without revealing actual values. The approach used by the example makes it impossible to guess the person's age and their income level is based on a 20,000 range. With this in mind, the following steps show how to combine the traits:

1.  Apply the average age to any `Age` column entries with a 0:

```
averageAge = sanitize_df['Age'].mean()
sanitize_df['Age'] = \
    [averageAge if x == 0 else x
    for x in sanitize_df['Age']]
print(sanitize_df)
```

It might be possible to argue that this step could skew the result, but not having a value at all will skew the result even more. There simply isn't a good solution when someone chooses not to provide input. The result of this step appears in *Figure 12.4*.

```
      Date    Age  Income Gender ProductColor  Price  Gadget1  Gadget2  Id
0  01/15/23  44.0   40000     M         Blue    199     True    False    0
1  01/15/23  32.0   80000     F          Red    169    False    False    1
2  01/15/23  49.0   60000     X       Purple    179     True     True    2
3  01/16/23  50.0  100000     N        Green    209     True     True    3
4  01/16/23  44.0   60000     M       Yellow    179    False     True    4
5  01/16/23  89.0   80000     N          Red    199     True     True    5
```

Figure 12.4 – The Age column with 0 values replaced with an average value

2.  Define a function for calculating the `AgeLevel` value for each record. The values are based on the first age value at each level, so the 20 to 29 age group begins with a value of 8:

```
def AgeLevel(Age):
    if Age >= 1 and Age <= 19:
        return 1
    elif Age >= 20 and Age <= 29:
        return 8
    elif Age >= 30 and Age <= 39:
        return 15
    elif Age >= 40 and Age <= 49:
        return 22
    elif Age >= 50 and Age <= 59:
        return 29
    elif Age >= 60 and Age <= 69:
        return 36
    elif Age >= 70:
        return 43
```

3.  Define a function for calculating the `IncomeLevel` value for each record. The income levels appear as values from 0 to 6 for each age group:

```
def IncomeLevel(Income):
    if Income == 0:
        return 0
    elif Income == 20000:
        return 1
    elif Income == 40000:
        return 2
    elif Income == 60000:
        return 3
    elif Income == 80000:
        return 4
    elif Income == 100000:
        return 5
    elif Income == 120000:
        return 6
```

4.  Define a function for calculating `GroupValue` for each record. To obtain `GroupValue`, it's only necessary to add `AgeLevel` to `IncomeLevel`:

```
def GroupValue(Age = 1, Income = 0):
    Group = AgeLevel(Age) + IncomeLevel(Income)
    return Group
```

5.  Create a list of group values:

```
GroupList = []
for Age, Income in \
    zip(sanitize_df['Age'], sanitize_df['Income']):
        GroupList.append(GroupValue(Age, Income))
print(GroupList)
```

This code processes each record in turn and determines the value of `GroupValue` for it. This value is added to the `GroupList` vector for later inclusion in the dataset. *Figure 12.5* shows the output of this step.

$$[24, 19, 25, 34, 25, 47]$$

Figure 12.5 – A vector of GroupValue entries for the dataset

6. Add GroupList to the Group column of the dataset:

```
sanitize_df['Group'] = GroupList
print(sanitize_df)
```

The sanitize_df dataset now has a combined trait column, Group, as shown in *Figure 12.6*.

```
      Date   Age  Income Gender ProductColor  Price  Gadget1  Gadget2  Id  \
0  01/15/23  44.0   40000      M         Blue    199     True    False   0
1  01/15/23  32.0   80000      F          Red    169    False    False   1
2  01/15/23  49.0   60000      X       Purple    179     True     True   2
3  01/16/23  50.0  100000      N        Green    209     True     True   3
4  01/16/23  44.0   60000      M       Yellow    179    False     True   4
5  01/16/23  89.0   80000      N          Red    199     True     True   5

   Group
0     24
1     19
2     25
3     34
4     25
5     47
```

Figure 12.6 – The sanitize_df dataset with the Group column

7. Drop the Age and Income columns:

```
sanitize_df.drop([
    "Age", "Income"],
    axis='columns', inplace=True)
print(sanitize_df)
```

The resulting dataset is now smaller, as shown in *Figure 12.7*, but the essence of the data remains intact so that it's possible to create a robust model. At this point, it would be very tough for someone to trace a particular record back to a specific individual and the resulting model will have a much smaller chance of being biased in any way.

```
      Date Gender ProductColor  Price  Gadget1  Gadget2  Id  Group
0  01/15/23      M         Blue    199     True    False   0     24
1  01/15/23      F          Red    169    False    False   1     19
2  01/15/23      X       Purple    179     True     True   2     25
3  01/16/23      N        Green    209     True     True   3     34
4  01/16/23      M       Yellow    179    False     True   4     25
5  01/16/23      N          Red    199     True     True   5     47
```

Figure 12.7 – The sanitize_df dataset with the Age and Income columns removed

It's possible that you could combine some steps, but using steps of this sort tends to reduce errors in combining the traits and makes the process more transparent. The issue of transparency in data manipulation is at the forefront of many privacy discussions today, so making your code as transparent and easy to follow as possible is going to save time and prevent frustration later.

## Eliminating unnecessary features

The final step in sanitizing a dataset is to remove unnecessary features. As presented in *Figure 12.7*, the example dataset no longer contains any PII, but it still retains all of the information that the original had in *Figure 12.1*. This particular step will remove some possibly useful information, but it really isn't necessary in this case. The idea is to discover which product features people like and including Gender in the model could still possibly bias it a little. As a consequence, the following code removes the Gender column:

```
sanitize_df.drop(["Gender"],
    axis='columns', inplace=True)
print(sanitize_df)
```

*Figure 12.8* shows the final dataset that is fully prepared for the analysis described at the outset of this example. It's now possible to determine which product features are likely to produce more sales. However, if you compare *Figure 12.8* with *Figure 12.1*, you will see that the latest iteration is considerably smaller and easier to work with. Plus, it's easier to explain to anyone outside the project who needs to know about it.

```
       Date ProductColor  Price Gadget1 Gadget2 Id  Group
0  01/15/23         Blue    199    True   False  0     24
1  01/15/23          Red    169   False   False  1     19
2  01/15/23       Purple    179    True    True  2     25
3  01/16/23        Green    209    True    True  3     34
4  01/16/23       Yellow    179   False    True  4     25
5  01/16/23          Red    199    True    True  5     47
```

Figure 12.8 – The final version of the sanitize_df dataset

As a final comment on this transition, the dataset in use now has a considerable number of security features added to it and it wasn't even necessary to program them by hand. For example, it's no longer possible to perpetrate identity theft because there is no PII to obtain. The data itself is less likely to suffer manipulation because of its structure. A hacker could possibly target the Price and Group columns, but it wouldn't be too hard to detect the manipulation in most cases. Overall, this dataset is a lot more secure for having been sanitized.

Now that you have some ideas on how to sanitize your dataset, it's time to consider just where the dataset comes from. It's essential to think about how data from various sources is collected and the intent behind the collection process. In addition, you need to give credit to data collectors where credit is due.

# Defining data source awareness

The internet makes it incredibly easy to locate many common kinds of dataset. There are so many datasets available for some purposes that sometimes it's hard to choose based on the content of the dataset alone. However, content isn't the only consideration. It's also important to consider the third party that collected it. In some cases, datasets are extremely biased or have special requirements that make them inappropriate to use for many kinds of analysis. Even if you were to ignore the issues with the dataset, the experimentation you perform with it would yield less-than-useful results.

## Validating user permissions

Part of data source awareness is to ensure that people using the dataset actually have the need and credentials to use it. This is especially true with datasets that deal with sensitive or confidential materials, or datasets that are controlled by government regulation, such as medical datasets that must follow **Health Insurance Portability and Accountability Act (HIPAA)** requirements. In short, you can't use a dataset that has restrictions placed on it unless you meet the requirements for using that dataset. Here are some considerations for validating user permissions from an ethical perspective when working with datasets:

- Ensure that you meet all the legal requirements

- Validate access to the data for all users that will engage with it

- Determine the limits of validation requirements, such as time issues

- Perform validation for all aspects of the dataset: actual manipulation, model building, data analysis, and result usage as a minimum

- Verify that the dataset will actually meet user needs and that the user has a need to know the information they provide

- Contact the dataset creator to ensure that everyone has the same ideas on what access means

- Train users to understand the responsibilities for using the dataset in an ethical manner and ensure that users sign off on dataset access requirements upon completion of training

It's important to realize that this list is an overview of common user validation requirements. You may have other requirements to consider that are specific to the dataset you want to use. For example, perhaps a dataset on chemical structures has some special government accountability requirement or a dataset used for statistical planning has a time requirement attached to it (*don't open until Christmas!*). This chapter can't detail those kinds of requirements, but you're still required to keep them in mind when validating users who will see them, the analysis provided by them, or the results created by any analysis.

## Using recognizable datasets

The communities of data scientists and researchers of the world create large numbers of datasets for specific purposes that are then peer-reviewed to ensure that they meet a variety of criteria. Generally, if you find a dataset associated with groups such as Kaggle (`https://www.kaggle.com/datasets`), organizations such as scikit-learn (`https://scikit-learn.org/stable/datasets/toy_dataset.html`), and institutions of higher learning where students create datasets for thesis work, you can be sure that the dataset is safe to use. If a dataset proves to be controversial in some way, such as the Boston Housing Dataset, these organizations will usually ensure that everyone knows about it and deprecate the dataset's use. These are the safe sort of datasets that you should seek out first. If a dataset doesn't quite meet your needs, you can always find ways to combine it with another using techniques such as those shown in the *Importing and combining the datasets* section of *Chapter 9*. The point is that safe is better than sorry. It's never a good idea to use the data now and ask for pardon later because data usage comes with consequences, many of them unpalatable.

A recognizable dataset may also be specific to your profession and not very useful to anyone else. In this case, you can find such datasets on websites specific to your organization, in articles, or even in printed material references. It's still important to vet such datasets to ensure that they meet your organization's ethical standards. Data collected in a manner that doesn't meet these standards can cause problems for you, your organization, and those that you serve (such as customers). The question should always come down to the following:

- Was the dataset collected in an ethical manner?

- Has the dataset been sanitized properly?

- Did the collector obtain all required permissions to use the data and are these permissions on open display?

- Can my organization use the dataset in an ethical manner?

- Is everyone who will touch the data in my organization cleared to do so?

- Does the data collector allow peer review?

The last bullet is especially important. Part of the process for creating a recognizable dataset is an open discussion about it in the form of peer review. Perhaps the dataset is really only a good start and it requires further manipulation to make it a better dataset, one that can tolerate full scrutiny by the public. Making sure that all stakeholders who are affected by the dataset have full access to it and can provide comments about it is critical.

> **Working with data ethically**
>
> There are groups that will create a dataset simply because they can. Of course, creating a dataset without any thought whatsoever as to the consequences is an extremely bad idea. In reading *Collating Hacked Data Sets* (`https://www.schneier.com/blog/archives/2020/01/collating_hacke.html`), it's momentarily inviting to think about the uses for such a dataset. For example, you might think that such a dataset would allow exploration of how to circumvent data hacking techniques. However, what the article is really telling you is that there is now a new technique available for collating hacked datasets that provides even more information for stealing money and performing various kinds of extortion, which aren't good uses for data. Whenever you hear the words dark web, it's time to think of looking somewhere else for the data for your project. This particular article is likely meant to make people aware that it's possible to do this sort of work with datasets, but it's still not well considered.

## Verifying third-party datasets

The two previous sections have looked at the elements needed to build awareness of datasets from the user and collection perspectives. It's also possible to obtain less-recognizable datasets that require you to jump through some hoops to obtain them. In most cases, you need to fill out a form to gain access to the dataset and in some cases, you must also pay a fee. These datasets are normally for special purposes and they're used by people who have a particular research or analysis need. However, before you decide to obtain a dataset that may or may not meet ethical standards, it's important to consider the third party that collected it using these criteria:

- Is the third party reputable?

- Has the data been collected in a manner that meets the requirements of the previous two sections?

- Is the data actually required for this project?

- Will restrictions placed on the dataset affect the ability to use the results of any analysis or output?

Depending on the third party and your profession, there are likely other questions you need to ask yourself. Data that is in any way secretive, hidden, restricted, policed, or otherwise out of the public view always carries some penalties with it, the consequences for using the data for any purpose. The biggest question that must be answered then is whether it's possible to live with the consequences of using the data. Do the pros of using the data outweigh the cons?

## Obtaining required third-party permissions

Many datasets come with various caveats. For example, you may have to provide credit to the dataset originator. The data may be licensed in some way, which means adhering to the requirements of the license. In rare circumstances, the data may have a time limit—you can use it today, but then the dataset must be deleted at some point. Whenever data, even public data from a recognizable source, has some sort of permissions restriction placed on it, you must be certain that you meet the permission requirement. Often, when looking at a dataset that appears to have no requirements, it's possible to think that it's free to use. However, unless the site hosting the data especially states that the data is in the public domain and has no permission requirement attached to it, you must assume that the data is copyrighted in some fashion and that you need permission to use it. When in doubt, follow these steps:

1. Contact the dataset originator to ensure that the dataset is free of encumbrance.

2. Obtain the dataset originator's permission in writing, electronic if necessary but written if possible.

3. Verify with your organization that the third-party dataset meets all legal, financial, professional, and policy-related requirements.

4. Keep all permission-related materials you have collected in a secure location that's accessible to your organization's legal, financial, and management staff who need to know about it.

5. Ensure that everyone on your team knows and understands the third-party requirements for working with the data.

Using these steps to ensure that you have proper permissions may seem like a lot of effort for little gain. However, it will almost always save you time and effort in the long run. Plus, following them will ensure that you meet any legal or financial requirements for working with the dataset.

Now that you have a sanitized dataset that meets all of the data awareness requirements, you need to ask whether the dataset is actually fair. In other words, even if the dataset is balanced, unbiased, and contains useful information gathered in an ethical manner, can you use it in a way that is fair to everyone involved? The next section answers this question and more.

## Understanding ML fairness

Aside from ethical concerns and ensuring that your dataset lacks issues such as bias, it's important that the dataset and its associated model deliver a fair result. It's possible for a dataset to lack any sort of PII, features that could be linked to particular groups, and unnecessary features, and yet remain unfair. One of the most controversial and well-known examples of ML unfairness is the models used to assess the recidivism risk of individuals seeking release from prison. *Fairness in Machine Learning – The Case of Juvenile Criminal Justice in Catalonia* (`https://blog.re-work.co/using-machine-learning-for-criminal-justice/`) tells of only one incidence. The problem is extremely widespread, leading many to ask whether ML is capable of being fair in this scenario. The following sections explore ML fairness in more detail.

# Determining what fairness means

The term fair isn't actually well understood in most contexts and is often based on people's biases when the truth of a matter is known to all. However, that's not what fair means in this case. A fair algorithm will output the same result for a given set of feature values regardless of race, sex, age, belief system, country of origin, or any of a number of other physical, locational, and emotional characteristics of the person involved. The problem with most models that exhibit an unfair result is that they were trained with too much of the wrong data, which is why the example in the *Sanitizing data correctly* section of this chapter stresses removing any data that isn't absolutely needed to train the model. A sanitized dataset that eliminates all of the unnecessary data is said to operate under the fairness by unawareness principle, which often doesn't work because being unaware doesn't mean the data itself doesn't hide some type of bias.

Unfairness can still creep into the model output in other ways. For example, if the police target specific groups so that those groups are arrested more often than other groups, the model will reflect that bias. The methodologies used by various members of the police force, such as the use of force, also play into the statistics used by the model and cause it to become unfair. The article *Policing Women: Race and gender disparities in police stops, searches, and use of force* (`https://www.prisonpolicy.org/blog/2019/05/14/policingwomen/`) provides some interesting statistics demonstrating biases along race and gender lines that end up influencing the datasets used for ML models and therefore make them unfair.

Note that even though this section has focused on law enforcement so far, the same problems exist in every other aspect of human interactions that are modeled by ML applications. For example, unfairness regularly creeps into hiring situations and the decisions managers make based on ML application evaluations of resumes, which the manager often doesn't read before making a decision. Even with the elimination of protected features, those that shouldn't appear in the dataset such as race or gender, the dataset will remain unfair because the manner in which the data is collected is unfair. This concept of an inability to be fair goes back to the discussion in the *Defining the human element* section of *Chapter 1, Defining Machine Learning Security*.

Part of the answer to the unfairness question is to ensure that data is normalized across all of those protected features before the protected features are removed. In this case, normalization means ensuring that the data fairly represents all parties involved. The data then needs to be sanitized so that the normalized dataset doesn't reveal any biases to the ML model. The normalized data must be configured such that there are no correlations between target features and protected features so that the ML model can learn the biases in the protected features through the target features. There are a number of techniques used to achieve the required normalization, which include demographic parity, equalized odds, and equal opportunity, none of which are completely effective. Humans must still validate that a dataset actually is fair before using it to train a production model.

From a security perspective, the concept of fairness is crucial because otherwise time, resources, and technical expertise are directed in the wrong place. If the result of a model is unfair, then it's not possible to come up with completely valid solutions for issues such as hacker attacks and their targets. What will happen is that the misdirected effort will provide a hole that the hacker can use to their advantage.

## Understanding Simpson's paradox

Simpson's paradox is a statistical phenomenon that occurs when the causal relationship between two categorical variables changes when modifications are made to non-associated variables. To make this easier to understand, say that there is a relationship between people who have red cars and drive fast 80% of the time, while people who have gray cars only drive fast 10% of the time. However, to better understand the relationship, the data is subdivided between people who drive city streets and those who drive on the highway. Suddenly the relationship changes. People who drive gray cars now drive fast nearly 60% of the time and people who drive red cars now only drive fast 40% of the time. Simpson's paradox presents three problems:

- People expect that causal relationships are immutable

- Simpson's paradox doesn't affect only small datasets of a particular type; it affects all datasets in some way

- Making inferences based on informal studies is extremely risky

In this case, the difference in driving speed is easy to understand because the scenario has introduced another variable that influences both the car's color and the rate of speed. The data is showing that people who drive gray cars are more likely to do so on the highway. Until this factor is considered, it's not possible to obtain a correct answer. However, the question that will boggle your mind is, "*How many other factors are there to consider?*"

The security implications of Simpson's paradox are straightforward. If a reliance on a causal relationship between two variables is too heavy, then the introduction of a new factor may disrupt assumptions made on the basis of this relationship. When working through security issues, it's important not to rely too heavily on any particular factor and to ensure that any assumptions are constantly examined for the potential for a new factor.

## Removing personal bias

Of all the kinds of bias that can enter a data analysis scenario, personal bias is the most difficult to see and mitigate. After looking at the example in the *Sanitizing data correctly* section, a person may decide that they don't really like the color red. Even if the color red becomes the most loved color for a new product, it doesn't appear on shelves because the personal bias of the person performing the survey analysis doesn't allow for the use of red. Of course, this is an extreme example. Generally, personal bias is a lot more subtle and incredibly hard to deal with. Getting past a like or dislike can prove nearly impossible for many people. To mitigate or avoid personal bias, it's important to try these approaches:

- Don't allow any single individual to perform an analysis; use a group instead

- Ensure that groups performing analysis have a healthy, constructive method of viewing the analysis critically

- Use the "blind taste test" method of detecting possible bias by removing that factor during analysis

- Test the results of an analysis with a group that doesn't have any vested interest in the outcome

Getting creative with personal bias is important. However, it's also important to know that it exists and that everyone on the planet has biases, so it's not as if a particular individual is alone in dealing with this issue. Realizing that bias is common and universal is the first step in really dealing with it.

## Defining algorithmic bias

Algorithmic bias occurs when an algorithm produces unfair results based on a number of factors that include training data, the manner used to code the algorithm, and the decision-making processes that appear as a result of executing the code. The determining factor of whether algorithmic bias is occurring is that the result is repeatable in a relatively precise manner. In other words, given a set of inputs, the algorithm will repeatedly produce an incorrect result for the same input values and that result will be precisely the same every time within a certain margin of error. Given that a model can consist of a number of algorithms, the model as a whole is affected, and locating the precise source of the problem can be difficult unless the algorithms are clearly understood.

# Addressing fairness concerns

The first step in solving a problem is to know the problem exists. Right now, most people who create models have no idea that a problem exists. This lack of knowledge and understanding is the reason that models such as those used in the **Correctional Offender Management Profiling for Alternative Sanctions (COMPAS)** software fail on a grand scale. If the software had been vetted with some type of fairness indicator, then there is a higher likelihood that it would have produced a fair result. The following sections look at methods of working with fairness indicators that make it possible to produce models that output a fairer result.

---

**Is a completely fair result possible?**

It isn't actually possible to create a completely fair result today. There is a paradox involved. Either a model can treat groups fairly or it can treat individuals fairly. You can find a wealth of white papers on the topic, such as *On the Apparent Conflict Between Individual and Group Fairness* (https://arxiv.org/pdf/1912.06883.pdf), but most of them simply restate that there is a problem; they don't offer an actual solution. This particular white paper does offer some interesting ideas, but there is a lack of examples that implement the ideas, so they remain an interesting hypothesis and nothing more. Following the guidelines in the white paper will help you produce models that are fairer, help you consider trade-offs between group and individual fairness better, and help overcome some misconceptions about the issue as a whole.

---

## Computing fairness indicators with TensorFlow

To ensure the fairness of your model, you need some type of tool to verify fairness in a manner that manual verification can't because humans aren't even-handed in this particular kind of check. The TensorFlow fairness indicators are a set of tools built on top of the **TensorFlow Model Analysis** (**TFMA**) (`https://www.tensorflow.org/tfx/model_analysis/get_started`) library. The TensorFlow fairness indicators are packaged with the **TensorFlow Data Validation** (**TFDV**) package (`https://www.tensorflow.org/tfx/data_validation/get_started`) and the What-If Tool (`https://pair-code.github.io/what-if-tool/`). The What-If Tool works similarly to the TensorBoard GUI shown in *Figure 10.14* in that it allows you to view and experiment with your data. The *Fairness Indicators* example at `https://www.tensorflow.org/tfx/guide/fairness_indicators` provides one of the best methods of determining whether a model is fair. The developers of the software emphasize that the product is still in beta testing, which is something to keep in mind when you use it.

All of this information is nice to know, but you may wonder what it's really all about. This really isn't a new form of analysis, but an extension of other kinds of analysis that you've already seen in this book, which focuses on slicing data across user types to see whether there is some type of analysis issue based on criteria such as race, gender, and belief system. So, the analysis performs targeted checks for the following:

- Descriptive statistics that perform the following tasks:

  - Summarize the characteristics of the dataset

  - Measure central tendency (the central position within the dataset using mean, median, and mode)

  - Measure variability (the spread of the dataset using variance and standard deviation)

  - Measure frequency distribution (the occurrence of data items within a dataset using count)

- Data schema (essentially a blueprint of the dataset's layout)

- Anomaly detection (see *Chapter 6, Detecting and Analyzing Anomalies*, for details)

Using the example dataset found in the *Sanitizing data correctly* section of the chapter, this kind of analysis might compare people who like the red product against those who like the blue product. The `ProductColor` column becomes the slice used to focus the statistics. The analysis would then answer questions such as the following:

- Is model performance the same for people who like red as those who like blue?

- Is the dataset distributed equally for people who like red or blue?

- Are there ways to improve the dataset to meet the needs of people who like red, versus those who like blue?

- What is the confidence level as to the effectiveness of the data analysis?

- Are there threshold values where inequities in responses are apparent?

You could possibly ask other questions as well. The point is to look for ways in which the data is unbalanced and then determine whether the unbalance is severe enough to cause issues with fairness. After looking at the examples in this book, we can see that the example in the *Building a fraud detection example* section of *Chapter 8, Locating Potential Fraud*, is unbalanced. A review of fairness indicators might show that the imbalance is enough to present fairness problems for groups who commit fraud (a check will show that this isn't the case because the dataset has been sanitized correctly).

## Solving fairness problems with TensorFlow-constrained optimization

A constrained optimization is one in which hard values are applied to some variables to determine what values they can contain. The other variables are optimized to match the requirements of the constraint, which reduces the number of possible outcomes. The following process is used to ensure fairness when using a constrained model:

1. Create an unconstrained model.

2. Evaluate the model using fairness metrics.

3. Relying on the results of the evaluation, apply constraints to focus features.

4. Retrain the now constrained model.

5. Evaluate the constrained model using fairness metrics.

6. Repeat *step 3* through *step 5* until the desired result is achieved.

This process may seem a bit like cheating, but it's one way to ensure that a model outputs a fair result despite the data fed to it (rather than because of it). The *TensorFlow Constrained Optimization Example Using CelebA Dataset* example at `https://www.tensorflow.org/responsible_ai/fairness_indicators/tutorials/Fairness_Indicators_TFCO_CelebA_Case_Study` demonstrates the techniques needed to use this form of fairness with a celebrity smile dataset.

## Mitigating privacy risks using federated learning and differential privacy

Some data is simply too hot to handle. You don't want to move it from its safe location unless absolutely necessary and if you do then you'll want to ensure that the data remains safe. Of course, it's still necessary to train a model in order to create the required application.

From a security perspective, creating some types of models would require the use of either federated learning or differential privacy techniques because the data concerning individuals of interest is locked down and inaccessible. To build a successful model, it may become necessary to create the model and then train it using these alternative techniques. The following sections discuss two common methods of mitigating privacy risks by keeping data safe and still providing a means to train a model.

## Distributing data and privacy risks using federated learning

Some situations call for obtaining data from multiple sources. A single source won't have enough data available to properly train a model so the use of that data is both ethical and fair. Avoiding all of the issues mentioned in the chapter so far means having a large enough data source to ensure that the model is correct. Using multiple data sources can also improve model accuracy and flexibility because each source is likely to have different collection methods (even if the methods vary only slightly, it's unlike they'll be precisely the same), and the environment in which the data is will differ. However, it may be that the required data is locked away and there isn't any way to transport it anywhere. This is a case where federated learning comes into play, which relies on the model moving to the data, rather than the data moving to the model. Consider the following process:

1. Company A creates a model based on a specific set of requirements.
2. Company A sends the model to Company B where the model is trained using Company B's data.
3. Company B sends the model back to Company A, which now sends it to Company C.
4. Company C trains the model using its data and sends it back again to Company A.
5. Company A shares the trained model with both Company B and Company C.

There are many advantages to this particular setup. For one thing, Company A could send the model to as many other companies as needed to ensure that training occurs on a large enough dataset. This means that the potential for a stable and accurate model increases greatly. In addition, the companies supplying data don't necessarily have to know about each other, which provides another level of security and privacy to the process.

The biggest problem with this approach is that it's a lot slower than training the model in one location all at one time. Consequently, you can't use this approach in situations where time is a factor. However, it does work exceptionally well for long-term research needs, especially when the data used for research purposes is of a nature where its distribution is prohibited by law or some other requirement.

## Relying on differential privacy

Differential privacy is a technique where patterns of groups of data are publicly shared, but individual records aren't. The shared data must be of such a structure that small changes to the dataset aren't enough to reveal individual records. This is a form of privacy often used by governments to ensure the good of the public while maintaining the privacy of individuals. For example, statistics published about a census would fall into this category. It's possible to obtain information about groups of people, but not possible to obtain information about an individual within the group.

In this scenario, a group is not included in the statistical dataset if it contains just one member. Consequently, unlike federated learning, the resulting model could be skewed because the information about the group in question is missing and there is no way to reconstruct the data using normal missingness techniques.

When working with this sort of data, the model must rely on grouped data techniques, except instead of grouping the data programmatically, the data is grouped at the outset. Consequently, you can quickly discover relationships between groups (as an example), but can't predict individual behavior. The white paper *Building prediction models with grouped data: A case study on the prediction of turnover intention* (`https://onlinelibrary.wiley.com/doi/full/10.1111/1748-8583.12396`) provides useful perspectives on how to accomplish this task.

An advantage of this approach is that it's faster than using federated learning because it's possible to obtain the data and train the model in one location. In addition, because the data is all in one location and already grouped, the need for most data cleaning and manipulation tasks is reduced or eliminated, saving time and effort.

## Summary

This chapter has discussed ethical issues surrounding ML, which include sanitizing data and ensuring that raw data remains secure. Many developers view this process as unnecessarily complicated and therefore avoid it at all costs. However, addressing ethical issues in data management also yields significant benefits to everyone involved in working with the data and associated ML models. The goal is to ensure that any analysis you make is both fair and secure.

Congratulations! You've made it to the end of the book. By now you've been introduced to a lot more than just security issues, and have addressed a wide range of data management and model creation issues that ultimately affect the results you receive from any data analysis. Ultimately, it doesn't matter whether you're working with text, graphics, sounds, or other data types; the result you obtain reflects the usefulness of the process you use to obtain it.

## Further reading

The following bullets provide you with some additional reading that you may find useful for understanding the materials in this chapter in greater depth:

- Read additional details about keeping PII out of your dataset: *Data publication: Removing identifiers from data* (https://libguides.library.usyd.edu.au/datapublication/desensitise-data)

- Understand the need for dimensionality reduction: *Machine learning: What is dimensionality reduction?* (https://bdtechtalks.com/2021/05/13/machine-learning-dimensionality-reduction/)

- Locate a dataset that's safe to use for your next experiment: *10 Great Places to Find Free Datasets for Your Next Project* (https://careerfoundry.com/en/blog/data-analytics/where-to-find-free-datasets/)

- Gain more insights into why machine learning applications are unlikely to be fair anytime in the near future: *Can Machine Learning Ever Be "Fair" — and What Does That Even Mean?* (https://towardsdatascience.com/can-machine-learning-ever-be-fair-and-what-does-that-even-mean-b84714dfae7e)

- Discover how even a well-designed model can be unfair: *How We Analyzed the COMPAS Recidivism Algorithm* (https://www.propublica.org/article/how-we-analyzed-the-compas-recidivism-algorithm)

- Learn about ways to mitigate algorithmic bias: *Algorithmic bias detection and mitigation: Best practices and policies to reduce consumer harms* (https://www.brookings.edu/research/algorithmic-bias-detection-and-mitigation-best-practices-and-policies-to-reduce-consumer-harms/)

- Gain a better understanding of federated learning: *PyGrid: A Peer-To-Peer Platform For Private Data Science And Federated Learning* (https://blog.openmined.org/what-is-pygrid-demo/)

- Get more insights into the use and application of differential privacy: *Harvard University Privacy Tools Project: Differential Privacy* (https://privacytools.seas.harvard.edu/differential-privacy)

# Index

## A

<packt>

Packt.com

Subscribe to our online digital library for full access to over 7,000 books and videos, as well as industry leading tools to help you plan your personal development and advance your career. For more information, please visit our website.

## Why subscribe?

- Spend less time learning and more time coding with practical eBooks and Videos from over 4,000 industry professionals

- Improve your learning with Skill Plans built especially for you

- Get a free eBook or video every month

- Fully searchable for easy access to vital information

- Copy and paste, print, and bookmark content

Did you know that Packt offers eBook versions of every book published, with PDF and ePub files available? You can upgrade to the eBook version at packt.com and as a print book customer, you are entitled to a discount on the eBook copy. Get in touch with us at customercare@packtpub.com for more details.

At www.packt.com, you can also read a collection of free technical articles, sign up for a range of free newsletters, and receive exclusive discounts and offers on Packt books and eBooks.

# Other Books You May Enjoy

If you enjoyed this book, you may be interested in these other books by Packt:

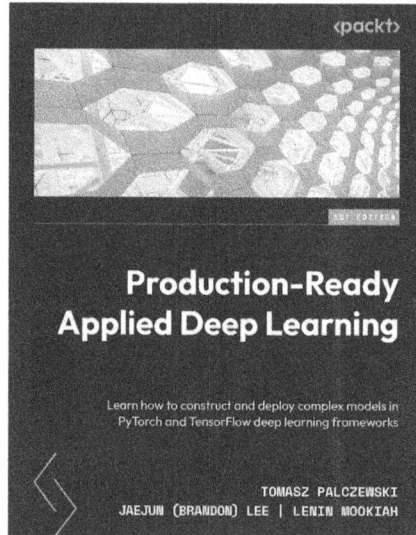

**Production-Ready Applied Deep Learning**

Tomasz Palczewski, Jaejun (Brandon) Lee, Lenin Mookiah

ISBN: 978-1-80324-366-5

- Understand how to develop a deep learning model using PyTorch and TensorFlow
- Convert a proof-of-concept model into a production-ready application
- Discover how to set up a deep learning pipeline in an efficient way using AWS
- Explore different ways to compress a model for various deployment requirements
- Develop Android and iOS applications that run deep learning on mobile devices
- Monitor a system with a deep learning model in production
- Choose the right system architecture for developing and deploying a model

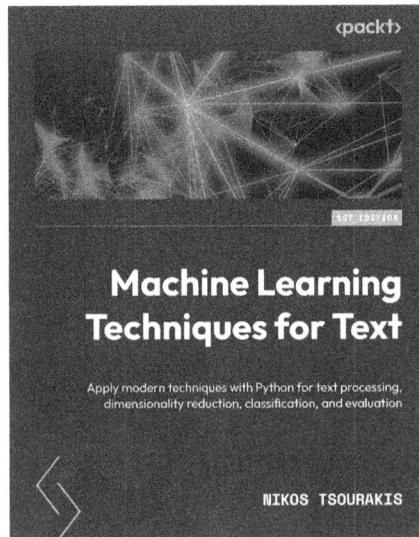

**Machine Learning Techniques for Text**

Nikos Tsourakis

ISBN: 978-1-80324-238-5

- Understand fundamental concepts of machine learning for text
- Discover how text data can be represented and build language models
- Perform exploratory data analysis on text corpora
- Use text preprocessing techniques and understand their trade-offs
- Apply dimensionality reduction for visualization and classification
- Incorporate and fine-tune algorithms and models for machine learning
- Evaluate the performance of the implemented systems
- Know the tools for retrieving text data and visualizing the machine learning workflow

## Packt is searching for authors like you

If you're interested in becoming an author for Packt, please visit `authors.packtpub.com` and apply today. We have worked with thousands of developers and tech professionals, just like you, to help them share their insight with the global tech community. You can make a general application, apply for a specific hot topic that we are recruiting an author for, or submit your own idea.

## Share Your Thoughts

Now you've finished *Machine Learning Security Principles*, we'd love to hear your thoughts! Scan the QR code below to go straight to the Amazon review page for this book and share your feedback or leave a review on the site that you purchased it from.

`https://packt.link/r/1-804-61885-3`

Your review is important to us and the tech community and will help us make sure we're delivering excellent quality content.

# Download a free PDF copy of this book

Thanks for purchasing this book!

Do you like to read on the go but are unable to carry your print books everywhere?

Is your eBook purchase not compatible with the device of your choice?

Don't worry, now with every Packt book you get a DRM-free PDF version of that book at no cost.

Read anywhere, any place, on any device. Search, copy, and paste code from your favorite technical books directly into your application.

The perks don't stop there, you can get exclusive access to discounts, newsletters, and great free content in your inbox daily

Follow these simple steps to get the benefits:

1. Scan the QR code or visit the link below

https://packt.link/free-ebook/9781804618851

2. Submit your proof of purchase
3. That's it! We'll send your free PDF and other benefits to your email directly

www.ingramcontent.com/pod-product-compliance
Lightning Source LLC
Chambersburg PA
CBHW081226220326
41598CB00037B/6887